VOLUME SEVEN HUNDRED AND EIGHT

METHODS IN ENZYMOLOGY

Carboxylases

METHODS IN ENZYMOLOGY

Editors-in-Chief

ANNA MARIE PYLE

Departments of Molecular, Cellular and Developmental Biology and Department of Chemistry Investigator, Howard Hughes Medical Institute Yale University

DAVID W. CHRISTIANSON

Roy and Diana Vagelos Laboratories Department of Chemistry University of Pennsylvania Philadelphia, PA

Founding Editors

SIDNEY P. COLOWICK and NATHAN O. KAPLAN

VOLUME SEVEN HUNDRED AND EIGHT

METHODS IN
ENZYMOLOGY
Carboxylases

Edited by

MARTIN ST. MAURICE
Department of Biological Sciences, Marquette University, Milwaukee, WI, United States

Academic Press is an imprint of Elsevier
50 Hampshire Street, 5th Floor, Cambridge, MA 02139, United States
525 B Street, Suite 1650, San Diego, CA 92101, United States
125 London Wall, London, EC2Y 5AS, United Kingdom

First edition 2024

Copyright © 2024 Elsevier Inc. All rights are reserved, including those for text and data mining, AI training, and similar technologies.

Publisher's note: Elsevier takes a neutral position with respect to territorial disputes or jurisdictional claims in its published content, including in maps and institutional affiliations.

No part of this publication may be reproduced or transmitted in any form or by any means, electronic or mechanical, including photocopying, recording, or any information storage and retrieval system, without permission in writing from the publisher. Details on how to seek permission, further information about the Publisher's permissions policies and our arrangements with organizations such as the Copyright Clearance Center and the Copyright Licensing Agency, can be found at our website: www.elsevier.com/permissions.

This book and the individual contributions contained in it are protected under copyright by the Publisher (other than as may be noted herein).

Notices
Knowledge and best practice in this field are constantly changing. As new research and experience broaden our understanding, changes in research methods, professional practices, or medical treatment may become necessary.

Practitioners and researchers must always rely on their own experience and knowledge in evaluating and using any information, methods, compounds, or experiments described herein. In using such information or methods they should be mindful of their own safety and the safety of others, including parties for whom they have a professional responsibility.

To the fullest extent of the law, neither the Publisher nor the authors, contributors, or editors, assume any liability for any injury and/or damage to persons or property as a matter of products liability, negligence or otherwise, or from any use or operation of any methods, products, instructions, or ideas contained in the material herein.

ISBN: 978-0-443-31310-3
ISSN: 0076-6879

> For information on all Academic Press publications
> visit our website at https://www.elsevier.com/books-and-journals

Publisher: Zoe Kruze
Editorial Project Manager: Saloni Vohra
Production Project Manager: James Selvam
Cover Designer: Gopalakrishnan Venkatraman
Typeset by MPS Limited, India

Contents

Contributors	*xi*
Preface	*xv*

1. Purification and characterization of mitochondrial biotin-dependent carboxylases from native tissues

Nicholas O. Schneider and Martin St. Maurice

1. Introduction	2
2. Before you begin	7
3. Key resources table	8
4. Materials and equipment	9
5. Step-by-step method details	10
5.1 Mitochondrial isolation from fresh pig liver	10
5.2 Mitochondrial lysis	13
5.3 Purification of biotin-dependent carboxylases from mitochondrial lysates	14
6. Expected outcomes	16
6.1 Kinetic assays of porcine pyruvate carboxylase	17
6.2 Assay procedure	20
7. Advantages	21
8. Limitations	21
9. Optimization and troubleshooting	22
9.1 Problems and potential solutions to optimize the procedure	22
10. Safety considerations and standards	23
11. Alternative methods/procedures	23
12. Summary and conclusions	24
References	25

2. Sample optimizations to enable the structure determination of biotin-dependent carboxylases

Jia Wei, Christine S. Huang, Yang Shen, Kianoush Sadre-Bazzaz, and Liang Tong

1. Introduction	32
2. Sample optimizations for structural studies of yeast ACC	34
2.1 Identifying a critical buffer component for crystallization and EM studies of yeast ACC	34

2.2 Using divide-and-conquer approach to study yeast ACC central domains – a cautionary tale		36
2.3 Preparation of phosphorylated ACC for structural studies		37
3. Sample optimization for crystal structure determination of PCC		38
4. Homologous domains are assembled to give highly distinct overall architectures of the biotin-dependent carboxylases		40
5. Summary		40
Acknowledgments		41
References		41

3. Purification of heteromeric acetyl-CoA carboxylases from *Escherichia coli* for structure solution

45

Amanda Silva de Sousa and Jeremy R. Lohman

1. Introduction		46
2. Expression construct design		49
2.1 Plasmid choice		50
2.2 Minioperon design		50
2.3 Notes		51
3. ACC complex overproduction in *E. coli*		51
3.1 Equipment		52
3.2 Strain and reagents		52
3.3 Procedure		53
3.4 Notes		53
4. Purification of the ACC complex		54
4.1 Equipment		55
4.2 Strain and reagents		55
4.3 Procedure		55
4.4 Notes		57
5. Activity assay by HPLC		59
5.1 Equipment		59
5.2 Reagents		59
5.3 Procedure		59
5.4 Notes		60
6. Summary and conclusions		62
Acknowledgments		63
References		63

4. Insights into the methodology of acetyl-CoA carboxylase inhibition

67

Mirela Tkalčić Čavužić, Brent A. Larson, and Grover L. Waldrop

1.	Background	69
	1.1 Reaction	69
	1.2 Protein structure	69
2.	Inhibitors for analysis of ACC structure and function	70
	2.1 Biotin carboxylase	70
	2.2 Enzymatic assay	70
	2.3 Product inhibitors	73
	2.4 ATP analogs	74
	2.5 Carboxyphosphate analogs	75
	2.6 Substrate inhibition	76
	2.7 Inhibition by BCCP	77
	2.8 Inhibition by inactive mutants of biotin carboxylase	77
	2.9 Negative feedback inhibition by palmitoyl-acyl carrier protein	78
	2.10 Carboxyltransferase	79
	2.11 Enzymatic Assay	79
	2.12 Biotin analogs	80
	2.13 Inhibition by BCCP	81
	2.14 Inhibition by DNA	82
	2.15 Inhibition by RNA	83
3.	Inhibitors of ACC with antibacterial properties	85
4.	Biotin carboxylase	85
	4.1 ATP site	85
	4.2 Optimization of pyridopyrimidines	87
	4.3 Amino-oxazole derivatives	88
	4.4 Benzimidazole carboxamide derivatives	89
5.	Biotin binding site	89
	5.1 Sulfoamidobenzamide	89
6.	Unknown binding site	91
	6.1 Pyrrolocin C and equisetin	91
7.	Carboxyltransferase	91
	7.1 Pyrrolidinediones	91
	7.2 Cinnamon derivatives	94
	7.3 Thailandamide A	96
	7.4 Dual-ligand and heterobivalent inhibitors	96
8.	Concluding remarks	98
	Author note	98
	References	98

viii Contents

5. Characterization of guanidine carboxylases 105

M. Sinn, J. Techel, A. Joachimi, and J.S. Hartig

1.	Introduction	106
2.	Before you begin	108
3.	Key resources table	109
4.	Materials and equipment	111
5.	Step-by-step method details	112
	5.1 Isolation of guanidine assimilating bacteria	112
	5.2 Sequencing of the guanidine assimilating bacteria	113
	5.3 Cloning of guanidine and urea carboxylase expression vectors	114
	5.4 Expression and purification of carboxylases	116
	5.5 Note	118
	5.6 Characterization of carboxylases	118
	5.7 Optional: Comparative proteome analysis	120
6.	Limitations	121
7.	Optimization and troubleshooting	121
	7.1 No colonies observed during isolation	121
8.	Safety considerations and standards	121
9.	Conclusion	122
	References	122

6. Methods to study prFMN-UbiD mediated (de)carboxylation 125

Dominic R. Whittall and David Leys

1.	Introduction	126
2.	*In vivo* production of *holo*-UbiD/*apo*-UbiD	129
	2.1 Equipment	130
	2.2 Buffers and reagents	130
	2.3 Molecular biology	131
	2.4 Heterologous production	131
	2.5 Purification of *holo*-UbiD	132
3.	*In vitro* reconstitution of *apo*-UbiD using prFMN^{reduced}	132
	3.1 Equipment	133
	3.2 Buffers and reagents	133
	3.3 Production of prFMN^{reduced}	134
	3.4 Reconstitution of *apo*-UbiD with prFMN^{reduced}	134
	3.5 Notes	135
4.	Synthesis of DMAP	135
	4.1 Equipment	135
	4.2 Reagents	135

4.3	Procedure	136
4.4	Preparing bis-triethylammonium phosphate (TEAP) solution	136
4.5	Preparing dimethylallyl monophosphate (DMAP)	136
4.6	Notes	136

5. *In vitro* oxidative maturation of prFMN containing UbiD — 137

5.1	Equipment	137
5.2	Buffers and reagents	137
5.3	Procedure	138

6. Characterisation of prFMN species via ESI-MS — 138

6.1	Equipment	138
6.2	Buffers and reagents	139
6.3	Procedure	139

7. ^{1}H NMR monitored enzyme catalysed deuterium exchange — 139

7.1	Equipment	140
7.2	Buffers and reagents	140
7.3	Procedure	140

8. (De)carboxylation assays monitored by HPLC/LC-MS — 140

8.1	Equipment	141
8.2	Buffers and reagents	141
8.3	Procedure	141

9. Preparation of whole-cell UbiD-biocatalysts — 142

9.1	Equipment	143
9.2	Buffers and reagents	143
9.3	Procedure	143

10. UbiD-substrate screening through monitoring flavin spectral shift — 144

10.1	Equipment	144
10.2	Buffers and reagents	144
10.3	Reconstitution of UbiD with FMN	144
10.4	Spectral shift assay	144

11.	Experimental strategies to drive carboxylation	145
12.	Conclusion	146
	References	147

7. Enzyme cascades for *in vitro* and *in vivo* FMN prenylation and UbiD (de)carboxylase activation under aerobic conditions — **151**

Anna N. Khusnutdinova, Khorcheska A. Batyrova, Po-Hsiang Wang, Robert Flick, Elizabeth A. Edwards, and Alexander F. Yakunin

1.	Introduction	152
2.	*In vitro* biosynthesis of prFMN and UbiD activation	156

2.1	Enzyme cascades for *in vitro* FMN prenylation and apoUbiD activation	157
2.2	Prenol phosphorylation to DMAP and FMN reduction using enzyme cascades	159
2.3	UbiD activation using prFMN cascades	160
2.4	Equipment	162
2.5	Materials	162
2.6	Procedure: recombinant protein expression and purification	163
2.7	Procedure: *in vitro* prFMN biosynthesis using enzyme cascades	164
2.8	Procedure: *in vitro* activation of purified UbiD using prFMN cascades	165
2.9	Notes	165
3.	Procedure: *in vivo* FMN prenylation and UbiD activation using recombinant *E. coli* cells	166
3.1	Equipment	167
3.2	Materials	168
3.3	Procedure	169
3.4	Notes	170
4.	Conclusions	170
Acknowledgments		171
References		171

8. Cellular and biochemical approaches to define GGCX carboxylation of vitamin K-dependent proteins

175

Kathleen L. Berkner, Kevin W. Hallgren, Mark A. Rishavy, and Kurt W. Runge

1.	Introduction	176
2.	Cellular analysis of VKD protein carboxylation	179
2.1	Analyzing VKD protein carboxylation in 293 cells expressing endogenous GGCX	179
2.2	Generating cell lines edited to eliminate endogenous GGCX	183
2.3	Expression of r-GGCX variants	184
3.	Biochemical analysis of VKD protein carboxylation	191
3.1	Assays to monitor GGCX catalysis	191
3.2	An assay to monitor the entire GGCX reaction	195
3.3	An assay to monitor GGCX processivity	198
4.	Considerations	199
References		201

9. Assessment of gamma-glutamyl carboxylase activity in its native milieu

207

Xuejie Chen, Darrel W. Stafford, and Jian-Ke Tie

1. Introduction	208
2. Cell-based assessment of VKD carboxylation activity using ELISA	211
2.1 Materials and equipment	213
2.2 Procedure	213
2.3 Notes	218
3. Functional study of naturally occurring GGCX mutations in the cellular milieu	219
3.1 Materials and equipment	221
3.2 Procedure	221
3.3 Notes	226
4. Assessment of vitamin K epoxidase activity in the cellular milieu	227
4.1 Materials and equipment	228
4.2 Procedure	228
4.3 Notes	231
5. Summary and conclusions	231
Acknowledgments	232
References	232

10. Expression, purification, and activation of one key enzyme in anaerobic CO_2 fixation: Carbon monoxide dehydrogenase II from *Carboxydothermus hydrogenoformans*

237

Kareem Aboulhosn and Stephen Wiley Ragsdale

1. Introduction	238
1.1 Wood-Ljungdahl pathway	238
1.2 Different types of CODH architecture	239
1.3 Relevance for producing CODH-II	240
2. Recombinant expression of CODH-II from *C. hydrogenoformans*	241
2.1 Cloning of CODH-II from *C. hydrogenoformans*	241
2.2 Use of fermenter to grow *E. coli* BL21 cells harboring pCooSII	242
2.3 Anaerobic harvest techniques	243
2.4 Lysis and purification of *Ch*CODH-II	244
2.5 Storage of *Ch*CODH-II	245
2.6 Use of anaerobic techniques in CODH characterization	245
3. Assay for CODH-catalyzed CO oxidation	246
3.1 Standard viologen-based CO oxidation assay	246

3.2	Reductive activation of CODH	247
3.3	EPR-based assay and metal quantification	248
4.	Optimization of growth conditions and lysis protocol	249
4.1	Lysis without the use of detergents leads to active *Ch*CODH-II	250
4.2	[Ni] during growth is critical for C-cluster loading	250
4.3	CooC1 and CooC3 do not mature *Ch*CODH-II	251
5.	Summary	253
Acknowledgments		254
References		254

11. Molybdenum-containing CO dehydrogenase and formate dehydrogenases
257

Russ Hille

1.	Overview	257
2.	CO Dehydrogenase	258
2.1	Introduction	258
2.2	Genetic context and protein structure	259
2.3	Mechanistic studies of the binuclear Mo/Cu center	261
2.4	Reaction with quinones	263
2.5	Experimental considerations	264
3.	Formate dehydrogenases	265
3.1	Introduction	265
3.2	Systematics	265
3.3	Reaction mechanism	269
3.4	Experimental considerations	270
4.	Summary	271
Acknowledgment		271
References		271

12. Assessing the role of redox carriers in the reduction of CO_2 by the oxo-acid:ferredoxin oxidoreductase superfamily
275

Sheila C. Bonitatibus, Mathew Walker, and Sean J. Elliott

1.	Introduction	276
2.	General considerations	279
3.	Materials and reagents	279
3.1	Molecular biology of the *Chlorobaculum* Fd proteins	279
3.2	Purification and characterization of the recombinant *Chlorobaculum* Fd proteins	280
3.3	Additional reagents	281

4.	Assay design and implementation	281
	4.1 The basic metronitazole assay	281
	4.2 Coupled biochemical assay with WT *Ct* PFOR	282
	4.3 Fd-mediated electrocatalytic assay	282
5.	Verification of redox potentials of Fd proteins	284
	5.1 General method	284
	5.2 Electrochemical characterization of Fd proteins	285
6.	Comparison of assay results	286
	6.1 Comparison of *Ct* Fd reactivity with *Ct* PFOR – *metronidazole assay*	286
	6.2 Comparison of *Ct* Fd reactivity with *Ct* PFOR – *reductive assay*	286
	6.3 *Ct* Fds as electron acceptors for *Ct* PFOR	288
7.	Conclusions and future directions	292
	Acknowlegments	293
	References	293

13. Measuring carbonic anhydrase activity in alpha-carboxysomes using stopped-flow   297

Nikoleta Vogiatzi and Cecilia Blikstad

1.	Introduction	298
2.	Cultivation of *H. neapolitanus*	301
	2.1 Equipment and materials	303
	2.2 Buffer and reagents	303
	2.3 Chemostat growth of *H. neapolitanus*	303
3.	Purification of carboxysomes	304
	3.1 Equipment and materials	305
	3.2 Buffers and reagents	306
	3.3 Purification of α-carboxysomes	306
4.	Measuring carbonic anhydrase activity using stopped-flow	309
	4.1 Equipment	311
	4.2 Reagents	311
	4.3 SX20 stopped-flow start-up and setting up the equipment	311
	4.4 Cleaning and preparing the SX20 stopped-flow before starting measurements	312
	4.5 Preparing assay buffer and making and diluting the CO_2 solution	313
	4.6 Making a standard measurement	313
	4.7 Measuring a saturation curve	315
	4.8 Measuring at different pH values	316

5. Data analysis	317
5.1 Calculate the experimental buffer factor	317
5.2 Calculate the kinetic constants k_{cat}, K_M and k_{cat}/K_M	317
6. Summary	319
Acknowledgment	319
References	319

14. Radiometric determination of rubisco activation state and quantity in leaves — **323**

Catherine J. Ashton, Rhiannon Page, Ana K.M. Lobo, Joana Amaral, Joao A. Siqueira, Douglas J. Orr, and Elizabete Carmo-Silva

1. Introduction	324
1.1 Method principles	326
1.2 Materials	328
1.3 Stock solutions to prepare in advance	331
2. Methods	333
2.1 Leaf sampling	333
2.2 Preparations for leaf extractions and rubisco activity assays	335
2.3 Protein extraction	336
2.4 Rubisco activity assays	337
2.5 Calculation of rubisco activation state	340
2.6 Rubisco quantification	342
3. Summary and conclusion	347
Acknowledgments	348
References	348

15. Computational methods for the study of carboxylases: The case of crotonyl-CoA carboxylase/reductase — **353**

Rodrigo Recabarren, Aharon Gómez Llanos, and Esteban Vöhringer-Martinez

1. Introduction	354
2. Molecular dynamics simulations identify conformational changes in crotonyl-CoA carboxylase/reductase increasing the local CO_2 concentration in the active site	357
2.1 Methods to quantify local CO_2 concentration and its dependence on protein conformation	359
2.2 Calculation of average CO_2 residence times	363
2.3 Simulation details and validation	366

3.	QM/MM molecular dynamics simulations reveal the reaction mechanism of the chemical CO_2 fixation step	369
	3.1 QM/MM calculations and free energy profiles	369
	3.2 The adaptive string method as an efficient tool for studying complex carboxylation reactions	372
	3.3 Corrections to free energy profiles	375
	3.4 Dissecting reactivity by comparing free energy profiles of different reaction mechanisms	376
	3.5 Model building for QM/MM MD simulations	379
	3.6 Simulation details for QM/MM MD simulations	379
4.	Summary	382
	References	382

16. Carboxylation in *de novo* purine biosynthesis — 389
Marcella F. Sharma and Steven M. Firestine

1.	Introduction	390
2.	Carboxylation in *de novo* purine biosynthesis	392
	2.1 AIR carboxylase	394
	2.2 N^5-CAIR synthetase and N^5-CAIR mutase	395
3.	CAIR synthesis as a drug target	398
4.	Expression and purification of *E. coli* $N^{5\text{-CAIR mutase}}$	400
	4.1 Equipment	400
	4.2 Reagents	401
	4.3 Procedure	401
5.	Expression and purification of *A. clavatus* $N^{5\text{-CAIR synthetase}}$	403
	5.1 Equipment	403
	5.2 Reagents	403
	5.3 Procedure	404
6.	Synthesis of CAIR and AIR	405
	6.1 Notes	406
7.	Activity assays	406
	7.1 CAIR decarboxylation assay (Note 1,2)	406
	7.2 AIR carboxylation assay	409
	7.3 SAICAR synthetase coupled assay for AIR carboxylase (Note 1)	410
	7.4 SAICAR synthetase coupled assay for N^5-CAIR mutase (Note 1)	412
	7.5 N^5-CAIR synthetase coupled assay	413

7.6	Discontinuous phosphate assay for N^5-CAIR synthetase	414
7.7	Isatin-fluorescein enzyme activity assay (Note 1)	416
7.8	Intrinsic Tryptophan Fluorescence Assay for N^5-CAIR Mutase	418
7.9	Thermal melting assay for N^5-CAIR mutase	419
8.	Summary and conclusions	420
References		421

Contributors

Kareem Aboulhosn
Program in Chemical Biology; Department of Biological Chemistry, University of Michigan, Ann Arbor, Michigan, United States

Joana Amaral
Lancaster Environment Centre, Lancaster University, Lancaster, United Kingdom

Catherine J. Ashton
Lancaster Environment Centre, Lancaster University, Lancaster, United Kingdom

Khorcheska A. Batyrova
Institute of Basic Biological Problems, Federal Research Center "Pushchino Scientific Center for Biological Research, Russian Academy of Sciences" (FRC PSCBR RAS), Pushchino, Moscow Region, Russia

Kathleen L. Berkner
Department of Cardiovascular and Metabolic Sciences, Lerner Research Institute, Cleveland Clinic Lerner College of Medicine at CWRU, Cleveland, OH, United States

Cecilia Blikstad
Department of Chemistry—Ångström Laboratory, Uppsala University, Uppsala, Sweden

Sheila C. Bonitatibus
Department of Chemistry, Boston University, Cummington Mall, Boston, MA, United States

Elizabete Carmo-Silva
Lancaster Environment Centre, Lancaster University, Lancaster, United Kingdom

Xuejie Chen
Department of Biology, University of North Carolina at Chapel Hill, Chapel Hill, NC, United States

Elizabeth A. Edwards
Department of Chemical Engineering and Applied Chemistry, University of Toronto, Toronto, Canada

Sean J. Elliott
Department of Chemistry, Boston University, Cummington Mall, Boston, MA, United States

Steven M. Firestine
Department of Pharmaceutical Sciences, Eugene Applebaum College of Pharmacy and Health Sciences, Wayne State University, Detroit, MI, United States

Robert Flick
Department of Chemical Engineering and Applied Chemistry, University of Toronto, Toronto, Canada

Kevin W. Hallgren
Department of Cardiovascular and Metabolic Sciences, Lerner Research Institute, Cleveland Clinic Lerner College of Medicine at CWRU, Cleveland, OH, United States

J.S. Hartig
Department of Chemistry; Konstanz Research School Chemical Biology (KoRS-CB), University of Konstanz, Germany

Russ Hille
Department of Biochemistry, University of California, Riverside, CA, United States

Christine S. Huang
Department of Biological Sciences, Columbia University, New York, NY, United States

A. Joachimi
Department of Chemistry, University of Konstanz, Germany

Anna N. Khusnutdinova
Centre for Environmental Biotechnology, School of Environmental and Natural Sciences, Bangor University, Bangor, United Kingdom; Institute of Basic Biological Problems, Federal Research Center "Pushchino Scientific Center for Biological Research, Russian Academy of Sciences" (FRC PSCBR RAS), Pushchino, Moscow Region, Russia

Brent A. Larson
Department of Biological Sciences, Louisiana State University, Baton Rouge, LA, United States

David Leys
Manchester Institute of Biotechnology, University of Manchester, Manchester, United Kingdom

Aharon Gómez Llanos
Departamento de Ciencias Biológicas y Químicas, Facultad de Medicina y Ciencia, Universidad San Sebastian, Lientur, Concepción, Chile

Ana K.M. Lobo
Lancaster Environment Centre, Lancaster University, Lancaster, United Kingdom

Jeremy R. Lohman
Department of Biochemistry and Molecular Biology, Michigan State University, East Lansing, MI, United States

Douglas J. Orr
Lancaster Environment Centre, Lancaster University, Lancaster, United Kingdom

Rhiannon Page
Lancaster Environment Centre, Lancaster University, Lancaster, United Kingdom

Stephen Wiley Ragsdale
Department of Biological Chemistry, University of Michigan, Ann Arbor, Michigan, United States

Rodrigo Recabarren
Departamento de Físico-Química, Facultad de Ciencias Químicas, Universidad de Concepción, Concepción, Chile

Contributors

Mark A. Rishavy
Department of Cardiovascular and Metabolic Sciences, Lerner Research Institute, Cleveland Clinic Lerner College of Medicine at CWRU, Cleveland, OH, United States

Kurt W. Runge
Department of Inflammation and Immunity, Lerner Research Institute, Cleveland Clinic Lerner College of Medicine at CWRU, Cleveland, OH, United States

Kianoush Sadre-Bazzaz
Department of Biological Sciences, Columbia University, New York, NY, United States

Nicholas O. Schneider
Department of Biological Sciences, Marquette University, Milwaukee, WI, United States

Marcella F. Sharma
Department of Pharmaceutical Sciences, Eugene Applebaum College of Pharmacy and Health Sciences, Wayne State University, Detroit, MI, United States

Yang Shen
Department of Biological Sciences, Columbia University, New York, NY, United States

Amanda Silva de Sousa
Department of Biochemistry and Molecular Biology, Michigan State University, East Lansing, MI, United States

M. Sinn
Department of Chemistry, University of Konstanz, Germany

Joao A. Siqueira
Lancaster Environment Centre, Lancaster University, Lancaster, United Kingdom

Martin St. Maurice
Department of Biological Sciences, Marquette University, Milwaukee, WI, United States

Darrel W. Stafford
Department of Biology, University of North Carolina at Chapel Hill, Chapel Hill, NC, United States

J. Techel
Department of Chemistry; Konstanz Research School Chemical Biology (KoRS-CB), University of Konstanz, Germany

Jian-Ke Tie
Department of Biology, University of North Carolina at Chapel Hill, Chapel Hill, NC, United States

Liang Tong
Department of Biological Sciences, Columbia University, New York, NY, United States

Nikoleta Vogiatzi
Department of Chemistry—Ångström Laboratory, Uppsala University, Uppsala, Sweden

Esteban Vöhringer-Martinez
Departamento de Físico-Química, Facultad de Ciencias Químicas, Universidad de Concepción, Concepción, Chile

Grover L. Waldrop
Department of Biological Sciences, Louisiana State University, Baton Rouge, LA, United States

Mathew Walker
Department of Chemistry, Boston University, Cummington Mall, Boston, MA, United States

Po-Hsiang Wang
Graduate Institute of Environmental Engineering, National Central University, Taoyuan, Taiwan

Jia Wei
Department of Biological Sciences, Columbia University, New York, NY, United States

Dominic R. Whittall
Manchester Institute of Biotechnology, University of Manchester, Manchester, United Kingdom

Alexander F. Yakunin
Centre for Environmental Biotechnology, School of Environmental and Natural Sciences, Bangor University, Bangor, United Kingdom; Department of Chemical Engineering and Applied Chemistry, University of Toronto, Toronto, Canada

Mirela Tkalčić Čavužić
Department of Biological Sciences, Louisiana State University, Baton Rouge, LA, United States

Preface

Carboxylases are a diverse group of enzymes that fix carbon dioxide into organic molecules through carbon–carbon bond formation. These reactions have profound implications in a wide array of biological processes, from fundamental metabolism to broader applications in health, biotechnology, agriculture, and global carbon cycling. The seemingly simple carboxylation reaction is, in fact, quite challenging; carboxylases must overcome significant thermodynamic and kinetic barriers to achieve catalysis. To accomplish this, carboxylases have evolved an array of strategies that have contributed to making them a challenging group of enzymes to study. Some, such as the classic example of rubisco, are notoriously slow and unstable catalysts. Others form multi-subunit complexes that must be properly assembled to catalyze the reaction. Most employ the aid of cofactors or cosubstrates such as biotin, reduced vitamin K, prenylated FMN, ferredoxin, or organometallic clusters to assist in overcoming the unfavorable thermodynamics of the reaction. Several, though not all, bind CO_2 directly, requiring them to overcome the challenge of weak binding by a sparingly soluble substrate. These features create a host of challenges to purifying and characterizing these enzymes.

This volume explores various methods for overcoming the challenges involved in obtaining, storing, and characterizing carboxylases that affect diverse biological systems. While the selection of carboxylases is not exhaustive, the featured enzymes undertake different catalytic strategies to achieve carboxylation, each with unique considerations. The volume covers a broad spectrum of approaches for studying these enzymes, from experimental techniques to computational analyses. The aim is to provide a set of practical reviews and guides that will be valuable to any laboratory interested in exploring these crucial enzymes.

Carboxylases contribute directly to human health through regulation of proteins and as essential components in key metabolic pathways that impact areas such as energy production, nucleotide biosynthesis, fat metabolism, and the synthesis of important biomolecules. Biotin-dependent carboxylases play a particularly important role in metabolism and are featured in Chapters 1–5 of this volume. Enzymes such as pyruvate carboxylase, which plays a crucial role in gluconeogenesis, or acetyl-CoA carboxylase, involved in fatty acid synthesis, highlight the indispensable nature of biotin-dependent carboxylases in maintaining metabolic balance in organisms ranging from bacteria to humans and reinforces the potential for these enzymes to serve as targets for therapeutic intervention. In Chapter 1, Schneider and St. Maurice describe revived methods for directly purifying biotin-dependent carboxylases from native tissues while, in Chapter 2, Tong and colleagues outline critical steps undertaken in the preparation of these enzymes for structural determination. In Chapter 3, de Sousa and Lohman extend this theme to the purification of heteromeric acetyl-CoA carboxylase from *E. coli* for structural and kinetic characterization. In Chapter 4, Waldrop and colleagues review a range of methods applied to discovering inhibitors of bacterial acetyl-CoA carboxylases. The exploration of biotin-dependent carboxylases concludes in Chapter 5, where Hartig and coworkers describe methods to identify and characterize newly discovered guanidine carboxylases from bacteria.

Several other carboxylases play important roles in human biological processes and metabolic pathways. Gamma-glutamyl carboxylase plays a critical role in carboxylating vitamin K dependent proteins that mediate blood clotting and regulate calcification. Assays for this enzyme are challenging and require a combination of appropriately prepared substrates and cofactors for both biochemical and cell-based methods, as outlined by Berkner and colleagues in Chapter 7 and Tie and colleagues in Chapter 8. Finally, in Chapter 16, Sharma and Firestine describe methods to characterize the carboxylase enzymes involved in the *de novo* purine biosynthesis pathway, where the carboxylation of 5-aminoimidazole ribonucleotide (AIR) by a single carboxylase (AIR carboxylase) in humans and higher eukaryotes differs from the combination of two enzymes, a synthetase and a mutase, that accomplish that same carboxylation in bacteria.

Beyond human health, carboxylases are gaining increasing attention in the realm of biotechnology. The ability of carboxylases to fix carbon dioxide into organic compounds makes them attractive tools for developing sustainable solutions to global challenges, such as climate change and the demand for renewable energy sources. The large family of prenylated FMN-dependent UbiD (de)carboxylases has attracted significant interest from researchers in this field. However, studies on these enzymes have been hindered by challenges in obtaining active enzyme properly reconstituted with the prenylated FMN cofactor. In Chapter 6, Whittall and Leys present methods for the *in vitro* reconstitution and characterization of active UbiD enzymes while, in Chapter 7, Yakunin and colleagues describe the use of enzyme cascades for FMN prenylation and UbiD (de)carboxylase activation. These two contributions will be of great value for those interested in exploring the growing potential of UbiD (de)carboxylases in the production of high-value chemicals.

Many other carboxylases can also be repurposed for biotechnological and synthetic purposes. In Chapter 12, Elliott and colleagues describe methods to characterize the ferredoxin redox carriers for the oxo-acid:-ferredoxin oxidoreductases that catalyze the reversible interconversion of CO_2 and oxo-acids. In Chapter 15, Vöhringer-Martinez and colleagues present computational methods for the study of crotonyl-CoA carboxylase/reductase that can be broadly applied to characterize the molecular determinants driving binding and catalysis in carboxylation reactions that employ CO_2 as a substrate. The carbon monoxide dehydrogenases and formate dehydrogenases also have great potential for the reversible storage of renewable energy in the production of biofuels, feedstocks and polymers. In Chapter 10, Aboulhosn and Ragsdale offer a detailed protocol for producing and characterizing the stand-alone, high-activity carbon monoxide dehydrogenase II in *Escherichia coli* under anaerobic conditions and, in Chapter 11, Hille reviews studies of the molybdenum-containing carbon monoxide dehydrogenase and formate dehydrogenases. Together, these chapters offer a modern collection of methods for working with many of the most promising carboxylases currently being employed for carbon fixation reactions in biotechnology.

Agriculture also benefits significantly from the study and application of carboxylases that contribute to plant growth and productivity. Rubisco, a well-known carboxylase, is responsible for carbon fixation in the Calvin cycle. It is highly abundant in plant tissues but it can be notoriously difficult to properly study. In Chapter 14, Carmo-Silva and colleagues carefully describe methods to sample and extract rubisco from plant leaves while minimizing its degradation and deactivation. In Chapter 13, Vogiatzi and Blikstad focus, instead, on the carbon dioxide concentration mechanism in the α-carboxysomes found in oceanic cyanobacteria and proteobacteria. They describe procedures to purify α-carboxysomes and to measure carbonic anhydrase enzymatic activity. Understanding and optimizing the function of such enzymes is critical for improving crop yields, especially in the context of a growing global population and changing environmental conditions.

This volume aims to provide a broad overview of methods that are being employed to study carboxylases in various biological and industrial contexts. By outlining methods to study carboxylases that play important roles in human health, biotechnology, and agriculture, this volume hopes to serve as a valuable resource for researchers, students, and professionals alike who are interested in employing similar studies in their own laboratories. As our understanding of carboxylases continues to evolve, so too will their applications in addressing some of the most pressing issues facing society today, from healthcare to environmental sustainability.

I would like to thank all the authors for agreeing to participate in this volume and for their excellent contributions. I gratefully acknowledge Siddharth Khattri and Zoe Cruze for their editorial assistance in bringing this volume to press. Finally, I extend my sincere thanks to David Christianson and Anna Pyle, Editors-in-Chief of *Methods in Enzymology*, for extending me the opportunity to organize this collection.

MARTIN ST. MAURICE
Department of Biological Sciences, Marquette University
Milwaukee, WI, United States

CHAPTER ONE

Purification and characterization of mitochondrial biotin-dependent carboxylases from native tissues

Nicholas O. Schneider and Martin St. Maurice[*]

Department of Biological Sciences, Marquette University, Milwaukee, WI, United States
*Corresponding author. e-mail address: martin.stmaurice@marquette.edu

Contents

1. Introduction	2
2. Before you begin	7
3. Key resources table	8
4. Materials and equipment	9
5. Step-by-step method details	10
5.1 Mitochondrial isolation from fresh pig liver	10
5.2 Mitochondrial lysis	13
5.3 Purification of biotin-dependent carboxylases from mitochondrial lysates	14
6. Expected outcomes	16
6.1 Kinetic assays of porcine pyruvate carboxylase	17
6.2 Assay procedure	20
7. Advantages	21
8. Limitations	21
9. Optimization and troubleshooting	22
9.1 Problems and potential solutions to optimize the procedure	22
10. Safety considerations and standards	23
11. Alternative methods/procedures	23
12. Summary and conclusions	24
References	25

Abstract

Biotin-dependent carboxylases catalyze the MgATP- and bicarbonate-dependent carboxylation of various acceptor substrates through a two-step carboxylation reaction. Biotin-dependent carboxylases play an essential role in the metabolism of key biomolecules and, therefore, they are the subject of ongoing drug discovery efforts, as well as of studies seeking to better characterize their structure and function. It has been an ongoing challenge to obtain high yields of mammalian biotin-dependent carboxylases for in vitro experimentation; these enzymes have not been successfully purified when recombinantly expressed from a bacterial expression host and only low yields of these recombinant, vertebrate enzymes have been obtained through expression in cell culture systems. Here, we describe a revived protocol to isolate mitochondrial biotin-dependent carboxylases (pyruvate carboxylase, propionyl-CoA carboxylase, and 3-methylcrotonyl-CoA carboxylase) from fresh pig liver. This serves as an inexpensive and effective alternative to using a recombinant expression system. This scalable protocol can be completed in less than 48 h and affords effective separation of mammalian, mitochondrial, biotin-dependent carboxylases from cellular components. The purified, mitochondrial biotin-dependent carboxylases can be used in downstream experiments focused on kinetic and structural characterization, as well as in initial drug discovery experiments.

1. Introduction

Biotin-dependent carboxylases are a diverse family of enzymes that use biotin as a cofactor to catalyze the ATP-dependent carboxylation of various substrates using bicarbonate as the carboxyl group donor. These enzymes serve important roles in the metabolism of carbohydrates, amino acids, and fatty acids (Jitrapakdee et al., 2008; Tong, 2013; Tong, 2017; Wakil et al., 1983; Waldrop et al., 2012). Dysregulation and aberrant activity of these enzymes drive a variety of metabolic, developmental, and neurological disorders (Grünert et al., 2012; Jitrapakdee et al., 2008; Wang et al., 2022; Wongkittichote et al., 2017). In mammals, the biotin-dependent carboxylases include acetyl-CoA carboxylase, propionyl-CoA carboxylase, 3-methylcrotonyl-CoA carboxylase, and pyruvate carboxylase. Urea carboxylase, guanidine carboxylase, acyl-CoA carboxylase, and geranyl-CoA carboxylase are also biotin-dependent carboxylases, but these enzymes are absent in mammals (Livieri et al., 2019; Schneider et al., 2020; Tong, 2013).

Mammalian biotin-dependent carboxylases utilize a multi-domain catalytic mechanism consisting of a MgATP and bicarbonate-dependent carboxylation of the biotin cofactor and subsequent carboxyl transfer from carboxybiotin to the acceptor substrate (Tong, 2013). All biotin-dependent carboxylases harbor three functionally homologous domains—the biotin carboxylase (BC) domain, the carboxyltransferase (CT) domain, and the biotin carboxyl carrier protein (BCCP) domain. The biotin cofactor is absolutely required for catalysis and is covalently tethered to a conserved lysine on the BCCP domain through a post-translational modification catalyzed by athe enzyme biotin protein ligase. The BC domain catalyzes the carboxylation of biotin from bicarbonate, enabled by the concomitant cleavage of MgATP to MgADP and inorganic phosphate. The BCCP-carboxybiotin then translocates to the CT domain, which catalyzes the carboxyl transfer from carboxybiotin to the acceptor substrate. The BC and

BCCP domains are evolutionarily conserved across the biotin-dependent carboxylases, reflecting a common origin and function. In contrast, the CT domains have evolved more distinctively to adapt to the specific substrates they act upon, leading to the functional differences among these enzymes.

Acetyl-CoA carboxylase (ACC) catalyzes the carboxylation of acetyl-CoA to form malonyl-CoA which represents the first committed step in fatty acid synthesis (Wakil et al., 1983). ACC continues to be extensively studied and has been shown to play a critical role in cancer and metabolic diseases. For example, ACC has been implicated in the progression of breast cancer, hepatocellular carcinoma, prostate cancer, and neck carcinoma (Brusselmans et al., 2005; Li et al., 2019; Milgraum et al., 1997; Wang et al., 2016; Yang et al., 2020). ACC has also been demonstrated to be important in pancreatic β-cell secretion (Cantley et al., 2019). Mammalian ACC is functionally a homodimer comprised of ~250 kDa subunits, with the primary dimerization interface maintained by contacts between an interaction domain (BT) and the CT domain (Hunkeler et al., 2018; Wei et al., 2016). Mammals have two isoforms of ACC: ACC1 and ACC2. ACC1 is expressed at a higher level than ACC2 and is located in the cytosol, while ACC2 is tethered to the cytosolic side of the outer mitochondrial membrane (Abu-Elheiga et al., 2000; Wang et al., 2022). ACC1 produces cytosolic malonyl-CoA to be used for fatty acid synthesis, while ACC2 produces malonyl-CoA for the regulation of fatty acid β-oxidation (Schreurs et al., 2010). As ACC1 and ACC2 are cytosolic enzymes, they are not effectively purified using the mitochondrial isolation method described in this chapter. However, some membrane-anchored ACC2 could be retained in mitochondrial preparations depending on the source of the mitochondria and the method by which the mitochondria have been prepared and processed.

Propionyl-CoA carboxylase (PCC) catalyzes the carboxylation of propionyl-CoA to methylmalonyl-CoA. PCC deficiency can result in propionic acidemia, a debilitating condition associated with metabolic acidosis and severe neurological issues that may lead to premature death (Baumgartner et al., 2014). PCC is important for amino acid and odd-chain fatty acid catabolism, leading to the downstream synthesis of succinyl-CoA, with its subsequent oxidation in the citric acid cycle (Wongkittichote et al., 2017). Aberrant PCC activity can alter the levels of a variety of citric acid cycle and citric acid cycle-derived metabolites such as succinate, glutamine, aspartate, alanine, oxaloacetate, pyruvate, and malate (Davison et al., 2011; Peuhkurinen, 1984; Wongkittichote et al., 2017). PCC is functionally a ~750 kDa heterododecamer consisting of 6 PCCα and 6 PCCβ subunits, forming an $\alpha_6\beta_6$ structure. The PCCα subunit is ~72 kDa in size and harbors the BC domain and the biotin-conjugated BCCP domain, while the ~56 kDa PCCβ subunit harbors the CT domain that binds propionyl-CoA (Wongkittichote et al., 2017). Like ACC, PCCα contains a BT domain which facilitates oligomerization with PCCβ (Huang et al., 2010). PCC is primarily located within the mitochondrial matrix but can also weakly associate with the matrix side of the inner mitochondrial membrane (Frenkel and Kitchens, 1975).

3-Methylcrotonyl-CoA carboxylase (MCC) localizes to the mitochondrial matrix and carboxylates 3-methylcrotonyl-CoA to form 3-methylglutaconyl-CoA, an essential step in the degradation of leucine (Chu and Cheng, 2007; Hu et al., 2022). MCC deficiency is one of the more prevalent inborn errors of metabolism, characterized by a spectrum of symptoms that can include neurological and metabolic disorders. Interestingly, it is estimated that only approximately 10 % of individuals with MCC deficiency exhibit symptoms (Forsyth et al., 2016; Grünert et al., 2012). MCC consists of MCCα and MCCβ subunits, with the MCCα subunit harboring the BC and biotin-conjugated BCCP domains, and MCCβ subunit harboring the 3-methylcrotonyl-CoA binding CT domain. MCC and PCC are homologous, but their overall structures are remarkably different (Huang et al., 2012). Most notably, while PCC oligomerizes through both BT-PCCβ and BC-PCCβ contacts, MCC oligomerizes primarily through BT-MCCβ contacts. This results in the MCC oligomer resembling a more "closed" conformation (Huang et al., 2012). Similar to PCC, MCC consists of a ~76 kDa MCCα and ~61 kDa MCCβ subunit. MCC has been observed to form both $\alpha_4\beta_4$ and $\alpha_6\beta_6$ structures and has recently been shown to form filaments consisting of stacked $\alpha_6\beta_6$ dodecamers (Díaz-Pérez et al., 2012; Hu et al., 2022; Huang et al., 2012).

The fourth mammalian biotin-dependent carboxylase, pyruvate carboxylase (PC), localizes to the mitochondrial matrix and catalyzes the carboxylation of pyruvate to form oxaloacetate. This represents the first committed step of gluconeogenesis. More generally, PC plays a crucial role in maintaining homeostatic levels of citric acid cycle intermediates and, consequently, serves as an early bottleneck in numerous biosynthetic pathways (Wallace et al., 1998). Because of this central metabolic role, PC has been implicated in the progression of various types of cancer and type-2 diabetes (Bahl et al., 1997; Cheng et al., 2011; Kumashiro et al., 2013; Liu et al., 2022; Ngamkham et al., 2022; Phannasil et al., 2015; Phannasil et al., 2017). PC also drives the pyruvate-malate shuttle, which contributes to mediating the cytosolic NADPH pool required for pancreatic insulin secretion (Xu et al., 2008). PC is unique among the biotin-dependent carboxylases because the acceptor substrate for this reaction is not an acyl-CoA, but rather a three-carbon alpha-keto acid (Tong, 2013). Consequently, the CT domain of PC is unique compared to the acyl-CoA carboxylases. Like mammalian ACC, the BC, CT, and BCCP domains of mammalian PC are all located on the same ~130 kDa subunit. PC is functionally a homotetramer, oligomerizing through a combination of the tetramerization (PT), CT, and BC domains (Chai et al., 2022; López-Alonso et al., 2022; St Maurice et al., 2007; Xiang and Tong, 2008).

Significant progress has been made in recent drug discovery efforts targeting ACC and PC, while advancements against PCC and MCC have been comparatively limited. Many potent inhibitors have been reported to target human ACCs, with mixed results on body weight reduction and fatty acid oxidation (reviewed in (Tong, 2013). Additionally, ACC1 and ACC2 have been implicated in several cancers and targeted inhibition of ACC in cancer cells have yielded promising anti-cancer effects. Numerous recent reports describe the significance of developing drugs targeting ACC1 and ACC2 for the treatment of cancer, and novel inhibitors of ACC have proven effective in preventing the growth of pancreatic cancer cells (Beckers et al., 2007; Chen et al., 2019; Petrova et al., 2016; Wu and Huang, 2020).

PC has also been validated as a therapeutic target for the treatment of various cancers and type-2 diabetes (Bahl et al., 1997; Cheng et al., 2011; Kumashiro et al., 2013; Ngamkham et al., 2022; Phannasil et al., 2015; Phannasil et al., 2017). Recent reports have reported potent inhibition for the compounds erianin and ZY-444 in several cancer cell lines, and ane-moside B4 in a colitis model. (Hong et al., 2022; Liang et al., 2023; Lin et al., 2020; Lv et al., 2024; Sheng et al., 2021; Shi et al., 2023). These compounds are proposed to inhibit cancer cell viability and inflammation by targeting PC, but this mechanism has not yet been confirmed through detailed in vitro studies. Additionally, there have been two novel classes of inhibitors that demonstrate potent inhibition of *Staphylococcocus aureus* PC, but they have not yet been confirmed to be effective against human PC (Burkett et al., 2019; Schneider et al., 2024). The CT domain of human PC is an attractive target for drug discovery because it lacks homology with the CT domains of the other biotin-dependent carboxylases.

Although human biotin-dependent carboxylases show promise as drug targets, challenges arise when attempting to express soluble forms of these recombinant enzymes in bacteria (Xiang and Tong, 2008). Eukaryotic expression systems have been the primary source for obtaining recombinant forms of vertebrate biotin-dependent carboxylases that can be purified using affinity tags (Chai et al., 2022; Chu and Cheng, 2007; Zhou et al., 2024a; Zhou et al., 2024b). One decades-old alternative to eukaryotic expression systems is the purification of biotin-dependent carboxylases directly from fresh animal tissue. Using the intrinsic physical and chemical properties of the enzymes, sometimes in combination with the application of reversible biotin-avidin affinity interactions, purification has been achieved from rat liver, rat brain, rat mammary glands, sheep kidney, bovine liver, porcine liver, and more (Ahmad et al., 1978; Ashman et al., 1972; Halenz and Lane, 1960; Lau et al., 1980; Mahan et al., 1975; Moss and Lane, 1971; Ochoa and Kaziro, 1965; Oei and Robinson, 1985; Thampy et al., 1988; Warren and Tipton, 1974). More recently, a commercially available streptavidin derivative, Streptactin®, has proven valuable for purifying biotin-dependent enzymes by utilizing their biotinylation as an affinity tag (Zhou et al., 2024a; Zhou et al., 2024b).

Purification from native tissue provides a relatively inexpensive, alternative method to obtain moderately concentrated and relatively pure preparations of vertebrate biotin-dependent enzymes. Recently, this approach has been used successfully to purify human biotin-dependent carboxylases for cryo-electron microscopy studies, where the requirements for high yield and high purity are significantly reduced relative to X-ray crystallography (Chai et al., 2022; Hu et al., 2022; Hunkeler et al., 2018; Zhou et al., 2024a; Zhou et al., 2024b). The purification of biotin-dependent carboxylases from fresh tissues and human cell culture also offers opportunities for kinetic characterization of the enzyme activities and can be a particularly useful tool for drug discovery, provided that the proper methods and controls are implemented in the kinetic studies. Working with purified samples offers major advantages over simply running assays on mitochondrial or cell lysates, where nonspecific interactions risk leading to the misinterpretation of results.

Here, we describe the purification of biotin-dependent carboxylases from native cells and tissues, which offers a quick, inexpensive, and scalable method to obtain relatively pure preparations of vertebrate enzymes (Fig. 1). This method provides concentrations and purity that are suitable for, but not limited to, structural studies using cryo-electron microscopy

Purification of biotin-dependent carboxylases from native tissue

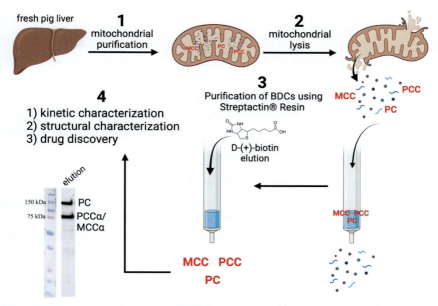

Fig. 1 Workflow for purifying biotin-dependent carboxylases (pyruvate carboxylase, PC; propionyl-CoA carboxylase, PCC; and 3-methylcrotonyl-CoA carboxylase, MCC) from the mitochondria of vertebrate animal cells or tissue. *Created with BioRender.com.*

and the screening and characterization of small molecule effectors for drug discovery. While the methods described here can be applied to any of the mitochondrially localized biotin-dependent carboxylases, our emphasis is rooted in our own experience working with pyruvate carboxylase.

2. Before you begin

Timing: 1–2h.
Critical: Store and adjust the pH of all buffers at 4 °C.
For the isolation of mitochondria from fresh pig liver:
1. Sucrose/Tris Buffer (ST buffer)
    a. 20 mM Tris-HCl pH 7.4, 250 mM sucrose, 10 mM potassium fluoride, 2 mM ethylene glycol-bis(β-aminoethyl ether)-N,N,N,N-tetraacetic acid (EGTA), 1 mM phenylmethylsulfonyl fluoride (PMSF).
        • **Note**: For PMSF, first prepare a 100 mM stock in ethanol or isopropanol. Dilute into the ST buffer to a final concentration of 1 mM.

For mitochondrial lysis:
2. Lysis buffer
    a. 100 mM Tris-HCl pH 7.4, 150 mM potassium chloride, 10% glycerol, 1% Triton X-100, 1 mM PMSF (prepared as described above), 1 μM pepstatin A, and 5 μM N-[N-(L-3-Trans-carboxirane-2-carbonyl)-L-leucyl]-agmatine (E-64).
    • **Note**: The protease inhibitors PMSF, pepstatin A, and E-64 should only be added immediately prior to use.
3. Resuspension buffer
    a. 100 mM Tris-HCl pH 7.4, 500 mM sucrose, 1 mM ethylenediaminetetraacetic acid (EDTA), 1 mM dithiothreitol (DTT)
    • **Note:** DTT is added immediately prior to use.

For the purification of biotin-dependent carboxylases from mitochondrial lysates:
4. Wash buffer
    a. 100 mM Tris pH 7.4, 150 mM KCl, 10% glycerol
5. Elution buffer
    a. Wash buffer with 50 mM D-(+)-biotin. To properly dissolve, readjust pH back to 7.4 after adding the biotin.
    • **Note**: ~2–5 mM desthiobiotin can be used in place of biotin.

For the kinetic assays of porcine PC using malate dehydrogenase:
6. Assay buffer
    a. 100 mM Tris pH 7.8, 150 mM KCl, 7 mM MgCl$_2$

3. Key resources table

Reagent or Resource	Source	Identifier
Biological Samples		
Fresh porcine liver	Wilson Farm Meats	
Chemicals, Peptides, and Recombinant Proteins		
Tris-HCl	Research Products International (RPI)	1185-53-1
Sucrose	RPI	S24065
Potassium fluoride	Sigma-Aldrich	229814

ethylene glycol-bis(β-aminoethyl ether)-N,N,N,N-tetraacetic acid (EGTA)	RPI	67-42-5
phenylmethylsulfonyl fluoride (PMSF)	Sigma-Aldrich	329-98-6
Glycerol	RPI	56-81-5
Triton X-100	Sigma-Aldrich	T8787
Pepstatin A	Sigma-Aldrich	26305-03-3
L-trans−3-Carboxyoxiran−2-carbonyl-L-leucylagmatine (E-64)	RPI	66701-25-5
dithiothreitol (DTT)	RPI	3483-12-3
d-biotin	Fisher Scientific	58-85-5
MgCl$_2$	RPI	7786-30-3
Critical Commercial Assays		
Malate dehydrogenase from porcine heart	Sigma-Aldrich	LMDH-RO

4. Materials and equipment

- Mitochondrial isolation from fresh pig liver
  - Food processing equipment: knife/scissors, meat mincer, food processor, cheesecloth filter, rubber spatulas.
  - Glass Dounce homogenizer (ideally > 50 mL)
  - Avanti J26XP centrifuge (Beckman Coulter)
  - JA-25.50 fixed-angle rotor (Beckman Coulter)
  - JA-8.1000 fixed-angle rotor (Beckman Coulter)
  - Large (250-1000 mL) and medium (∼50 mL) centrifuge bottles
  - UV-1800 spectrophotometer (Shimadzu Scientific Instruments)
- Mitochondrial lysis
  - Eppendorf tube rotator
  - Ammonium sulfate (Thermo Fisher Scientific)
  - pH meter
  - JA-25.50 fixed-angle rotor (Beckman Coulter)
  - Avanti J26XP centrifuge (Beckman Coulter)
- Purification of biotin-dependent carboxylases from mitochondrial lysates
  - Streptactin® Sepharose® resin (IBA Life Sciences, Göttingen, Germany)

- **Note**: This has recently been discontinued and replaced with Strep-Tactin®XT 4Flow® high capacity resin
- Microplate spectrophotometer (Spectramax i3x, Molecular Devices)
- 96 well plates (Santa Cruz Biotechnology)
- Eppendorf tube rotator
- Centrifuge (tabletop or Avanti J26XP centrifuge (Beckman Coulter))
- Liquid nitrogen

5. Step-by-step method details
5.1 Mitochondrial isolation from fresh pig liver
Timing: 8–14 h.

This method describes the preparation of mitochondria from a single fresh pig liver. The result of this procedure should be >15 mL of relatively pure and concentrated mitochondria that can be frozen and stored for >6 months without losing enzyme activity (as determined by measuring the activity of pyruvate carboxylase). The yield and relative purity of the mitochondrial extracts will impact the downstream purification. The degree of mitochondrial purity can be evaluated by Western blot using antibodies against cytosolic and mitochondrial protein markers to compare the starting tissue with the purified mitochondria, as previously described (Chen et al., 2013).

Sucrose/Tris Buffer (ST buffer).
- 20 mM Tris-HCl pH 7.4, 250 mM sucrose, 10 mM potassium fluoride, 2 mM EGTA, 1 mM PMSF.
  - **Note**: For PMSF, first prepare a 100 mM stock in ethanol or isopropanol. Once fully dissolved, dilute into the ST buffer to a final concentration of 1 mM.

Critical: All steps are performed at room temperature, but all solutions (with and without liver components) are always kept at 4 °C or on ice.
Critical: All centrifuge steps are performed at 4 °C.
1. Harvest a whole, fresh liver from an adult pig (~1 kg).
   a. We obtained our fresh pig livers through generous donations from Wilson Farm Meats (Elkhorn, WI, USA).
   b. **Critical**: For the best results, the liver should be processed immediately (1-2 h) after harvesting.
   c. **Critical**: After harvest, the liver should immediately be placed and constantly maintained on ice until the start of the mitochondrial isolation.

2. Cut the liver into small pieces, approximately $4 \times 4 \times 4$ cm.
3. Feed the liver pieces into a meat mincer and collect the minced liver in a beaker with 500 mL ST buffer on ice. A medium-sized pig liver, approximately 1 kg, should yield 0.75–1.5 L of minced liver.
4. Add cold ST buffer to the minced liver to a final ratio of at least 1:1 of ST buffer to minced liver. This will create 2–3 L of a heterogenous solution.
5. Add the mixture to a food processor, and blend in 10–15 s intervals. Between intervals, allow the unblended pieces to settle to the bottom. Repeat until the total blend time reaches approximately 1 min
   a. **Note**: At this stage, unblended pieces of minced liver will often remain visible.
6. Centrifuge the blended mixture using large, 1 L centrifuge bottles and a JA-8.1000 fixed-angle rotor at 700 g for 10 min to pellet large debris.
   a. **Note**: smaller centrifuge bottles (e.g. 500 mL) and rotors can be used but this will increase the total processing time.
7. Filter the supernatant through cheesecloth and collect in a beaker placed on ice to keep for step 9. This is the filtered supernatant.
8. Dislodge and resuspend the pellet with 1 to 5 equivalents of ST buffer, and repeat the blending, centrifugation, and filtering steps (steps 5–7) once more.
   a. **Note**: Two cycles of steps 5–7 should result in ~1 L of filtered supernatant that is entirely free of large debris. If a larger volume is used in steps 4 and 8, the final volume may exceed 1 L of filtered supernatant.

Pause Point: Due to our centrifuge and homogenization capacity, we typically advance a 200–400 mL portion of the filtered supernatant at a time to process through steps 9–16, maintaining the remaining filtered supernatant in a 4 °C refrigerator or cold room. Once these processing steps have been completed, another 200–400 mL portion of filtered supernatant can be processed through steps 9–16. This is repeated until all the filtered supernatant has been processed. Processing 200–400 mL of filtered supernatant through steps 9-16 typically takes approximately 2 h.

9. Using the JA-25.50 fixed-angle rotor, centrifuge the filtered supernatant at 4 °C in 50 mL centrifuge tubes at 15,000 g for 20 min
   a. **Note**: A beige pellet will be observed following centrifugation. This corresponds to the crude mitochondrial fraction.

b. Note: A dark red pellet is typically observed at the center of the beige pellet – this corresponds to a mixture of debris and red blood cells. Avoid resuspending this part of the pellet, and discard after step 10.

10. Using a spatula, scrape the beige portion of the pellet to loosen it without significantly disturbing the red pellet.
    a. Note: The beige pellet from 50 mL of centrifuged filtered supernatant is typically suspended in approximately 15 mL ST buffer.
    b. Critical: Avoid disturbing or resuspending the dark red portion of the pellet. This is important to the overall quality of the mitochondrial purification.

11. Transfer the suspended, loose beige pellet to a glass Dounce homogenizer that has been pre-chilled on ice.

12. Fully homogenize the mitochondria until no visible pellet remains.
    a. Note: We used a hand-powered homogenizer. ~5 homogenization cycles is typically sufficient for complete homogenization (pushing the pestle down and then pulling it up is considered to be one cycle).

13. Centrifuge the homogenized mitochondria at 1,000 g for 10 min to separate the debris from the mitochondrial suspension.
    a. Note: The loosely pelleted debris can be difficult to see and work around. It will be beige in color (similar to the suspension), loosely floating, and "mucus-like".

14. Slowly and carefully aspirate the mitochondrial suspension while minimizing aspiration of the mucus-like debris. Transfer the aspirated suspension to a fresh 50 mL centrifuge tube and discard the loosely pelleted debris.

15. Centrifuge the transferred mitochondrial suspension in 50 mL centrifuge tubes at 15,000 g for 20 min. If any of the dark red precipitate remains, repeat steps 10 - 14. Repeat these steps until a clean, beige pellet is observed after this final centrifugation step (this will typically require at least 2-3 repetitions of steps 10–15).

16. Once a uniform beige pellet is observed after the centrifugation in step 15, discard the supernatant and resuspend the pellet with ~15 mL ST buffer. Dispense the resuspended mitochondria in 500 μL aliquots into microcentrifuge tubes (or a volume appropriate for future use).
    a. Note: A ~1 kg liver should yield several hundred 500 μL aliquots.
    b. Critical: The final resuspension volume (after several rounds of purification) will dictate the final concentration of the mitochondrial

Purification of biotin-dependent carboxylases from native tissue 13

isolate. This should be a consideration when deciding on the final resuspension volume prior to aliquoting.

17. Flash freeze the microcentrifuge tubes with the aliquoted mitochondria in liquid nitrogen and store at −80 °C.

Pause Point: A typical purified mitochondrial preparation from pig liver will yield hundreds of frozen aliquots of isolated mitochondria. Stored at −80 °C, these preparations can yield active pyruvate carboxylase after > 6 months of storage.

5.2 Mitochondrial lysis

Timing: 2–3 h.

This procedure describes the lysis of the purified mitochondria and a subsequent crude purification of biotin-dependent carboxylases using ammonium sulfate precipitation. The lysis and ammonium sulfate precipitation steps are performed at room temperature. This procedure assumes a starting material equivalent to ~35 × 500 µL aliquots of isolated mitochondria, and a total lysis volume of 100 mL, but the procedure is scalable for those interested in obtaining a higher overall yield.

Lysis buffer.

- 100 mM Tris-HCl pH 7.4, 150 mM potassium chloride, 10 % glycerol, 1 % Triton X-100, 1 mM PMSF (prepared as described above), 1 µM pepstatin A, and 5 µM E-64.
  - **Note**: The protease inhibitors PMSF, pepstatin A, and E-64 are added immediately prior to use.

    Resuspension buffer
- 100 mM Tris-HCl pH 7.4, 500 mM sucrose, 1 mM EDTA, 1 mM DTT
  - **Note**: DTT is added immediately prior to use.

Critical: Consider keeping samples from the following steps for analysis by SDS-PAGE and Western Blotting: whole mitochondria (step 1), the supernatant after centrifugation of the ammonium sulfate precipitation (step 10), the ammonium sulfate resuspension (step 11), and the insoluble pellet (step 12).

1. Thaw the desired number of aliquots of isolated mitochondria on ice.

2. Pellet the mitochondria by centrifuging at 15,000g for 3 min and discard the supernatant.

3. Resuspend the pelleted mitochondria in each aliquot by adding a sufficient volume of lysis buffer (+ protease inhibitors) to thoroughly resuspend by pipetting up and down (~250 µL).

4. Add the resuspended mitochondria from each aliquot to ~70 mL lysis buffer (+ protease inhibitors).

5. Bring the total volume to 100 mL by adding lysis buffer (+ protease inhibitors).
   a. **Critical**: Monitor the pH through step 8. The pH must remain above 6.7 and should be kept within a pH range of 7.0–7.4. If needed, add 1 M Tris Base to increase the pH. We do not typically see the pH drop below 7.0.
6. Stir slowly using a magnetic stir bar for 1 h at room temperature.
7. Slowly add solid ammonium sulfate to 35 % saturation (~1.4 M) over approximately two min.
8. Continue to stir slowly for an additional 30 min
9. Centrifuge the 35 % saturated ammonium sulfate solution at 4 °C in two 50 mL centrifuge bottles at a speed of 15,000 g for 30 min
10. A white ammonium sulfate pellet will be observed, likely on the side of the centrifuge tube. A beige pellet is frequently observed floating on the surface—this is debris. Discard the beige pellet and the supernatant.
    a. **Note**: If a beige pellet is observed on the side of the centrifuge tube (~1/3 up the side), and no white pellet is observed, the white protein pellet may be intermixed with the beige debris. If this occurs, resuspend the entire pellet as in step 11 and continue to step 12. Pay close attention to the absorbance at 280 nm recorded in step 13 to ensure that a high protein concentration was still obtained.
11. Resuspend the ammonium sulfate pellet using 3 mL resuspension buffer (+DTT) for every 50 mL centrifugation from step 9.
12. Centrifuge the resuspension at 15,000 g for 5 min to pellet any remaining insoluble material.
13. Collect the supernatant and measure the absorbance at 280 nm to ensure that the protein concentration is sufficiently high.
    a. **Note**: The absorbance at 280 nm should be > 2 in a 1 cm pathlength quartz cuvette. The resuspension may have a yellow tint.

5.3 Purification of biotin-dependent carboxylases from mitochondrial lysates

Timing: 3–5 h.

This procedure describes the purification of biotin-dependent carboxylases using a Streptactin® resin. The protocol is designed to yield ~500 µL of relatively pure biotin-dependent enzymes.

Wash buffer.
- 100 mM Tris pH 7.4, 150 mM potassium chloride, 10 % glycerol
  Elution buffer

- Wash buffer with 50 mM D-(+)-biotin. To properly dissolve, readjust pH back to 7.4 after adding the biotin.

  Critical: All steps are performed at 4 °C.

  Critical: Consider keeping the following steps for analysis by SDS-PAGE and Western Blotting: unbound protein (step 5), washing steps (step 6), elution(s) (step 9), and a sample of the beads to evaluate the elution competence (a.k.a. the uneluted protein that remains bound to the resin).

1. For relatively small-scale purifications, ~500 µL of Streptactin® Sepharose® (IBA Lifesciences GmBH, Göttingen, Germany) resin (50 % slurry) may be used in a batch-style purification.
2. Equilibrate resin with 20 column volumes (CV) of wash buffer.
3. Pellet the resin by centrifuging at 300 g for ~30 s and remove the supernatant by careful aspiration.
   a. **Critical**: Pellet the resin at low speeds. High centrifugation speeds risk damaging the integrity of the resin.
   b. **Note**: At these low centrifugation speeds, the resin will form a very loose pellet. Take care not to disturb or aspirate the resin when aspirating the supernatant. Pay close attention to the presence of resin in the aspirated supernatants—there should be very little.
4. Load the mitochondrial lysate resuspension onto the resin and gently rock/rotate to constantly mix for 1 h.
5. Pellet the resin by centrifuging at 300 g for ~30 s and remove the unbound protein by gently aspirating and discarding the supernatant.
6. Wash the resin with 20 CV of wash buffer.
   a. **Note**: Considering the use of the batch purification method, the wash may be performed in consecutive, smaller washing steps as demonstrated in Figs. 2 and 3.
7. Add ~500 µL elution buffer to the resin.
8. Rock/rotate for 1 h.
9. Pellet the resin by centrifuging at 300g for ~30 s and collect the eluted, purified protein by gently aspirating the supernatant.
   a. **Note**: Avoid aspirating the resin. It is not unusual to aspirate a small amount of resin. Should that occur, the collected elution should be centrifuged again at 300g for 1 min to remove this small amount of resin from the eluant.
   b. **Note**: Repeated elution steps should be considered. A significant amount of biotin-dependent enzyme remains bound to the Streptactin® resin after the first elution (Fig. 2).

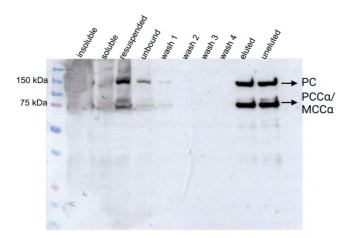

Fig. 2 Representative Western blot of the purification of biotin-dependent carboxylases from pig liver mitochondria. Biotin-dependent carboxylases were purified by small-scale batch purification on Streptactin® Sepharose as described. Following 8 % SDS-PAGE and blotting, biotinylated proteins were visualized by Streptavidin-HRP following the method described by (Cui and Ma, 2018). Individual purification steps are noted at the top of each well. The "resuspended" fraction is the resuspended ammonium sulfate precipitation. Each wash step represents a wash with ~5 column volumes of wash buffer. PC (~130 kDa), PCCα and MCCα (~80 kDa) are all present in the purified fraction.

    c. **Note**: Recent reports describing the cryo-EM structures of biotin-dependent enzymes from mitochondrial isolates include one or more size-exclusion chromatographic steps following the Streptactin® purification step (Zhou et al., 2024b; Zhou et al., 2024a). The use of DEAE-Sepharose to further separate the individual biotin-dependent enzymes has also been described (Oei and Robinson, 1985).
10. For long-term storage of the eluted purified biotin-dependent carboxylases, 40 μL aliquots are flash frozen in liquid nitrogen and stored at −80 °C.

6. Expected outcomes

The presence of biotin-dependent carboxylases in each purification step can be monitored using Western blot or SDS-PAGE (Coomassie stain/silver stain). Biotin-dependent carboxylases can be visualized on a Western blot using streptavidin conjugated horseradish peroxidase (Streptavidin-HRP).

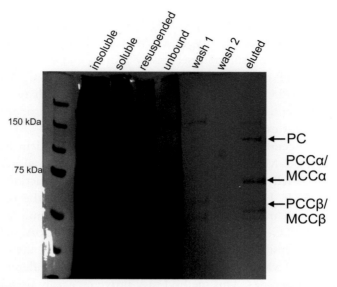

Fig. 3 Purification of biotin-dependent carboxylases from pig liver mitochondria. Biotin-dependent carboxylases were purified by small-scale batch purification on Streptactin® Sepharose as described. Individual purification steps are noted at the top of each well. Samples were subjected to 8 % SDS-PAGE and were visualized by silver staining. The "resuspended" fraction is the resuspended ammonium sulfate precipitation. Each wash step represents a wash with ~20 column volumes of wash buffer. PC (~130 kDa), PCCα and MCCα (~80 kDa), PCCβ and MCCβ (56 and 61 kDa, respectively) are all present in the purified fraction.

Streptavidin-HRP detects all biotinylated proteins on the Western blot without requiring a secondary antibody for signal amplification (Cui and Ma, 2018). Antibodies directed against the individual enzymes can also be employed, but Streptavidin-HRP offers a quick and inexpensive approach to detect the biotin-dependent carboxylases. Fig. 2 is a representative Western blot performed using Streptavidin-HRP to detect biotinylated proteins throughout the purification process. Fig. 3 is a representative silver-stained SDS-PAGE gel displaying the protein fractions through the individual purification steps. Since the purification protocol exploits the natural biotin affinity tag of biotin-dependent carboxylases, the biotin-dependent carboxylases are the predominant proteins present in the final elution (Fig. 3).

6.1 Kinetic assays of porcine pyruvate carboxylase

The enzyme activity of biotin-dependent carboxylases can be determined in a variety of ways. When using the above protocol to purify a collection of biotin-dependent carboxylases, ATP cleavage assays are not recommended to assess specific activity. PCC, MCC, and ACC all act on similar substrates with varying substrate selectivity profiles and they all will likely contribute to ATP cleavage in the presence of a given acyl–CoA substrate. Additionally, HPLC alone may not offer sufficiently high resolution to accurately differentiate between the individual CoA esters produced by the biotin-dependent carboxylases (Liu et al., 2020).

Liu et al. recently reported a strategy using HPLC-tandem mass spectrometry (HPLC-MS/MS) to assess PCC activity (Liu et al., 2016; Liu et al., 2020). Briefly, this method involves initiating the reaction with the addition of ATP, HCO_3^- and propionyl-CoA, allowing the reaction to proceed for two min, terminating the reaction with formic acid, then performing HPLC-MS/MS to detect methylmalonyl-CoA. This assay provides high sensitivity and is expected to be generally useful for the detection of the various CoA esters produced by the different biotin-dependent carboxylases. This method may prove especially useful to assay MCC, where limited coupled enzyme assays have been established.

In addition to ^{14}C-bicarbonate fixation, ACC activity has been determined using a variety of methods (see, also, Chapters 3 and 4 of this volume): coupling ACC activity with fatty acid synthase and monitoring ^{3}H-palmitic acid production from ^{3}H–acetyl-CoA, coupling ACC activity with citrate synthase and 5,5′-dithiobis-(2-nitrobenzoic acid) and monitoring 5-thio-2-nitrobenzoic acid production, and coupled assays with citrate synthase and malate dehydrogenase (Santoro et al., 2006; Seethala et al., 2006; Willis et al., 2008; Zempleni et al., 1997). An important note is that ACC is the only biotin-dependent carboxylase that is primarily located in the cytosol. ACC2 localizes to the outer mitochondrial membrane, but ACC2 is not typically observed in the elution of purified, mitochondrial biotin-dependent carboxylases. Therefore, a modification to this protocol that purifies ACC from the cytosol should result primarily in the purification of ACC, facilitating the use of activity assays that measure ATP cleavage.

Relative to the acyl-CoA biotin-dependent carboxylases, PC can be more reliably assayed due to its unique 3-carbon substrate, pyruvate. A common method to assay for PC activity couples the formation of oxaloacetate with its reduction to malate by malate dehydrogenase (MDH), where the oxidation of NADH can be followed spectrophotometrically (Utter and Keech, 1963). This assay has sometimes been used to assess the inhibition of PC activity in cell or mitochondrial lysates, but without sufficient controls to ensure that the activity being assessed is due to PC

(Hong et al., 2022; Liang et al., 2023; Lv et al., 2024; Sheng et al., 2021; Shi et al., 2023). Vertebrate PC is highly dependent on the allosteric activator acetyl-CoA for activity (Utter and Keech, 1963). Given the potential presence of contaminating proteins in crude extracts and, potentially, in the more highly purified preparations from tissue, it is strongly recommended to always include two negative controls: one that excludes acetyl-CoA and the other that excludes pyruvate from the reaction conditions. Mitochondrial lysates, in particular, will yield a relatively high background rate of NADH oxidation in the presence of pyruvate, likely catalyzed by mitochondrial lactate dehydrogenase (Ketchum et al., 1988; Paventi et al., 2017). Therefore, the inclusion of *both* of these negative controls is necessary to ensure that the measured activity can be attributed to PC (Fig. 4). As demonstrated in Fig. 4, mitochondrial lysates will often display a statistically insignificant PC activity over background. Therefore, any conclusions drawn from such MDH-coupled assays with mitochondrial extracts should be treated with great caution. To improve the signal over background, it is highly recommended that the purification

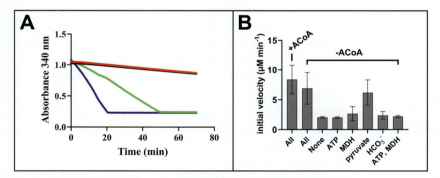

Fig. 4 Streptactin® purification is required to accurately assess the activity of PC from native tissue. (A) Assessing the activity of PC purified from pig liver using the malate dehydrogenase coupled assay. Initial velocity traces of a negative enzyme control (no enzyme added; black), a negative activator control (no acetyl-CoA added; red), purified PC from pig liver (green), and 10 μg mL^{-1} of recombinantly expressed and purified *Staphylococcus aureus* PC from an *E. coli* expression host (blue). (B) Assessing PC activity directly in mitochondrial lysates. The x-axis describes the substrates (HCO$_3^-$, ATP, pyruvate, or MDH) that were included in the assay, either in the presence or absence of acetyl-CoA (ACoA). Mitochondrial lysate is present in each condition. In this example, the rate of the reaction with pyruvate alone is equivalent to the reaction with all PC substrates, strongly suggesting that the predominant activity in the lysate results from mitochondrial lactate dehydrogenase.

protocol outlined above be followed prior to using the MDH-coupled assay to characterize PC activity directly from cellular systems.

The protocol outlined below can be used to assay PC activity following its partial purification from mitochondrial isolates.

6.2 Assay procedure

Timing: 1–3 h.

Assay buffer.

- 100 mM Tris (pH 7.8), 150 mM potassium chloride, 7 mM $MgCl_2$

Critical: Kinetic assays of PC can be performed on samples purified from any vertebrate cell or tissue source using a standard malate dehydrogenase coupled enzyme assay. However, it is important to include control reactions that are performed both in the absence of acetyl-CoA as well as in the absence of pyruvate. These controls are critical to ensure that the activity being measured is being catalyzed by PC. If appreciable product formation is observed in the absence of **either** pyruvate **or** acetyl-CoA, this is very likely a result of contaminating enzyme(s)/substrate(s) that contribute to the background oxidation of NADH. The background oxidation observed in the presence of both acetyl-CoA and pyruvate should be subtracted from the measured rate in the presence of acetyl-CoA and pyruvate. For example, in Fig. 4, the rate in the absence of acetyl-CoA (red trace) should be subtracted from the rate in the presence of acetyl-CoA (blue trace) to determine the initial velocity for the reaction catalyzed by pyruvate carboxylase.

Note: The following protocol is optimized for a 96-well plate. The volumes can be adjusted to accommodate assays in individual cuvettes or 384-well plates, as appropriate.

1. Prepare stocks of assay components:
   a. Acetyl-Coenzyme A, 20x stock: 5 mM
   b. Substrates, 10x stocks: 250 mM $NaHCO_3^-$, 120 mM pyruvate, 10 mM ATP, 2.5 mM NADH
   c. Malate dehydrogenase, 10x stock: 200 U mL^{-1}
2. Prepare a substrate master mix by combining equal volumes of the 10x stocks of 250 mM $NaHCO3^-$, 120 mM pyruvate, 10 mM ATP, 2.5 mM NADH, and 200 U mL-1 malate dehydrogenase.
3. For the experimental condition, add the following to three replicate wells in a 96-well plate:
   a. 20 µL of the purified enzyme

Purification of biotin-dependent carboxylases from native tissue

      i. **Note**: Consider running a range of volumes of the enzyme to account for variation in PC concentration between different purifications.
    b. 10 μL of 5 mM acetyl–CoA
    c. 70 μL assay buffer
4. Initiate the reaction by adding 100 μL substrate master mix.
5. Monitor the reaction at 340 nm using a microplate spectrophotometer.
    a. Note: An example initial velocity trace, in the presence and absence of acetyl–CoA, is shown in Fig. 4.
6. To calculate the specific activity (e.g. μmol min^{-1} mg^{-1}):
    a. Determine the slope for the initial velocity phase of the reaction, in Absorbance sec^{-1}.
    b. Convert to appropriate units of concentration time^{-1} using the extinction coefficient for NADH ($6220\,M^{-1}cm^{-1}$) (e.g. μM min^{-1}).
    c. Measure the total protein concentration in the sample using the bicinchoninic acid (BCA) assay, or an equivalent assay (e.g. μg mL^{-1}).
    d. Divide the value obtained in b by the value obtained in c.

7. Advantages

The purification of biotin-dependent carboxylases from native tissue provides a straightforward means to acquire native, vertebrate biotin–dependent enzymes directly from animal tissue. While this general approach preceded the advent of recombinant DNA technology, the development of commercially available Streptactin® resin has greatly facilitated the reversible binding of biotinylated proteins. In this method, the biotin cofactor effectively serves as an affinity tag for purifying biotin-dependent enzymes from tissue. This method is particularly useful for overcoming the difficulties associated with expressing soluble, full-length vertebrate biotin-dependent carboxylases in bacterial systems. The method we have described here is quick and cost-effective, especially when compared to the recombinant expression of these enzymes in eukaryotic expression systems.

8. Limitations

Purifying biotin-dependent carboxylases from native tissue does not yield large amounts of protein and the steps outlined in this protocol do not effectively purify the individual mitochondrially-located, biotin-dependent enzymes from one another. The procedure may be scaled up, offering a yield that should be more than sufficient for many downstream applications, including cryo-EM. However, the purified enzymes will be consumed quickly when used in enzyme assays, especially when downstream applications are centered on high-throughput small molecule screening for drug discovery. It must also be noted that this approach does not facilitate the separation of the three mitochondrial biotin-dependent carboxylases (PC, MCC, PCC). Downstream steps such as size-exclusion chromatography and ion-exchange have been reported to enhance the purification of the individual enzymes (Oei and Robinson, 1985). If the desired protein must be purified away from all other mitochondrial biotin-dependent carboxylases, recombinant DNA technology is the preferred approach. For example, affinity tags have been used to recombinantly express and purify human PC from human cell culture (Chai et al., 2022; Jitrapakdee et al., 1999). An additional limitation is the availability and willingness to work with fresh animal tissue. We obtained our fresh pig livers through generous donations from Wilson Farm Meats (Elkhorn, WI, USA), but access to a processing facility and the willingness of lab personnel to work with animal tissues may present a barrier.

9. Optimization and troubleshooting

9.1 Problems and potential solutions to optimize the procedure

If problems are encountered with purity, it can be beneficial to assess the overall purity of the mitochondrial extracts. The degree of mitochondrial purity can be evaluated by Western blot using antibodies against cytosolic and mitochondrial protein markers to compare the starting tissue with the purified mitochondria, as previously described (Chen et al., 2013).

The ammonium sulfate precipitation step can be optimized to enhance the capture of specific biotin-dependent carboxylases. While our protocol has been optimized for PC, a different concentration of ammonium sulfate may be optimal in preferentially precipitating other biotin-dependent carboxylases. We have not fully optimized this step for each of the individual biotin-dependent carboxylases. A titration of ammonium sulfate concentrations should be included in the initial troubleshooting steps to find the optimal concentration for the desired enzyme. Ammonium sulfate concentrations between 10–40% saturation result in different levels of protein precipitation.

The elution of biotin-dependent carboxylases from the Streptactin® resin is often incomplete, with a significant amount of protein remaining bound to the resin. This reduces the overall yield. The elution step can be optimized to increase the overall yields. In our experience, overnight incubation with elution buffer does not increase the yield. However, consecutive elution steps have provided a modest increase in the total protein yield. Elution may also be further optimized by modifying the pH or the concentration of desthiobiotin/biotin in the elution buffer.

10. Safety considerations and standards

When handling animal tissues in the lab, the OSHA Laboratory Safety Guidelines should be meticulously followed. The transmission of animal borne diseases is possible, and it is highly recommended to obtain fresh animal tissue from a source that performs common pathogen screens following slaughter. Lab personnel should wear nitrile or other non-permeable gloves when working with raw and processed animal tissue to reduce exposure to allergens and potentially zoonotic agents. Gloves should be removed before leaving the work area, and hands must be thoroughly washed with soap and water. Personal protective equipment (PPE) such as lab coats, face masks, and goggles are also recommended to protect against splashes and contamination. All work with tissue should be conducted in designated areas like biosafety cabinets to minimize exposure. Proper labeling, storage, and disposal of tissues and waste in appropriate biohazard containers is essential. All personnel should be trained in proper handling procedures, be aware of biohazard risks, and be familiar with incident response and ethical guidelines.

11. Alternative methods/procedures

Recently, some groups have used Streptactin® resin to purify biotin-dependent carboxylases for structural characterization by cryo-EM (Hu et al., 2022; Zhou et al., 2024a; Zhou et al., 2024b). Zhou et al. performed downstream size exclusion chromatography to further purify the desired biotin-dependent carboxylases (Zhou et al., 2024a; Zhou et al., 2024b). It is important to note that using size exclusion as an additional purification step is effective in further purifying the biotin-dependent enzymes from contaminating proteins, but it is less effective at separating PCC, MCC, and PC due to their similar overall molecular weights (PCC: ~750 kDa, MCC: ~850 kDa, PC: ~520 kDa).

In the methods described in this chapter, only the mitochondrial enzymes PCC, MCC, and PC were purified. The protocol might be adjusted to purify ACC. Rather than discarding the cytosolic components and isolating the mitochondria, the cytosolic components should be retained. Further purification of ACC from the cytosolic components by ammonium sulfate precipitation and affinity purification via Streptactin® resin is expected to yield relatively pure ACC.

12. Summary and conclusions

Dysregulation and aberrant activity of biotin-dependent carboxylases are implicated in various metabolic, developmental, and neurological disorders. Therefore, they have been the subject of many recent efforts aimed at drug discovery and structure determination. Given that expression of recombinant vertebrate biotin-dependent carboxylases in a bacterial host has not been successful, alternative approaches such as recombinant expression in a eukaryotic host or purification from native cells/tissues has become a necessary step to obtain sufficient yields for downstream drug discovery efforts and kinetic/structural characterization.

This protocol describes a straightforward method to purify biotin-dependent carboxylases from fresh porcine liver. Considering the high sequence identity between pig and human proteins (i.e. the porcine PC protein sequence is 97 % identical and 98 % similar to human PC), this can serve as an inexpensive and valuable tool for studies aimed at impacting human health. The protocol is also amenable to purifying human PC from cultured human liver cells. Biotin-dependent carboxylases are purified to a high degree using this approach, and the purified fraction demonstrates considerable PC activity (Figs. 2–4). While this protocol is effective in purifying the three mitochondrial biotin-dependent carboxylases, downstream steps are needed to further purify the individual carboxylases. Depending on the application, many experimentalists will find that working with the purified fraction of biotin-dependent carboxylases is a great improvement over working with mitochondrial extracts. In such instances, the protocol we have described should be sufficient for their needs.

References

Abu-Elheiga, L., Brinkley, W. R., Zhong, L., Chirala, S. S., Woldegiorgis, G., & Wakil, S. J. (2000). The subcellular localization of acetyl-CoA carboxylase 2. *Proceedings of the National Academy of Sciences, 97*(4), 1444–1449. https://doi.org/10.1073/pnas.97.4.1444.

Ahmad, F., Ahmad, P. M., Pieretti, L., & Watters, G. T. (1978). Purification and subunit structure of rat mammary gland acetyl coenzyme A carboxylase. *The Journal of Biological Chemistry, 253*(5), 1733–1737.

Ashman, L. K., Keech, D. B., Wallace, J. C., & Nielsen, J. (1972). Sheep kidney pyruvate carboxylase: Studies on its activation by acetyl coenzyme A and characteristics of its acetyl coenzyme A independent reaction. *Journal of Biological Chemistry, 247*(18), 5818–5824. https://doi.org/10.1016/S0021-9258(19)44831-X.

Bahl, J. J., Matsuda, M., DeFronzo, R. A., & Bressler, R. (1997). In vitro and in vivo suppression of gluconeogenesis by inhibition of pyruvate carboxylase. *Biochemical Pharmacology, 53*(1), 67–74. https://doi.org/10.1016/S0006-2952(96)00660-0.

Baumgartner, M. R., Hörster, F., Dionisi-Vici, C., Haliloglu, G., Karall, D., Chapman, K. A., ... Chakrapani, A. (2014). Proposed guidelines for the diagnosis and management of methylmalonic and propionic acidemia. *Orphanet Journal of Rare Diseases, 9*, 130. https://doi.org/10.1186/s13023-014-0130-8.

Beckers, A., Organe, S., Timmermans, L., Scheys, K., Peeters, A., Brusselmans, K., ... Swinnen, J. V. (2007). Chemical inhibition of acetyl-CoA carboxylase induces growth arrest and cytotoxicity selectively in cancer cells. *Cancer Research, 67*(17), 8180–8187. https://doi.org/10.1158/0008-5472.CAN-07-0389.

Brusselmans, K., De Schrijver, E., Verhoeven, G., & Swinnen, J. V. (2005). RNA interference-mediated silencing of the acetyl-CoA-carboxylase-alpha gene induces growth inhibition and apoptosis of prostate cancer cells. *Cancer Research, 65*(15), 6719–6725. https://doi.org/10.1158/0008-5472.CAN-05-0571.

Burkett, D. J., Wyatt, B. N., Mews, M., Bautista, A., Engel, R., Dockendorff, C., ... St. Maurice, M. (2019). Evaluation of α-hydroxycinnamic acids as pyruvate carboxylase inhibitors. *Bioorganic & Medicinal Chemistry, 27*(18), 4041–4047. https://doi.org/10.1016/j.bmc.2019.07.027.

Cantley, J., Davenport, A., Vetterli, L., Nemes, N. J., Whitworth, P. T., Boslem, E., ... Biden, T. J. (2019). Disruption of beta cell acetyl-CoA carboxylase-1 in mice impairs insulin secretion and beta cell mass. *Diabetologia, 62*(1), 99–111. https://doi.org/10.1007/s00125-018-4743-7.

Chai, P., Lan, P., Li, S., Yao, D., Chang, C., Cao, M., ... Fan, X. (2022). Mechanistic insight into allosteric activation of human pyruvate carboxylase by acetyl-CoA. *Molecular Cell, 82*(21), 4116–4130.e6. https://doi.org/10.1016/j.molcel.2022.09.033.

Chen, X., Cui, Z., Wei, S., Hou, J., Xie, Z., Peng, X., ... Yang, F. (2013). Chronic high glucose induced INS-1β cell mitochondrial dysfunction: A comparative mitochondrial proteome with SILAC. *Proteomics, 13*(20), 3030–3039. https://doi.org/10.1002/pmic.201200448.

Chen, L., Duan, Y., Wei, H., Ning, H., Bi, C., Zhao, Y., ... Li, Y. (2019). Acetyl-CoA carboxylase (ACC) as a therapeutic target for metabolic syndrome and recent developments in ACC1/2 inhibitors. *Expert Opinion on Investigational Drugs, 28*(10), 917–930. https://doi.org/10.1080/13543784.2019.1657825.

Cheng, T., Sudderth, J., Yang, C., Mullen, A. R., Jin, E. S., Matés, J. M., & DeBerardinis, R. J. (2011). Pyruvate carboxylase is required for glutamine-independent growth of tumor cells. *Proceedings of the National Academy of Sciences of the United States of America, 108*(21), 8674–8679. https://doi.org/10.1073/pnas.1016627108.

Chu, C.-H., & Cheng, D. (2007). Expression, purification, characterization of human 3-methylcrotonyl-CoA carboxylase (MCCC). *Protein Expression and Purification, 53*(2), 421–427. https://doi.org/10.1016/j.pep.2007.01.012.

Cui, Y., & Ma, L. (2018). Sequential use of milk and bovine serum albumin for streptavidin-probed Western blot. *Biotechniques, 65*(3), 125–126. https://doi.org/10.2144/btn-2018-0006.

Davison, J. E., Davies, N. P., Wilson, M., Sun, Y., Chakrapani, A., McKiernan, P. J., ... Peet, A. C. (2011). MR spectroscopy-based brain metabolite profiling in propionic acidaemia: Metabolic changes in the basal ganglia during acute decompensation and effect of liver transplantation. *Orphanet Journal of Rare Diseases, 6*, 19. https://doi.org/10.1186/1750-1172-6-19.

Díaz-Pérez, C., Rodríguez-Zavala, J. S., Díaz-Pérez, A. L., & Campos-García, J. (2012). Co-expression of α and β subunits of the 3-methylcrotonyl-coenzyme A carboxylase from Pseudomonas aeruginosa. *World Journal of Microbiology and Biotechnology, 28*(3), 1185–1191. https://doi.org/10.1007/s11274-011-0921-1.

Forsyth, R., Vockley, C. W., Edick, M. J., Cameron, C. A., Hiner, S. J., Berry, S. A., ... Arnold, G. L. (2016). Outcomes of cases with 3-methylcrotonyl-CoA carboxylase (3-MCC) deficiency—Report from the Inborn Errors of Metabolism Information System. *Molecular Genetics and Metabolism, 118*(1), 15–20. https://doi.org/10.1016/j.ymgme.2016.02.002.

Frenkel, E. P., & Kitchens, R. L. (1975). Intracellular localization of hepatic propionyl-CoA carboxylase and methylmalonyl-CoA mutase in humans and normal and vitamin B12 deficient rats. *British Journal of Haematology, 31*(4), 501–513. https://doi.org/10.1111/j.1365-2141.1975.tb00885.x.

Grünert, S. C., Stucki, M., Morscher, R. J., Suormala, T., Bürer, C., Burda, P., ... Baumgartner, M. R. (2012). 3-methylcrotonyl-CoA carboxylase deficiency: Clinical, biochemical, enzymatic and molecular studies in 88 individuals. *Orphanet Journal of Rare Diseases, 7*, 31. https://doi.org/10.1186/1750-1172-7-31.

Halenz, D. R., & Lane, M. D. (1960). Properties and purification of mitochondrial propionyl carboxylase. *Journal of Biological Chemistry, 235*(3), 878–884. https://doi.org/10.1016/S0021-9258(19)67953-6.

Hong, J., Xie, Z., Yang, F., Jiang, L., Jian, T., Wang, S., ... Huang, X. (2022). Erianin suppresses proliferation and migration of cancer cells in a pyruvate carboxylase-dependent manner. *Fitoterapia, 157*, 105136. https://doi.org/10.1016/j.fitote.2022.105136.

Hu, J. J., Lee, J. K. J., Liu, Y.-T., Yu, C., Huang, L., Aphasizheva, I., ... Zhou, Z. H. (2022). Discovery, structure, and function of filamentous 3-methylcrotonyl-CoA carboxylase. *Structure (London, England: 1993)*, S0969-2126(22)00486-5. https://doi.org/10.1016/j.str.2022.11.015.

Huang, C. S., Ge, P., Zhou, Z. H., & Tong, L. (2012). An unanticipated architecture of the 750-kDa α6β6 holoenzyme of 3-methylcrotonyl-CoA carboxylase. *Nature, 481*(7380), 219–223. https://doi.org/10.1038/nature10691.

Huang, C. S., Sadre-Bazzaz, K., Shen, Y., Deng, B., Zhou, Z. H., & Tong, L. (2010). Crystal structure of the alpha(6)beta(6) holoenzyme of propionyl-coenzyme A carboxylase. *Nature, 466*(7309), 1001–1005. https://doi.org/10.1038/nature09302.

Hunkeler, M., Hagmann, A., Stuttfeld, E., Chami, M., Guri, Y., Stahlberg, H., & Maier, T. (2018). Structural basis for regulation of human acetyl-CoA carboxylase. *Nature, 558*(7710), 470–474. https://doi.org/10.1038/s41586-018-0201-4.

Jitrapakdee, S., Maurice, M., St., Rayment, I., Cleland, W. W., Wallace, J. C., & Attwood, P. V. (2008). Structure, mechanism and regulation of pyruvate carboxylase. *The Biochemical Journal, 413*(3), 369–387. https://doi.org/10.1042/BJ20080709.

Jitrapakdee, S., Walker, M. E., & Wallace, J. C. (1999). Functional expression, purification, and characterization of recombinant human pyruvate carboxylase. *Biochemical and Biophysical Research Communications, 266*(2), 512–517. https://doi.org/10.1006/bbrc.1999.1846.

Ketchum, C. H., Robinson, A., Hall, L. M., & Grizzle, W. E. (1988). Lactate dehydrogenase isolated from human liver mitochondria: its purification and partial biochemical characterization. *Clinical Biochemistry, 21*(4), 231–237. https://doi.org/10.1016/S0009-9120(88)80006-7.

Kumashiro, N., Beddow, S. A., Vatner, D. F., Majumdar, S. K., Cantley, J. L., Guebre-Egziabher, F., ... Samuel, V. T. (2013). Targeting pyruvate carboxylase reduces gluconeogenesis and adiposity and improves insulin resistance. *Diabetes, 62*(7), 2183–2194. https://doi.org/10.2337/db12-1311.

Lau, E. P., Cochran, B. C., & Fall, R. R. (1980). Isolation of 3-methylcrotonyl-coenzyme A carboxylase from bovine kidney. *Archives of Biochemistry and Biophysics, 205*(2), 352–359. https://doi.org/10.1016/0003-9861(80)90117-4.

Li, K., Zhang, C., Chen, L., Wang, P., Fang, Y., Zhu, J., ... Liu, Y. (2019). The role of acetyl-coA carboxylase2 in head and neck squamous cell carcinoma. *PeerJ, 7*. https://doi.org/10.7717/peerj.7037.

Liang, Q., Li, Q., Chen, Z., Lv, L., Lin, Y., Jiang, H., ... Xu, Q. (2023). Anemoside B4, a new pyruvate carboxylase inhibitor, alleviates colitis by reprogramming macrophage function. *Inflammation Research*. https://doi.org/10.1007/s00011-023-01840-x.

Lin, Q., He, Y., Wang, X., Zhang, Y., Hu, M., Guo, W., ... Chen, Y. (2020). Targeting pyruvate carboxylase by a small molecule suppresses breast cancer progression. *Advanced Science (Weinheim, Baden-Wurttemberg, Germany), 7*(9), 1903483. https://doi.org/10.1002/advs.201903483.

Liu, X., Feng, X., Ding, Y., Gao, W., Xian, M., Wang, J., & Zhao, G. (2020). Characterization and directed evolution of propionyl-CoA carboxylase and its application in succinate biosynthetic pathway with two CO_2 fixation reactions. *Metabolic Engineering, 62*, 42–50. https://doi.org/10.1016/j.ymben.2020.08.012.

Liu, Y.-N., Liu, T.-T., Fan, Y.-L., Niu, D.-M., Chien, Y.-H., Chou, Y.-Y., ... Chiu, Y.-H. (2016). Measuring propionyl-CoA carboxylase activity in phytohemagglutinin stimulated lymphocytes using high performance liquid chromatography. *Clinica Chimica Acta, 453*, 13–20. https://doi.org/10.1016/j.cca.2015.11.023.

Liu, Y., Liu, C., Pan, Y., Zhou, J., Ju, H., & Zhang, Y. (2022). Pyruvate carboxylase promotes malignant transformation of papillary thyroid carcinoma and reduces iodine uptake. *Cell Death Discovery, 8*(1), 423. https://doi.org/10.1038/s41420-022-01214-y.

Livieri, A. L., Navone, L., Marcellin, E., Gramajo, H., & Rodriguez, E. (2019). A novel multidomain acyl-CoA carboxylase in Saccharopolyspora erythraea provides malonyl-CoA for de novo fatty acid biosynthesis. *Scientific Reports, 9*(1), 6725. https://doi.org/10.1038/s41598-019-43223-5.

López-Alonso, J. P., Lázaro, M., Gil-Cartón, D., Choi, P. H., Dodu, A., Tong, L., & Valle, M. (2022). CryoEM structural exploration of catalytically active enzyme pyruvate carboxylase. *Nature Communications, 13*(1), 6185. https://doi.org/10.1038/s41467-022-33987-2.

Lv, L., Li, Q., Wang, K., Zhao, J., Deng, K., Zhang, R., ... Xu, Q. (2024). Discovery of a new anti-inflammatory agent from anemoside B4 derivatives and its therapeutic effect on colitis by targeting pyruvate carboxylase. *Journal of Medicinal Chemistry, 67*(9), 7385–7405. https://doi.org/10.1021/acs.jmedchem.4c00222.

Mahan, D. E., Mushahwar, I. K., & Koeppe, R. E. (1975). Purification and properties of rat brain pyruvate carboxylase. *Biochemical Journal, 145*(1), 25–35. https://doi.org/10.1042/bj1450025.

Milgraum, L. Z., Witters, L. A., Pasternack, G. R., & Kuhajda, F. P. (1997). Enzymes of the fatty acid synthesis pathway are highly expressed in in situ breast carcinoma. *Clinical Cancer Research: An Official Journal of the American Association for Cancer Research, 3*(11), 2115–2120.

Moss, J., & Lane, M. D. (1971). The biotin-dependent enzymes. In In. A. Meister (Vol. Ed.), (1st ed.). *Advances in enzymology—And related areas of molecular biology: 35*, (pp. 321–442). Wiley. https://doi.org/10.1002/9780470122808.ch7.

Ngamkham, J., Siritutsoontorn, S., Saisomboon, S., Vaeteewoottacharn, K., & Jitrapakdee, S. (2022). CRISPR Cas9-mediated ablation of pyruvate carboxylase gene in colon cancer cell line HT-29 inhibits growth and migration, induces apoptosis and increases sensitivity to 5-fluorouracil and glutaminase inhibitor. *Frontiers in Oncology, 12*, 966089. https://doi.org/10.3389/fonc.2022.966089.

Ochoa, S., & Kaziro, Y. (1965). Chapter V—Carboxylases and the role of biotin. In M. Florkin, & E. H. Stotz (Vol. Eds.), *Comprehensive biochemistry: 16*, (pp. 210–249). Elsevier. https://doi.org/10.1016/B978-1-4831-9710-4.50012-X.

Oei, J., & Robinson, B. H. (1985). Simultaneous preparation of the three biotin-containing mitochondrial carylases from rat liver. *Biochimica et Biophysica Acta (BBA)—General Subjects, 840*(1), 1–5. https://doi.org/10.1016/0304-4165(85)90154-0.

Paventi, G., Pizzuto, R., & Passarella, S. (2017). The occurence of l-lactate dehydrogenase in the inner mitochondrial compartment of pig liver. *Biochemical and Biophysical Research Communications, 489*(2), 255–261. https://doi.org/10.1016/j.bbrc.2017.05.154.

Petrova, E., Scholz, A., Paul, J., Sturz, A., Haike, K., Siegel, F., ... Liu, N. (2016). Acetyl-CoA carboxylase inhibitors attenuate WNT and Hedgehog signaling and suppress pancreatic tumor growth. *Oncotarget, 8*(30), 48660–48670. https://doi.org/10.18632/oncotarget.12650.

Peuhkurinen, K. J. (1984). Regulation of the tricarboxylic acid cycle pool size in heart muscle. *Journal of Molecular and Cellular Cardiology, 16*(6), 487–495. https://doi.org/10.1016/s0022-2828(84)80637-9.

Phannasil, P., Ansari, I. H., El Azzouny, M., Longacre, M. J., Rattanapornsompong, K., Burant, C. F., ... Jitrapakdee, S. (2017). Mass spectrometry analysis shows the biosynthetic pathways supported by pyruvate carboxylase in highly invasive breast cancer cells. *Biochimica et Biophysica Acta (BBA)—Molecular Basis of Disease, 1863*(2), 537–551. https://doi.org/10.1016/j.bbadis.2016.11.021.

Phannasil, P., Thuwajit, C., Warnnissorn, M., Wallace, J. C., MacDonald, M. J., & Jitrapakdee, S. (2015). Pyruvate carboxylase is up-regulated in breast cancer and essential to support growth and invasion of MDA-MB-231 cells. *PLoS One, 10*(6), e0129848. https://doi.org/10.1371/journal.pone.0129848.

Santoro, N., Brtva, T., Roest, S. V., Siegel, K., & Waldrop, G. L. (2006). A high-throughput screening assay for the carboxyltransferase subunit of acetyl-CoA carboxylase. *Analytical Biochemistry, 354*(1), 70–77. https://doi.org/10.1016/j.ab.2006.04.006.

Schneider, N. O., Gilreath, K., Henriksen, N. M., Donaldson, W. A., Chaudhury, S., & St. Maurice, M. (2024). Synthesis and evaluation of 1,3-disubstituted imidazolidine-2,4,5-triones as inhibitors of pyruvate carboxylase. *ACS Medicinal Chemistry Letters, 15*(7), 1088–1093. https://doi.org/10.1021/acsmedchemlett.4c00183.

Schneider, N. O., Tassoulas, L. J., Zeng, D., Laseke, A. J., Reiter, N. J., Wackett, L. P., & Maurice, M. S. (2020). Solving the conundrum: Widespread proteins annotated for urea metabolism in bacteria are carboxyguanidine deiminases mediating nitrogen assimilation from guanidine. *Biochemistry, 59*(35), 3258–3270. https://doi.org/10.1021/acs.biochem.0c00537.

Schreurs, M., Kuipers, F., & Van Der Leij, F. R. (2010). Regulatory enzymes of mitochondrial β-oxidation as targets for treatment of the metabolic syndrome. *Obesity Reviews, 11*(5), 380–388. https://doi.org/10.1111/j.1467-789X.2009.00642.x.

Seethala, R., Ma, Z., Golla, R., & Cheng, D. (2006). A homogeneous scintillation proximity assay for acetyl coenzyme A carboxylase coupled to fatty acid synthase. *Analytical Biochemistry, 358*(2), 257–265. https://doi.org/10.1016/j.ab.2006.07.037.

Sheng, Y., Chen, Y., Zeng, Z., Wu, W., Wang, J., Ma, Y., ... Wang, F. (2021). Identification of pyruvate carboxylase as the cellular target of natural bibenzyls with potent anticancer activity against hepatocellular carcinoma via metabolic reprogramming. *Journal of Medicinal Chemistry*. https://doi.org/10.1021/acs.jmedchem.1c01605.

Shi, H., Yang, J., Qiao, Z., Li, L., Liu, G., Dai, Q., ... Ma, X. (2023). Design, synthesis and structure–activity relationship studies on erianin analogues as pyruvate carboxylase inhibitors in hepatocellular carcinoma cells. *Organic & Biomolecular Chemistry, 21*(34), 7005–7017. https://doi.org/10.1039/D3OB01114C.

St Maurice, M., Reinhardt, L., Surinya, K. H., Attwood, P. V., Wallace, J. C., Cleland, W. W., & Rayment, I. (2007). Domain architecture of pyruvate carboxylase, a biotin-dependent multifunctional enzyme. *Science (New York, N. Y.), 317*(5841), 1076–1079. https://doi.org/10.1126/science.1144504.

Thampy, K. G., Huang, W.-Y., & Wakil, S. J. (1988). A rapid purification method for rat liver pyruvate carboxylase and amino acid sequence analyses of NH2-terminal and biotin peptide. *Archives of Biochemistry and Biophysics, 266*(1), 270–276. https://doi.org/10.1016/0003-9861(88)90258-5.

Tong, L. (2013). Structure and function of biotin-dependent carboxylases. *Cellular and Molecular Life Sciences: CMLS, 70*(5), 863–891. https://doi.org/10.1007/s00018-012-1096-0.

Tong, L. (2017). Chapter Five—Striking diversity in holoenzyme architecture and extensive conformational variability in biotin-dependent carboxylases. In T. Karabencheva-Christova (Vol. Ed.), *Advances in protein chemistry and structural biology: 109*, (pp. 161–194). Academic Press. https://doi.org/10.1016/bs.apcsb.2017.04.006.

Utter, M. F., & Keech, D. B. (1963). Pyruvate carboxylase. *Journal of Biological Chemistry, 238*(8), 2603–2608. https://doi.org/10.1016/S0021-9258(18)67873-1.

Wakil, S. J., Stoops, J. K., & Joshi, V. C. (1983). Fatty acid synthesis and its regulation. *Annual Review of Biochemistry, 52*(1), 537–579. https://doi.org/10.1146/annurev.bi.52.070183.002541.

Waldrop, G. L., Holden, H. M., & St. Maurice, M. S. (2012). The enzymes of biotin dependent CO_2 metabolism: What structures reveal about their reaction mechanisms. *Protein Science: A Publication of the Protein Society, 21*(11), 1597–1619. https://doi.org/10.1002/pro.2156.

Wallace, J. C., Jitrapakdee, S., & Chapman-Smith, A. (1998). Pyruvate carboxylase. *The International Journal of Biochemistry & Cell Biology, 30*(1), 1–5. https://doi.org/10.1016/S1357-2725(97)00147-7.

Wang, M.-D., Wu, H., Fu, G.-B., Zhang, H.-L., Zhou, X., Tang, L., ... Wang, H.-Y. (2016). Acetyl-coenzyme A carboxylase alpha promotion of glucose-mediated fatty acid synthesis enhances survival of hepatocellular carcinoma in mice and patients. *Hepatology (Baltimore, Md.), 63*(4), 1272–1286. https://doi.org/10.1002/hep.28415.

Wang, Y., Yu, W., Li, S., Guo, D., He, J., & Wang, Y. (2022). Acetyl-CoA carboxylases and diseases. *Frontiers in Oncology, 12*, 836058. https://doi.org/10.3389/fonc.2022.836058.

Warren, G. B., & Tipton, K. F. (1974). Pig liver pyruvate carboxylase. Purification, properties and cation specificity. *Biochemical Journal, 139*(2), 297–310. https://doi.org/10.1042/bj1390297.

Wei, J., Zhang, Y., Yu, T.-Y., Sadre-Bazzaz, K., Rudolph, M. J., Amodeo, G. A., ... Tong, L. (2016). A unified molecular mechanism for the regulation of acetyl-CoA carboxylase by phosphorylation. *Cell Discovery, 2*, 16044. https://doi.org/10.1038/celldisc.2016.44.

Willis, L. B., Omar, W. S. W., Sambanthamurthi, R., & Sinskey, A. J. (2008). *Non-radioactive assay for acetyl-CoA carboxylase activity, 2*.

Wongkittichote, P., Mew, N. A., & Chapman, K. A. (2017). Propionyl-CoA carboxylase—A review. *Molecular Genetics and Metabolism, 122*(4), 145–152. https://doi.org/10.1016/j.ymgme.2017.10.002.

Wu, X., & Huang, T. (2020). Recent development in Acetyl-CoA carboxylase inhibitors and their potential as novel drugs. *Future Medicinal Chemistry, 12*(6), 533–561. https://doi.org/10.4155/fmc-2019-0312.

Xiang, S., & Tong, L. (2008). Crystal structures of human and Staphylococcus aureus pyruvate carboxylase and molecular insights into the carboxyltransfer reaction. *Nature Structural & Molecular Biology, 15*(3), 295–302. https://doi.org/10.1038/nsmb.1393.

Xu, J., Han, J., Long, Y. S., Epstein, P. N., & Liu, Y. Q. (2008). The role of pyruvate carboxylase in insulin secretion and proliferation in rat pancreatic beta cells. *Diabetologia, 51*(11), 2022–2030. https://doi.org/10.1007/s00125-008-1130-9.

Yang, J. H., Kim, N. H., Yun, J. S., Cho, E. S., Cha, Y. H., Cho, S. B., ... Yook, J. I. (2020). Snail augments fatty acid oxidation by suppression of mitochondrial ACC2 during cancer progression. *Life Science Alliance, 3*(7), https://doi.org/10.26508/lsa.202000683.

Zempleni, J., Trusty, T. A., & Mock, D. M. (1997). Lipoic acid reduces the activities of biotin-dependent carboxylases in rat liver. *The Journal of Nutrition, 127*(9), 1776–1781. https://doi.org/10.1093/jn/127.9.1776.

Zhou, F., Zhang, Y., Zhu, Y., Zhou, Q., Shi, Y., & Hu, Q. (2024a). Filament structures unveil the dynamic organization of human acetyl-CoA carboxylase (p. 2024.06.16. 599241). bioRxiv. https://doi.org/10.1101/2024.06.16.599241.

Zhou, F., Zhang, Y., Zhu, Y., Zhou, Q., Shi, Y., & Hu, Q. (2024b). Structural insights into human propionyl-CoA carboxylase (PCC) and 3-methylcrotonyl-CoA carboxylase (MCC). *eLife, 13*. https://doi.org/10.7554/eLife.98885.1.

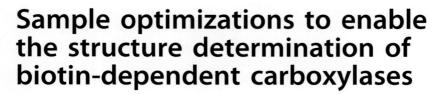

CHAPTER TWO

Sample optimizations to enable the structure determination of biotin-dependent carboxylases

Jia Wei, Christine S. Huang, Yang Shen, Kianoush Sadre-Bazzaz, and Liang Tong[*,1]

Department of Biological Sciences, Columbia University, New York, NY, United States
*Corresponding author. e-mail address: ltong@columbia.edu

Contents

1. Introduction	32
2. Sample optimizations for structural studies of yeast ACC	34
2.1 Identifying a critical buffer component for crystallization and EM studies of yeast ACC	34
2.2 Using divide-and-conquer approach to study yeast ACC central domains – a cautionary tale	36
2.3 Preparation of phosphorylated ACC for structural studies	37
3. Sample optimization for crystal structure determination of PCC	38
4. Homologous domains are assembled to give highly distinct overall architectures of the biotin-dependent carboxylases	40
5. Summary	40
Acknowledgments	41
References	41

Abstract

Biotin-dependent carboxylases have central roles in the metabolisms of fatty acids, amino acids and other compounds. Their functional importance is underscored by their strong conservation from bacteria to humans. These enzymes are large, multi-domain or multi-subunit complexes, and can have molecular weights of 500 to 750 kDa. Despite their large sizes, the first structures of most of these enzymes were determined using X-ray crystallography. This chapter presents various technical challenges that were overcome during their structure determination, which involves extensive optimization of the protein samples and their crystals. The cryo electron microscopy resolution revolution has made it easier to study these large complexes at the atomic level.

[1] Present addresses: Structural and Protein Sciences, Johnson & Johnson Innovative Medicine, Spring House, PA (J.W.); Department of Protein Chemistry, Genentech Inc., South San Francisco, CA (C.S.H.); Department of Bispecifics, Regeneron Pharmaceuticals, Tarrytown, NY (Y.S.); ARUP Laboratories, University of Utah, Salt Lake City, UT (K.S-B.).

1. Introduction

Biotin-dependent carboxylases are central metabolic enzymes (Tong, 2013). Perhaps the best-known metabolic role is in fatty acid biosynthesis, which is mediated by acetyl-CoA carboxylase (ACC) (Wakil & Abu-Elheiga, 2009). ACC catalyzes the carboxylation of the acetyl group in acetyl-CoA, converting it to malonyl-CoA, which provides the building blocks for the synthesis of long-chain fatty acids. In humans and other mammals, a second isoform of ACC (ACC2) has a crucial role in fatty acid oxidation, where the malonyl-CoA product inhibits the transport of long-chain acyl-CoAs into the mitochondria for oxidation. Biotin-dependent carboxylases that are active toward other CoA esters also exist, such as propionyl-CoA carboxylase (PCC), 3-methylcrontonyl-CoA carboxylase (MCC), geranyl-CoA carboxylase (GCC), and long-chain acyl-CoA carboxylase (LCC) (Tran et al., 2015). PCC and MCC are critical for the metabolism of selected amino acids and several other compounds, and mutations in these two enzymes lead to serious metabolic diseases such as propionic acidemia. Biotin-dependent carboxylases can also act on simple organic compounds, such as pyruvate carboxylase (PC) and urea carboxylase (UC). Mutations in PC are also linked to metabolic diseases.

Biotin-dependent carboxylases share the common prosthetic group biotin, linked covalently to a biotin carboxyl carrier protein (BCCP) through an amide bond with a conserved lysine side chain. Biotin is the carrier of an activated CO_2, mediating its transfer from bicarbonate to the acceptor molecule. These enzymes therefore contain two distinct catalytic activities (Knowles, 1989), with the biotin carboxylase (BC) catalyzing the carboxylation of biotin with bicarbonate as the CO_2 donor, a process that also requires ATP hydrolysis. The carboxybiotin then translocates to the active site of the carboxyltransferase (CT), where it is transferred to the acceptor molecule.

The functional importance of these enzymes is underscored by the fact that many of them are found in all domains of life, from bacteria to humans. In some organisms, the BC, CT and BCCP activities exist as separate protein subunits, while in other organisms they may be fused together. For example, E. coli ACC contains four subunits, BC (50 kDa), BCCP (17 kDa), and two subunits for the CT activity (70 kDa total for the two subunits) (Cronan & Waldrop, 2002). In comparison, eukaryotic ACC is a single-chain, multi-domain enzyme with molecular weight of ∼250 kDa, with the BC domain close to the N terminus and the two CT

Sample optimizations to enable the structure determination of biotin-dependent carboxylases 33

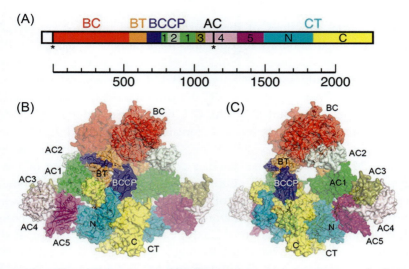

Fig. 1 Overall structure of yeast ACC dimer. (A) Domain organizations of eukaryotic ACCs. The domains are given different colors and labeled. A scale bar for the number of amino acid residues is shown at the bottom. The two phosphorylation sites that inhibit ACC are indicated with the black lines and the asterisks, one just prior to the BC domain and the other in the middle of the protein. Yeast ACC contains only the latter phosphorylation site. BC: biotin carboxylase; BT: BC-CT interaction; BCCP: biotin carboxyl carrier protein; AC: ACC central; CT: carboxyltransferase. The ACC central region has 5 domains, named AC1 through AC5. (B) Schematic drawing of the yeast ACC dimer structure. The 10 domains in each monomer are colored as in panel A and labeled. The BC domain dimer is located at the top (red). The CT domain has two subdomains, N and C. AC: ACC central. (C) Structure of the yeast ACC dimer after a 45° rotation around the vertical axis from panel A, showing a hole in the center of the structure. Dissociation of the BC domain dimer at the top leads to the linear and bent conformations of the yeast ACC.

domains, homologous to the two CT subunits of *E. coli* ACC, at the C terminus (Fig. 1A). Moreover, biotin-dependent carboxylases function as large oligomers. For example, yeast ACC is a dimer of 500 kDa, while human ACC dimers can form polymer filaments. PCC, MCC and GCC are 750 kDa, $\alpha_6\beta_6$ dodecamers, with the α subunit containing the BC and BCCP activities and the β subunit the CT activity. PC is a 500 kDa homotetramer in most organisms (Jitrapakdee et al., 2008), although two-subunit PCs are found in some bacteria (Choi et al., 2016).

The large sizes of these enzymes make them very challenging to study by crystallographic methods, although they are now well suited for studies by cryo electron microscopy (cryo-EM). Using the divide and conquer approach, initial structures of these enzymes were obtained by crystallography for their subunits or domains, for example *E. coli* ACC BC subunit (Waldrop, Rayment, & Holden, 1994), *E. coli* ACC BCCP subunit (Athappilly & Hendrickson, 1995), yeast ACC CT domain (Zhang, Yang, Shen, & Tong, 2003), yeast ACC BC domain (Shen, Volrath, Weatherly, Elich, & Tong, 2004), and *E. coli* ACC CT subunits (Bilder et al., 2006). The structures of these subunits and domains have provided great molecular insights into the enzymes (Tong, 2013; Waldrop, Holden, & St. Maurice, 2012). On the other hand, a complete understanding of these enzymes requires structural information on the entire oligomer. Remarkably, the first structures of many of these large enzymes were determined by crystallography, before the resolution revolution in cryo-EM, including the PCs (Choi et al., 2016; Choi et al., 2017; Lietzan et al., 2011; St. Maurice et al., 2007; Sureka et al., 2014; Xiang & Tong, 2008), bacterial PCC (Huang et al., 2010), bacterial MCC (Huang, Ge, Zhou, & Tong, 2012), yeast UC (Fan, Chou, Tong, & Xiang, 2012), yeast ACC (Wei & Tong, 2015), bacterial LCC (Tran et al., 2015), and bacterial GCC (Jurado, Huang, Zhang, Zhou, & Tong, 2015). The technological advancements in X-ray crystallography were important for the successful data collection on crystals of such large complexes. For example, crystals of yeast ACC have a unit cell dimension of 614 Å along the *c* axis (space group *P*4$_3$), and highly focused beams and large detectors at synchrotron beamlines allowed the diffraction spots to be resolved for data collection. More importantly, sample optimizations were critical for the production of crystals that were of sufficient quality for structure determination, and some of these optimization protocols are described in this chapter. These protocols, and some of the observations on these large enzymes, should be relevant for studying their structures by the cryo-EM technique as well as for understanding protein structures in general.

2. Sample optimizations for structural studies of yeast ACC

2.1 Identifying a critical buffer component for crystallization and EM studies of yeast ACC

We initially attempted to express full-length yeast ACC (2233 residues) in *E. coli* using the pET system (Novagen), with the strong T7 promoter, but found that the recombinant protein was only in the insoluble fraction.

We reasoned that this might be due to *E. coli* having difficulty folding such a large protein, and it might be advantageous to use a weaker promoter. This was prompted by our experience with expressing human mitochondrial NAD^+-dependent malic enzyme in *E. coli*. Using the pET system only gave us insoluble protein, while using the weaker *trp* promoter gave us a large amount of soluble protein (Loeber, Infante, Maurer-Fogy, Krystek, & Dworkin, 1991; Yang, Lanks, & Tong, 2002). We therefore tried to express full-length yeast ACC using the *trp* promoter, and were able to purify a small amount of soluble protein (Wei & Tong, 2015).

We successfully produced crystals with the purified full-length yeast ACC sample, but the diffraction quality of the crystals was poor, extending only to around 8 Å resolution. We then optimized the sample by removing some flexible residues at the N terminus of the protein, prior to the BC domain (Fig. 1A), and found that a sample lacking the first 21 residues produced crystals that diffracted much better. This ultimately led to a structure of yeast ACC 500 kDa homodimer at 3.1 Å resolution (Wei & Tong, 2015) (Fig. 1B and C).

The structure determination of this enzyme by crystallography occurred near the inception of the resolution revolution in electron microscopy. We examined the purified sample of yeast ACC by negative stain EM, and to our surprise, we found a highly conformationally heterogeneous sample. The particles assume linear or bent conformation in the images (Wei et al., 2016), highly distinct from the conformation in the crystal structure, which has the overall shape of a quarter of a disk (Fig. 1B). Such a sample would not be of sufficient quality for EM studies. In fact, it is highly unlikely that such a heterogeneous sample can be crystallized. The linear or bent conformation arises due to the dissociation of the relatively unstable BC domain dimers into monomers, while the CT domain dimer is highly stable and maintains the overall dimer of the yeast ACC (Fig. 1C).

Noting that this was exactly the same sample that produced crystals diffracting to 3.1 Å resolution, we wondered what could be the cause of the distinct conformations observed in negative stain EM. We hypothesized that the crystallization buffer may have stabilized the conformation of the protein, allowing good quality crystals to form. We then examined the yeast ACC sample in the crystallization buffer under negative stain EM, and satisfyingly found a much more conformationally homogeneous sample, with the particles assuming a shape that is similar to what we observed in the crystal structure. The crystallization buffer contained 14% (w/v) PEG3350, 4% (v/v) *tert*-butanol, and 0.2 M sodium citrate (Wei & Tong, 2015). We then examined the effect of each of these compounds on the conformation of yeast ACC by negative stain EM, and found that citrate was the critical component. Using a sample in the presence of only citrate as an additive, we were able to produce a 3D reconstruction of yeast ACC from negative stain images, and the crystal structure could be readily docked into the reconstruction (Wei et al., 2016).

Remarkably, citrate is an activator of human ACCs, although it is not known to have an effect on the catalytic activity of yeast ACC. Citrate is converted to acetyl-CoA, the substrate of ACC, by the enzyme ATP-citrate lyase (ACLY), and therefore is an upstream compound in this metabolic pathway (feed forward activation). Moreover, citrate was in the crystallization buffer of yeast CT domain (Zhang et al., 2003) and yeast ACC central (AC) domains (Wei & Tong, 2015) as well. However, we did not observe citrate in any of these crystal structures, suggesting that it might have a nonspecific effect on the structure of yeast ACC and its domains.

The yeast ACC sample produced a sharp peak on the gel filtration column, but our EM experiments showed that it is still conformationally highly heterogeneous. Such a sample would not be of sufficient quality for structure determination by EM or crystallography. Fortunately, we were able to identify citrate as a compound that can stabilize the conformation of yeast ACC through crystallization screening and follow up EM studies. This may have general implications for studying other samples. A sample that appears to be heterogeneous in EM studies could be stabilized by compound(s), which would facilitate or enable their structure determination. Moreover, crystallization screening could be a method for examining a large collection of buffer conditions and identifying such compound(s), in finding conditions that can give (micro) crystals (or possibly some ordered aggregates as visualized by a UV microscope) (Wei et al., 2016). Other methods of screening for compounds or buffer conditions may also be possible, for example thermal shift assays.

2.2 Using divide-and-conquer approach to study yeast ACC central domains – a cautionary tale

Eukaryotic ACCs are large, multi-domain enzymes (Fig. 1A), while bacterial ACCs are multi-subunit. The divide-and-conquer approach was highly successful in determining the structures of various subunits and domains of these enzymes. Between the BCCP and CT domains, in the central region of eukaryotic ACCs, there is a large segment, with ∼700 amino acid residues, that bears no clear sequence homology to other proteins. Therefore, this segment appears to be unique to eukaryotic ACCs, and its structure was not known.

Through much trial and error, we were able to express, purify and crystallize several segments of the central region of yeast ACC (Wei & Tong, 2015). These segments are all dimeric in the crystal, with extensive interfaces between the two monomers. Although some of them are monomeric in solution, the sample containing the entire central region is dimeric in solution as well. This appears to be consistent with the observation that ACCs are only active as dimers, and the dimers of the central region could possibly represent their structures in the full-length enzyme.

However, after we determined the structure of the full-length enzyme, it became clear that the ACC central segment is actually monomeric in it, as a scaffold holding the BC dimer and CT dimer together (Wei & Tong, 2015) (Fig. 1B and C). Therefore, the dimers of the ACC central domains observed when they are in the crystals on their own appear to be artifacts, because they had been cut away from the other domains of ACC. This serves as a cautionary tale for the divide-and-conquer approach, in that observations from the fragments may not always correspond to the conformation in the full-length protein.

Human ACCs are known to form polymers, and it was shown that the polymer interface primarily involves the ACC central domains (Hunkeler et al., 2018). Therefore, this region of ACC can be involved in homotypic interactions. On the other hand, yeast ACC is not known to form polymers. A polymer in the crystal, due to the presence of a 4_3 screw axis, involves contacts between the AC3–AC5 domains (Fig. 1A) of neighboring molecules. This is reminiscent of the AC3-AC5 domain dimer, but is different in many details.

2.3 Preparation of phosphorylated ACC for structural studies

Eukaryotic ACCs are inhibited by phosphorylation, at a site just before the BC domain (in animal ACCs, Fig. 1A) and another site in the central region. However, the site just before the BC domain is not present in yeast ACC, and the phosphorylation site in the central region was believed to be Ser1157 (Hunkeler, Stuttfeld, Hagmann, Imseng, & Maier, 2016), but the mechanism of this regulation was not clear.

We expressed and purified the yeast SNF1 protein kinase heterotrimer that phosphorylates ACC (Amodeo, Rudolph, & Tong, 2007) as well as the constitutively active upstream kinase Tos3 that activates SNF1 (Wei et al., 2016). We used these kinases to phosphorylate a purified sample of domains AC3-AC5 of yeast ACC *in vitro*. The phosphorylation reactions were carried out at room temperature, and contained 4 μM ACC, 0.8 μM SNF1, 0.2 μM Tos3, 2 mM ATP and 5 mM $MgCl_2$. We observed a shift of the protein band in the SDS gel, indicating complete phosphorylation. We then incubated a solution of domains AC3-AC5 at 2.9 mg/ml concentration with the protein kinases for 20 min, at a molar ratio of 225:5:1 for AC3-AC5:SNF1:Tos3, and set up the mixture for crystallization. The crystallographic analysis revealed clear electron density for the phosphate group on Ser1157 (Wei et al., 2016). Moreover, we used SNF1 to phosphorylate full-length yeast ACC, and examined the sample by negative stain EM. We observed the inactive linear and bent conformations even in the presence of citrate. Therefore, these studies demonstrate that SNF1 can directly phosphorylate Ser1157 in the central region, and this phosphorylation leads to a reorganization of the yeast ACC dimer and the inhibition of its catalytic activity.

3. Sample optimization for crystal structure determination of PCC

We were able to express human PCC (HsPCC) in *E. coli* by constructing a bicistronic plasmid for its two subunits with the pET system, and produced crystals (Huang et al., 2010). They have the shape of a large portion of a disk, with a round edge. They diffracted X-rays poorly, and a data set could be collected to 5.5 Å resolution only. The apparent space group is *R*32, but the intensity statistics suggested extensive twinning. Efforts at solving this structure by molecular replacement were not successful, due to the twinning and the low resolution.

We then turned to bacterial homologs of HsPCC, and after screening through several different organisms, we were able to express, purify and crystallize the enzyme from *Ruegeria pomeroyi* (RpPCC). RpPCC shares 54% and 65% sequence identity with the α and β subunits of HsPCC, respectively. The crystals diffracted X-rays better, and a data set to 3.3 Å resolution was collected. They were in space group *P*3, but exhibited perfect hemihedral twinning. We were able to solve the structure by molecular replacement using the structures of BC and CT domains as the model, but we were not able to observe additional electron density for other parts of the protein, likely due to the twinning. Therefore, it is important to break the twining habit of these crystals.

Sample optimizations to enable the structure determination of biotin-dependent carboxylases

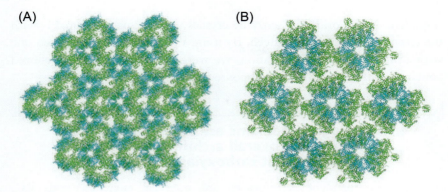

Fig. 2 Crystal packing of PCC. (A) Crystal packing of RpPCC, viewed down the 3-fold axis of the enzyme, which is also the crystallographic 3-fold axis. The α subunits are colored in green, and β subunits in cyan. (B) Crystal packing of RpPCCα-RdPCCβ hybrid, also viewed down the 3-fold axis of the enzyme.

We reasoned that if we could change the crystal packing, it might break the twinning. In the RpPCC crystal, the 3-fold symmetry of the PCC enzyme coincides with the crystallographic 3-fold axis, and there are three $\alpha_2\beta_2$ heterodimers in the asymmetric unit, giving a rather distinct packing pattern (Fig. 2A). The packing interactions are mediated by contacts between the α subunits as well as between the α and β subunits of neighboring dodecamers. In order to disrupt this packing arrangement, we came up with the idea of using a hybrid PCC enzyme, with the two subunits from different organisms, since we had already screened through PCCs where both subunits came from the same organism. We found that crystals of a PCC containing the α subunit of RpPCC and the β subunit of *Roseobacter denitrificans* PCC (RdPCC) were no longer twinned (space group *P*1). Glycerol (16 % (v/v)) was included in the crystallization buffer to slow down crystal growth. Most of the crystals diffracted X-rays poorly and were highly mosaic, and many crystals were screened to find ones of sufficient quality for data collection (which was also the case for RpPCC crystals). We were able to solve the structure by molecular replacement, and readily located the electron density for the BT (BC-CT interaction) domain in the α subunit (Huang et al., 2010). This was the first time that the BT domain was recognized in the structures of biotin-dependent carboxylases.

The RdPCC β subunit shares 88 % sequence identity with that of RpPCC. Despite this high sequence conservation, the packing of the hybrid PCC enzyme crystal is entirely different from that of RpPCC

(Fig. 2B). The hybrid PCC enzyme crystals were in space group P1, with the entire enzyme in the unit cell. By using a different β subunit, even one with a high sequence conservation, we were able to change the preferred contacts among the subunits and consequently the crystal packing.

4. Homologous domains are assembled to give highly distinct overall architectures of the biotin-dependent carboxylases

Biotin-dependent carboxylases are modular enzymes, with BC, BCCP, CT, and BT (or PT for pyruvate carboxylase) domains. The domains among these enzymes share high sequence conservation. Therefore, it would have been expected that the enzymes should share similar three-dimensional structures. Remarkably, the overall architectures of these enzymes are strikingly different (Tong, 2017). The homology among these enzymes only ensures that their domains have similar structures, but the domains are assembled in distinct fashions to give different overall architectures. This is analogous to using the same building blocks to build entirely different buildings. These observations with the biotin-dependent carboxylases should have general implications for other proteins as well.

The assembly of the β subunits of PCC and MCC represents another example of the dramatic structural difference seen in these enzymes. The β subunits of both enzymes form $β_6$ hexamers, and their overall architecture is rather similar. The hexamers have a 3-fold symmetry, and pseudo 6-fold symmetry due to the homology between the two domains (N and C domains) in each subunit. However, the positions of the N and C domains in each subunit are swapped in MCC compared to PCC (Huang et al., 2012). The linker between the two domains runs in the opposite direction in the two enzymes and connects different domains in this pseudo hexamer. This again highlights the remarkable diversity in how the biotin-dependent carboxylases are assembled from their homologous domains.

5. Summary

With the resolution revolution, cryo-EM is now the preferred technique for determining structures of such large complexes as the biotin-dependent carboxylases. In fact, cryo-EM has already been used to show the filament structure of human ACC (Hunkeler et al., 2018), domain movements in the catalysis by PC (López–Alonso et al., 2022), the assembly pathway for PCC (Lee et al., 2023), a filament structure of *Leishmania tarentolae* MCC (Hu et al., 2023), and a structure of trypanosome MCC (Plaza-Pegueroles, Aphasizheva, Aphasizhev, Fernandez-Tornero, & Ruiz, 2024). These enzymes can be readily purified with avidin affinity resins, taking advantage of their biotin prosthetic group. Therefore, it can be expected that many additional structures of biotin-dependent carboxylases determined by cryo-EM will be reported in the coming years. Nonetheless, the experience and lessons learned from determining the structures of these large enzymes using crystallography would be beneficial for studying these enzymes by cryo-EM, and for studying other proteins in general.

Acknowledgments

We thank Hailong Zhang, Zhiru Yang, Song Xiang, Chi-Yuan Chou, Linda P.C. Yu, Philip Choi, Timothy H. Tran, Yu-Shan Hsiao, Ben Tweel, Ashley Paulson, Jiang Li, and Martin Bush for carrying out structural studies on biotin-dependent carboxylases in the laboratory of LT over the years; Yixiao Zhang and Thomas Walz for help with the EM studies on yeast ACC; Binbin Deng, Peng Ge and Z. Hong Zhou for the EM studies on PCC and MCC. This research was supported by NIH grants R01DK67238 and R35GM118093 (to LT).

References

Amodeo, G. A., Rudolph, M. J., & Tong, L. (2007). Crystal structure of the heterotrimer core of Sacharyomyces cerevisiae AMPK homolog SNF1. *Nature, 449*, 492–495.

Athappilly, F. K., & Hendrickson, W. A. (1995). Structure of the biotinyl domain of acetyl-coenzyme A carboxylase determined by MAD phasing. *Structure (London, England: 1993), 3*, 1407–1419.

Bilder, P., Lightle, S., Bainbridge, G., Ohren, J., Finzel, B., Sun, F., ... Waldrop, G. L. (2006). The structure of the carboxyltransferase component of acetyl-CoA carboxylase reveals a zinc-binding motif unique to the bacterial enzyme. *Biochem, 45*, 1712–1722.

Choi, P. H., Jo, J., Lin, Y. C., Lin, M. H., Chou, C.-Y., Dietrich, L. E. P., & Tong, L. (2016). A distinct holoenzyme organization for two-subunit pyruvate carboxylase. *Nature Communications, 7*, 12713.

Choi, P. H., Vu, T. M. N., Pham, H. T., Woodward, J. J., Turner, M. S., & Tong, L. (2017). Structural and functional studies of pyruvate carboxylase regulation by cyclic-di-AMP in lactic acid bacteria. *Proceedings of the National Academy of Sciences of the United States of America, 114*, E7226–E7235.

Cronan, J. E., Jr., & Waldrop, G. L. (2002). Multi-subunit acetyl-CoA carboxylases. *Progress in Lipid Research, 41*, 407–435.

Fan, C., Chou, C.-Y., Tong, L., & Xiang, S. (2012). Crystal structure of urea carboxylase provides insights into the carboxyltransfer reaction. *The Journal of Biological Chemistry, 287*, 9389–9398.

Hu, J. J., Lee, J. K. J., Liu, Y. T., Yu, C., Huang, L., Aphasizheva, I., ... Zhou, Z. H. (2023). Discovery, structure, and function of filamentous 3-methylcrotonyl-CoA carboxylase. *Structure (London, England: 1993), 31*(1), 100–110.e104.

Huang, C. S., Ge, P., Zhou, Z. H., & Tong, L. (2012). An unanticipated architecture of the 750-kDa a6b6 holoezyme of 3-methylcrotonyl-CoA carboxylase. *Nature, 481*, 219–223.

Huang, C. S., Sadre-Bazzaz, K., Shen, Y., Deng, B., Zhou, Z. H., & Tong, L. (2010). Crystal structure of the a6b6 holoenzyme of propionyl-coenzyme A carboxylase. *Nature, 466*, 1001–1005.

Hunkeler, M., Hagmann, A., Stuttfeld, E., Chami, M., Guri, Y., Stahlberg, H., & Maier, T. (2018). Structural basis for regulation of human acetyl-CoA carboxylase. *Nature, 558*, 470–474.

Hunkeler, M., Stuttfeld, E., Hagmann, A., Imseng, S., & Maier, T. (2016). The dynamic organization of fungal acetyl-CoA carboxylase. *Nature Communications, 7*, 11196.

Jitrapakdee, S., St. Maurice, M., Rayment, I., Cleland, W. W., Wallace, J. C., & Attwood, P. V. (2008). Structure, mechanism and regulation of pyruvate carboxylase. *The Biochemical Journal, 413*, 369–387.

Jurado, A. R., Huang, C. S., Zhang, X., Zhou, Z. H., & Tong, L. (2015). Structure and substrate selectivity of the 750-kDa a6b6 holoenzyme of geranyl-CoA carboxylase. *Nature Communications, 6*, 8986.

Knowles, J. R. (1989). The mechanism of biotin-dependent enzymes. *Annual Review of Biochemistry, 58*, 195–221.

Lee, J. K. J., Liu, Y. T., Hu, J. J., Aphasizheva, I., Aphasizhev, R., & Zhou, Z. H. (2023). CryoEM reveals oligomeric isomers of a multienzyme complex and assembly mechanics. *Journal of Structural Biology: X, 7*, 100088.

Lietzan, A. D., Menefee, A. L., Zeczycki, T. N., Kumar, S., Attwood, P. V., Wallace, J. C., ... St. Maurice, M. (2011). Interaction between the biotin carboxyl carrier domain and the biotin carboxylase domain in pyruvate carboxylase from Rhizobium etli. *Biochem, 50*, 9708–9723.

Loeber, G., Infante, A. A., Maurer-Fogy, I., Krystek, E., & Dworkin, M. B. (1991). Human NAD^+-dependent mitochondrial malic enzyme. *The Journal of Biological Chemistry, 266*, 3016–3021.

López-Alonso, J. P., Lázaro, M., Gil-Cartón, D., Choi, P. H., Dodu, A., Tong, L., & Valle, M. (2022). CryoEM structural exploration of catalytically active enzyme pyruvate carboxylase. *Nature Communications, 13*(1), 6185.

Plaza-Pegueroles, A., Aphasizheva, I., Aphasizhev, R., Fernandez-Tornero, C., & Ruiz, F. M. (2024). The cryo-EM structure of trypanosome 3-methylcrotonyl-CoA carboxylase provides mechanistic and dynamic insights into its enzymatic function. *Structure (London, England: 1993)*.

Shen, Y., Volrath, S. L., Weatherly, S. C., Elich, T. D., & Tong, L. (2004). A mechanism for the potent inhibition of eukaryotic acetyl-coenzyme A carboxylase by soraphen A, a macrocyclic polyketide natural product. *Molecular Cell, 16*, 881–891.

St. Maurice, M., Reinhardt, L., Surinya, K. H., Attwood, P. V., Wallace, J. C., Cleland, W. W., & Rayment, I. (2007). Domain architecture of pyruvate carboxylase, a biotin-dependent multifunctional enzyme. *Science (New York, N. Y.), 317*, 1076–1079.

Sureka, K., Choi, P. H., Precit, M., Delince, M., Pensinger, D. A., Huynh, T. N., ... Woodward, J. J. (2014). The cyclic dinucleotide c-di-AMP is an allosteric regulator of metabolic enzyme function. *Cell, 158*, 1389–1401.

Tong, L. (2013). Structure and function of biotin-dependent carboxylases. *Cellular and Molecular Life Sciences: CMLS, 70*, 863–891.

Tong, L. (2017). Striking diversity in holoenzyme architecture and extensive conformational variability in biotin-dependent carboxylases. *Advances in Protein Chemistry and Structural Biology, 109*, 161–194.

Tran, T. H., Hsiao, Y.-S., Jo, J., Chou, C.-Y., Dietrich, L. E. P., Walz, T., & Tong, L. (2015). Structure and function of a single-chain, multi-domain long-chain acyl-CoA carboxylase. *Nature, 518*, 120–124.

Wakil, S. J., & Abu-Elheiga, L. A. (2009). Fatty acid metabolism: Target for metabolic syndrome. *Journal of Lipid Research, 50,* S138–S143.

Waldrop, G. L., Holden, H. M., & St. Maurice, M. (2012). The enzymes of biotin dependent CO_2 metabolism: What structures reveal about their reaction mechanisms. *Prot. Sci. 21,* 1597–1619.

Waldrop, G. L., Rayment, I., & Holden, H. M. (1994). Three-dimensional structure of the biotin carboxylase subunit of acetyl-CoA carboxylase. *Biochem, 33,* 10249–10256.

Wei, J., & Tong, L. (2015). Crystal structure of the 500-kDa yeast acetyl-CoA carboxylase holoenzyme dimer. *Nature, 526,* 723–727.

Wei, J., Zhang, Y., Yu, T.-Y., Sadre-Bazzaz, K., Rudolph, M. J., Amodeo, G. A., ... Tong, L. (2016). A unified molecular mechanism for the regulation of acetyl-CoA carboxylase by phosphorylation. *Cell Disc, 2,* 16044.

Xiang, S., & Tong, L. (2008). Crystal structures of human and Staphylococcus aureus pyruvate carboxylase and molecular insights into the carboxyltransfer reaction. *Nature Structural & Molecular Biology, 15,* 295–302.

Yang, Z., Lanks, C. W., & Tong, L. (2002). Molecular mechanism for the regulation of human mitochondrial NAD-dependent malic enzyme by fumarate and ATP. *Structure (London, England: 1993), 10,* 951–960.

Zhang, H., Yang, Z., Shen, Y., & Tong, L. (2003). Crystal structure of the carboxyltransferase domain of acetyl-coenzyme A carboxylase. *Science (New York, N. Y.), 299,* 2064–2067.

CHAPTER THREE

Purification of heteromeric acetyl-CoA carboxylases from *Escherichia coli* for structure solution

Amanda Silva de Sousa and Jeremy R. Lohman[*]

Department of Biochemistry and Molecular Biology, Michigan State University, East Lansing, MI, United States
*Corresponding author. e-mail address: jlohman@msu.edu

Contents

1.	Introduction	46
2.	Expression construct design	49
	2.1 Plasmid choice	50
	2.2 Minioperon design	50
	2.3 Notes	51
3.	ACC complex overproduction in *E. coli*	51
	3.1 Equipment	52
	3.2 Strain and reagents	52
	3.3 Procedure	53
	3.4 Notes	53
4.	Purification of the ACC complex	54
	4.1 Equipment	55
	4.2 Strain and reagents	55
	4.3 Procedure	55
	4.4 Notes	57
5.	Activity assay by HPLC	59
	5.1 Equipment	59
	5.2 Reagents	59
	5.3 Procedure	59
	5.4 Notes	60
6.	Summary and conclusions	62
	Acknowledgments	63
	References	63

Abstract

The primary role of acetyl-CoA carboxylases (ACCs) is to generate malonyl-CoA for use in fatty acid and lipid biosynthesis. However, malonyl-CoA is also used in other various metabolic processes such as secondary metabolite biosynthesis. The diverse utilization

Methods in Enzymology, Volume 708
ISSN 0076-6879, https://doi.org/10.1016/bs.mie.2024.10.024
Copyright © 2024 Elsevier Inc. All rights are reserved, including those for text and data mining, AI training, and similar technologies.

of malonyl-CoA makes ACCs targets for the development of inhibitors and also a target for engineering allosteric regulation for biofuel and secondary metabolite production. The ACC from *Escherichia coli* is representative most of bacterial systems, and is heteromeric, being comprised of four proteins encompassing three distinct subunits. Historically the purification of active *E. coli* ACC complexes has been problematic due to the reported facile dissociation into subunits. Most studies on heteromeric ACCs study the isolated subunits, which are active on their own. Nevertheless, in reconstituted systems, the subunits appear to have allosteric interactions. In this chapter, we provide methods to generate, purify and characterize these heteromeric ACCs complexes. We have used these methods to solve cryogenic electron microscopy structures of active *E. coli* ACC complexes. Purification of active ACC complexes represents a significant step forward in our ability to characterize how allosteric interactions and effectors alter catalytic activity. We expect future studies on the heteromeric ACC complexes will enable rational engineering of new antibiotics and biofuel production.

1. Introduction

In bacteria like *Escherichia coli*, acetyl-CoA carboxylases (ACCs) generate malonyl-CoA primarily for use in fatty acid and lipid biosynthesis (Cronan, 2021), with a bit directed toward lipoic acid and biotin biosynthesis (Cronan , 2014). In other organisms, malonyl-CoA has many other roles besides lipid biosynthesis, such as incorporation into polyketides and other secondary metabolites or in the 3-hydroxypropionate pathway for CO_2 fixation (Berg, Kockelkorn, Buckel, & Fuchs, 2007; Zarzycki, Brecht, Muller, & Fuchs, 2009). A balance between energy production via acetyl-CoA oxidation or incorporation into fatty acids and other metabolites must be maintained. As such, ACCs are highly regulated. For example, the *E. coli* ACC is feedback regulated by acyl carrier proteins (ACPs) bearing acylthioester lipid precursors (Davis & Cronan, 2001). In addition, overexpression of the *E. coli* ACC subunits must be coordinated. Otherwise, growth is compromised (Davis, Solbiati, & Cronan, 2000). Since all *E. coli* ACC subunits are essential, ACC is an attractive target for the development of antibiotics. Furthermore, the *E. coli* ACC is representative of the system in ESKAPE pathogens. Due to the function of ACCs in broad contexts, there is a desire to alter their activity both positively and negatively. Thus, having structures of ACCs bound to effectors would be extremely helpful for rational engineering and antibiotic discovery.

The *E. coli* ACC is heteromeric, consisting of four proteins (AccA, AccB, AccC, AccD) making up three functional subunits, Fig. 1 (Tong, 2013). The AccC subunit is a biotin carboxylase (BC) and AccB is a carboxy biotin

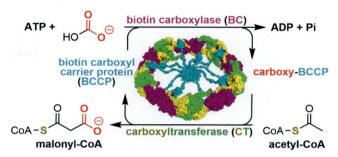

Fig. 1 Catalytic activity of the ACC complex and structure. The two distinct catalytic activities of the ACC system are shown. Surface representation for a model of the *E. coli* ACC complex is shown in the middle of the reaction scheme. AccC (BC) colored magenta, AccB (BCCP) colored cyan, AccA/AccD (CT) colored green/yellow, respectively.

carrier protein (BCCP). Based on AlphaFold the AccB protein has an extended N-terminus which is predicted to form an αββ-motif that oligomerizes into an extended β-sheet if 2 to 7 copies or a β-barrel if more than 10 copies are included (both extended β-sheets and β-barrels are predicted for 8 and 9 copies)(Xu, de Sousa, Boram, Jiang, & Lohman, 2024). The AccA:AccD proteins form the carboxyltransferase (CT) subunit. The AccA protein (sometimes called CT-α) has an N-terminal coiled coil domain and the AccD subunit (sometimes called CT-β) has a 4 cysteine zinc-binding domain. The AccA N-terminal coiled-coil domain, AccD N-terminal zinc-binding domain, and AccB extended N-terminal domain differentiates the *E. coli* like ACCs from propionyl-CoA carboxylases and other acyl-CoA carboxylases that are also heteromeric but built around a hexameric CT core. In humans and most other eukaryotes, ACCs are homodimeric, with most plants having homodimeric ACC in the cytosol and heteromeric ACC in the plastid to support fatty acid biosynthesis (Salie & Thelen, 2016).

Early efforts to study the prokaryotic heteromeric complex were challenging due to problematic disassociation of the complex upon purification from wild-type cells (Cronan, 2021; Fall, 1976; Guchhait et al., 1974; Lane, Moss, & Polakis, 1974). Initial purifications of *E. coli* ACC using ammonium sulfate fractionation (activity precipitated between 25–45%) followed by chromatography on alumina yielded two fractions (Alberts & Vagelos, 1968). The fraction containing the BC (AccC) and BCCP (AccB) subunits bound to the alumina, while the CT subunit was not retained. This study provided the first hints that the BC and BCCP form a somewhat stable subcomplex. Of note, *E. coli* ACC subunit purifications published in Methods of Enzymology Volume 35 started with 1–2 kg of wet *E. coli* cells and yielded ~5–10 mg of each subunit per kg of cell paste (Fall & Vagelos, 1975; Guchhait, Polakis, & Lane, 1975a, 1975b). Subsequently, an ACC complex from *Pseudomonas citronellolis* was purified using the same protocol of ammonium sulfate fractionation (precipitated between 0–45%) and alumina chromatography (400 mM potassium phosphate pH 7.7 elution) (Fall, 1976). The use of a high salt concentration (400 mM ammonium sulfate) allowed purification of a *P. citronellolis* complex (~400 kDa) over a Sepharose 4B column. These early studies gave clues for how to purify heteromeric ACC complexes from wild-type cells. However, introduction of overexpression plasmids with affinity purification tags has largely led to the abandonment of traditional ammonium sulfate fractionation as a purification technique. The use of overexpression is especially attractive since the first purifications required kilograms of *E. coli* cells to obtain milligrams of purified subunits (Ras B. Guchhait et al., 1974). In turn, difficulties in reconstituting *E. coli* ACC complexes from the tagged subunits, may be due to interactions of the tagged termini being disrupted.

Nevertheless, the early studies combined with characterization of over-expressed subunits established that, in the first catalytic step, the ~100 kDa homodimeric BC domain uses ATP and bicarbonate to generate carboxy-biotin linked to the ~17 kDa BCCP. In the second catalytic step, the car-boxy-BCCP translocates to the ~135 kDa heterodimeric CT, where the carboxy group is transferred to acetyl-CoA to generate malonyl-CoA. Using careful enzymology it was established that protein:protein interactions (PPI) within the (sub)complexes alter reaction kinetics and prevent the reverse reaction, establishing a regulatory role for complex formation (Broussard, Price, Laborde, & Waldrop, 2013; Ras B. Guchhait et al., 1974; Soriano et al., 2006). Further studies characterizing isolated domains provided puzzling information about the state of the complex. While the CT is always reported to behave as an $AccA_2$-$AccD_2$ heterotetramer (Bilder et al., 2006), the BC, AccC, can be mutated to generate a monomer with a couple of mutations at the dimer interface (Shen, Chou, Chang, & Tong, 2006). Yet, BC dimerization is required for activity in vivo (Smith & Cronan, 2012). Furthermore, the BC dimer displays "half-site" reactivity, where only one active site in the dimer undergoes catalysis at a time (Blanchard, Lee, Frantom, & Waldrop, 1999; de Queiroz & Waldrop, 2007; Janiyani, Bordelon, Waldrop, & Cronan, 2001; Mochalkin et al., 2008). The BCCP, AccB, is even more enigmatic as it can be proteolyzed in *E. coli* grown to stationary phase (Fall & Vagelos, 1972; Fall, Nervi, Alberts, & Vagelos, 1971). The full-length BCCP is composed of an N-terminal region, which prior to AlphaFold, escaped structural prediction (Choi-Rhee & Cronan, 2003). In vivo proteolysis of the BCCP generates a monomeric C-terminal domain (residues 80–156) that is structured and carries the attached biotin (Athappilly & Hendrickson, 1995). The full-length BCCP alone can behave as a dimer, or higher ordered aggregate depending on the conditions (Choi-Rhee & Cronan, 2003; Fall & Vagelos, 1972; Nenortas & Beckett, 1996). The BCCP and BC can be co-purified, and depending on conditions, the ratio of BC:BCCP ranges from 1:2 to 1:3 (Choi-Rhee & Cronan, 2003), with a structure of the BC:BCCP complex having a ratio of 1:1 (Broussard, Kobe, et al., 2013). The simplest form of ACC obeying the known dimer stoichiometry of the BC and CT would be a 2:2:2:2 complex of AccA:AccD:AccC:AccB (~270 kDa, ~300 kDa if 4 AccB). Reconstitution from tagged subunits has yielded a complex larger than ~640 kDa, corresponding to at least 4 of each protein (Broussard, Price, et al., 2013).

To clarify how the *E. coli* subunits interact in the rate limiting step of fatty acid biosynthesis (Davis et al., 2000), as potential antibacterial target sites (Radka & Rock, 2022), or to stabilize the complex for biofuel production, we set out to develop methods to purify a complex. Inspired by a report of co-overexpression from plasmids leading to increased fatty acid levels (Davis et al., 2000), we generated a plasmid that co-expressed *accA/accD/accB/accC* along with *birA* from a single transcript and successfully purified an active complex without the use of affinity tags. Traditional ammonium sulfate precipitation yielded very active preparations that we could further purify by size exclusion chromatography. Finally, cryo-EM analysis revealed a complex of 10 of each subunit that forms rings which stack into fibers. These preparations pave the way to further explore biophysical communication between the BC and CT active sites. Furthermore, our methods are likely applicable to study other somewhat unstable enzyme complexes that are assembled from moderate hydrophobic interfaces.

2. Expression construct design

Due to the relatively low expression levels of ACC subunits in wild-type cells, we needed to either use large amounts of cells or use overexpression. We chose overexpression since it would also allow us to explore the roles of various subunit appendages via mutagenesis. There are multiple options to consider for the co-expression of proteins. Plasmids have variable copy numbers that directly influence the amount of protein produced, for example pRSFDuet-1 is around 200 copies and pCDFDuet-1 or pCOLADuet-1 are around 20 copies. The proteins can be expressed behind promoters individually or as operons with ribosome binding sites (RBS) between each open reading frame.(Kim et al., 2004) The addition of promoters in front of each gene and appropriate spacers would include another ~120 bp per gene increasing plasmid size. To minimize the overall plasmid size and take advantage of the native RBS between *accB* and *accC,* we chose instead to include RBS between the other genes. Another component, the *birA* gene was also included to ensure that expression of the chromosomal copy wasn't limiting for the generation of biotinylated BCCP (Abdel-Hamid & Cronan, 2007).

2.1 Plasmid choice

We were inspired by a previous study where active *E. coli* ACC complex was generated from a low copy plasmid expressing all four subunits under T7 control (Davis et al., 2000). An important observation was that overexpression of the ACC system slowed growth and an imbalance in ACC subunit ratio was toxic (Davis et al., 2000). Nevertheless, due to their success with expression behind T7 promoters we followed their lead rather than exploring the tightly controlled pBAD system which is induced with arabinose. Ultimately, the use of a pRSFDuet-1 based plasmid leads to very high complex production. However, when we mutated subunit interaction regions, the plasmids become unstable. A medium copy plasmid, pCDFDuet-1, ultimately allowed us to generate and characterize mutants, while retaining much of the high overexpression levels (Xu et al., 2024).

2.2 Minioperon design

Some biosynthetic pathways have been overexpressed through the use of minioperons with the introduction of synthetic RBSs between open reading frames.(McGoldrick et al., 2005) In our case, we introduced the sequence 5′-TAAAGaggagaATACTAGATG-3′ between the *accA/accD, accD/accB,* and *accC/birA* open reading frames (the lowercase nucleotides indicate a RBS, underlined nucleotides are stop and start codons). We included the native sequence, 5′-TAACGAGGCGAACATG-3′ between the *accB/accC* open reading frames. The final *accA/accD/accB/accC/birA* minioperon construct was synthesized by GenScript and cloned into both pRSFDuet-1 and pCDFDuet-1, with the sequence 5′-*TTTAATAAGG AGATATACA*|minioperon|TCCGTAATCT*CCTCGAGTCTGGTA AAGA*-3′ (nucleotides corresponding to the vectors are italicized).

2.3 Notes

Termini of the proteins was taken into consideration for the potential inclusion of affinity tags. The activity of aminopeptidases is strongly influenced by amino acids downstream of the initiating N-formyl methionine, requiring amino acids with small side chains (Xiao, Zhang, Nacev, Liu, & Pei, 2010). The retainment of an N-terminal methionine is indicative that the methionine is functional and likely to be disrupted by affinity tags. Furthermore, our cryo-EM structure revealed important considerations for potential inclusions of affinity tags (Xu et al., 2024). The AccA N-terminus begins with M-S-L, suggesting retainment of the N-terminal methionine. A tag here is likely to disrupt ACC complex filament formation. The AccA C-terminus ends in a cleft at the AccA:AccD interface, and a tag here might also be detrimental for proper oligomerization. The AccB protein begins with M-D-I, indicating the N-terminal methionine is retained, and since mutants with deletions of the first four amino acids abolish complex formation between AccB:AccC the addition of a tag is likely detrimental to overall complex formation (Choi-Rhee & Cronan, 2003). The AccC protein begins with M-L-D, again indicating the N-terminal methionine is important, which is involved in hydrophobic interactions near the surface of the protein. In the complex, the AccC N-terminus faces the exterior of the ring and is a promising site for an affinity tag. The AccC C-terminus appears to interact with the AccA N-terminal coiled coil domain and a tag here is likely problematic. The AccD gene begins with M-S-W-I and the W-I residues are buried in a hydrophobic pocket on AccC. A tag here is likely very problematic. The C-terminus of AccD contains a long unstructured segment that can be tagged. Thus, the only places where tags are unlikely to be problematic are at the N-terminus of AccC and C-terminus of AccD and potentially the C-terminus of AccA. We have successfully purified active ACC tagged on the C-terminus of AccD with either a tetracysteine tag or a mCitrine fluorescent protein. Also of note, constructs lacking the *birA* open reading frame give preparations with greatly reduced activity, presumably due to incomplete BCCP biotinylation.

3. ACC complex overproduction in *E. coli*

The plasmids pRSFDuet-1 or pCDFDuet-1 bearing the *accA/D/B/C/ birA* minioperon, carrying kanamycin and streptomycin resistance, respectively, were transformed into electrocompetent *E. coli* BL21(DE3) cells. This strain has the lacUV5 gene controlling inducible T7 RNA polymerase expression by lactose or isopropyl β-D-1-thiogalactopyranoside (IPTG) (Studier & Moffatt, 1986). Expression of the ACC complex is reported to slow growth and overexpression of individual subunits is toxic to *E. coli* growth (Davis et al., 2000). Therefore, we sought a way to repress the basal expression from the T7 promoter during initial cultures (Grossman, Kawasaki, Punreddy, & Osburne, 1998). To mitigate the negative consequences of ACC basal expression we included 1 % glucose for catabolite repression in overnight cultures. Those overnight *E. coli* cultures were used as inoculum for over-expression in Terrific Broth at 18 or 37 °C. During overexpression we included zinc (in the form of trace metal solution)(Studier, 2005) and biotin as additives. The presence of zinc is important for the activity of AccD, which exhibits a zinc finger (Bilder et al., 2006), and biotin is needed for generating holo-BCCP by BirA.

3.1 Equipment
- Electroporator (Gene Pulser Xcell Total System, Bio-Rad).
- Electroporation cuvettes (1 mm).
- Shaker (New Brunswick Innova 44, 2 in. orbit).
- Incubator.
- Sterile spreading device.
- High-speed centrifuge.
- 50 mL conical tubes capable of centrifugation at 11,000g.

3.2 Strain and reagents
- *E. coli* BL21(DE3) electrocompetent cells.
- SOC medium (2 % tryptone, 0.5 % yeast extract, 10 mM NaCl, 2.5 mM KCl, 10 mM $MgCl_2$, 10 mM $MgSO_4$, and 20 mM glucose).
- LB agar medium (1 % sodium chloride, 1 % tryptone, 0.5 % yeast extract, 1.5 % agar) with appropriate antibiotic at 50 μg/mL.
- LB medium (1 % sodium chloride, 1 % tryptone, 0.5 % yeast extract).
- Kanamycin or streptomycin, 50 mg/mL stock solution.
- Filter sterilized glucose (20 % glucose) stock solution.
- 2 M $MgCl_2$
- Terrific Broth (12 g/L peptone, 24 g/L yeast extract, 2.2 g/L potassium phosphate monobasic, 9.4 g/L potassium phosphate dibasic, 10 g/L glycerol).
- 1000 × Trace metals (50 mM $FeCl_3$ in 120 mM HCl, 20 mM $CaCl_2$, 10 mM $MnCl_2$, 10 mM $ZnSO_4$, 2 mM each $CoCl_2$, $CuCl_2$, $NiCl_2$, Na_2MoO_4, Na_2SeO_3, H_3BO_3).

Purification of ACC 53

- D-biotin.
- Isopropyl β-D-1-thiogalactopyranoside (IPTG), 400 mM.
- Lysis buffer (50 mM tris HCl, 50 mM NaCl and 10 % glycerol pH 8.0).

3.3 Procedure

1. 500 ng of plasmid pRSF or pCDF with the *E. coli* ACC minioperon is added to 25 μL of electrocompetent cells and transferred to an electroporation cuvette.
2. The bacterial suspension is electroporated using manufacturer recommend parameters (2.1 kV, 100 Ω, and 25 μF, time constant ~2.6 ms).
3. Immediately after electroporation, 975 μL of warm SOC medium is added to the cuvette, and the culture is transferred to 1.5 mL tubes and incubated in a shaker at 37 °C for 1 h.
4. 100 μL of the recovered culture is streaked on LB agar plates containing 50 μg/mL of kanamycin or streptomycin and 10 g/L of glucose, and then incubated overnight at 37 °C.
5. A single colony is picked and added to 50 mL LB supplemented with 10 g/L glucose and 50 μg/mL of kanamycin or streptomycin and grown overnight at 37 °C with shaking at 180 rpm.
6. 10 mL of the pre-culture (note 1) is added to 1 L of Terrific Broth supplemented with 100 μL of 1000 × trace metals (0.1 × final concentration) and 5 mL 2 M $MgCl_2$ (10 mM final concentration), and 1 mL of kanamycin or streptomycin stock (50 μg/mL final concentration).
7. The 1 L cultures are grown to an OD A_{600} of ~1.2 at 37 °C with shaking at 180 rpm (increase rpm for incubators with 1″ throw), at which point the temperature is reduced to 18 °C.
8. Upon reaching thermal equilibrium, 1 mL of IPTG (0.4 mM final concentration), and 24 mg of solid biotin (100 μM final concentration) is added to the cultures and shaking is continued at 18 °C overnight.
9. Cells are harvested by centrifugation at 5000g, the pellet weight is measured (note 2) and pellets are resuspended in ~ 35 mL of lysis buffer then frozen at −20 °C in 50 mL conical tubes capable of centrifugation at 11,000g.

3.4 Notes

1. We have generated glycerol stocks of *E. coli* BL21 (DE3) with the pRSF or pCDF plasmids by adding equal volumes of 50 % glycerol to overnight cultures grown in the presence of glucose and freezing at −80 °C. Some of these stocks can be used to inoculate fresh batches of overnight

cultures and successfully lead to overexpressed ACC. However, if the overnight cultures are grown for more than 20 h before preservation, they are likely to lead to a lack of overexpression upon induction in large scale cultures.

2. Typical cell pellets from 1 L of Terrific Broth for growth at 37 °C after 4 h of induction are around 3 g, whereas for overnight at 18 °C pellets are around 4 g. Since both temperatures generate preparations with similar specific activities, overnight cultures were preferred due to the larger pellet. However, the proportion of intact to cleaved BCCP has not been quantitated between the two different preparations, and it is known that increasing incubation times can lead to BCCP degradation (Fall & Vagelos, 1972).

4. Purification of the ACC complex

Classical ammonium sulfate precipitation preserves the structure of proteins while driving them from solution (Wingfield, 1998), a method outlined in the first volume of Methods in Enzymology (Green & Hughes, 1955). Ammonium sulfate was also used in the initial purifications of *E. coli* ACC subunits (Alberts & Vagelos, 1968). Our overexpressed ACC complex was precipitated by 30–45 % ammonium sulfate, similar to the natively expressed complex (Alberts & Vagelos, 1968). Highly active ACC is recovered upon resuspension of the ammonium sulfate pellets, however further purification is influenced by the buffer used for resuspension and for initial resuspension of cells for lysis. Size exclusion chromatography of active complex is most successful if 10 % glycerol is maintained during ammonium sulfate precipitation and resuspension. For example, our lysis buffer has 10 % glycerol and traditional ammonium sulfate precipitation is done by adding solid ammonium sulfate, which does not dilute the glycerol. If solutions of ammonium sulfate are used that lack glycerol we recover less active ACC activity after size exclusion chromatography. The composition of the lysis buffer and resuspension buffers also alter the amount of activity recovered, for example the addition of 0.4 M potassium chloride can increase recovered activity. The procedure outlined below is likely an effective starting point but should be further optimized to suit ACC complexes from other organisms.

4.1 Equipment

- Sonicator.
- High speed centrifuge
- Rotor for 50 mL conical tubes
- Benchtop swinging bucket centrifuge for 50 mL conical tubes.
- Minisart® NML Plus 0.7 µm GF syringe filters.
- Nanodrop UV/Vis spectrophotometer.
- ÄKTA pure™ protein purification system.
- Size exclusion chromatography columns, Superose™ 6 Increase 10/300 GL (Cytiva), HiPrep 16/60 Sephacryl S-500 HR (Cytiva).
- Bio-Rad Mini-PROTEAN system for SDS-PAGE electrophoresis.
- Vivaspin® 5 kDa MWCO polyethersulfone concentrator.

4.2 Strain and reagents

- Lysozyme.
- DNAse 2 mg/mL stock solution.
- RNAse 1 mg/mL stock solution.
- 500 mM EDTA pH 8.0.
- 1 M $MnCl_2$
- Saturated ammonium sulfate with 10 % glycerol.
- Resuspension buffer (10 mM tris, 200 mM NaCl, pH 8.0).
- Any kD Mini-PROTEAN TGX precast gels.
- Flamingo™ fluorescent protein gel stain.

4.3 Procedure

1. The frozen resuspended cell pellets are thawed in a room temperature water bath and lysozyme is added to 1 mg/mL along with 50 µL each of DNAse and RNAse stocks. A 100 µL sample is saved for activity and SDS-PAGE analysis.
2. Sonication is performed on ice at 40 % amplitude with a program of 10 s on and 10 s off for a total pulse time of 3 min.
3. Cell lysates are centrifuged at 11,000g for 45 min at 4 °C to remove insoluble materials. Optional, add nucleic acid precipitation/ribosome denaturation such as 50 mM EDTA or 50 mM $MnCl_2$ and centrifuge again. A 100 µL sample is saved for activity and SDS-PAGE analysis.
4. Ammonium sulfate precipitation is performed by slow addition (4 to 5 portions with gentle mixing between over the course of 4–5 min) of saturated ammonium sulfate with 10 % glycerol to cell lysates on ice to

final values of 10%, 20%, and 30% (note 1) with centrifugation in a swinging bucket rotor at 5000g at 4 °C for 20 min between additions to generate fractionated pellets (note 2).
5. The ammonium sulfate pellets are slurried and frozen at −80 °C in 200 μL aliquots.
6. Aliquots of the slurried pellets are resuspended in resuspension buffer to obtain sample solutions with an A_{280} of ~5–10 which are syringe filtered. The samples are applied to either a Superose™ 6 Increase 10/300 GL (100–500 μL of sample solution) or HiPrep 16/60 Sephacryl S-500 HR (500–1000 μL of sample solution) size exclusion chromatography (note 3) column at 4 °C pre-equilibrated and resolved with resuspension buffer at half the recommended flow rate for room temperature, and fractions are collected for further analysis, Fig. 2.
7. The presence of the ACC subunits is determined by SDS-PAGE on small portions of the pellets using Flamingo staining (note 5) and activity assays described below, Fig. 3.
8. Fractions containing the active ACC complex are pooled and concentrated using centrifugal concentrators with 5 kDa cutoffs to ensure retention of the BCCP component and frozen in liquid nitrogen.

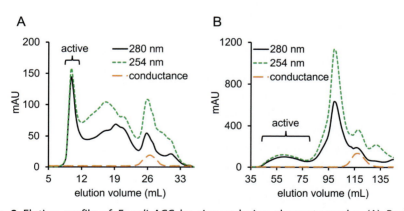

Fig. 2 Elution profile of *E. coli* ACC by size exclusion chromatography. (A) Profile obtained using a Superose™ 6 Increase 10/300 GL column. (B) Profile obtained using a HiPrep 16/60 Sephacryl S-500 HR column. The black solid line represents absorbance at 280 nm, the green dashed line corresponds to absorbance at 254 nm, and the orange long-dashed line indicates conductivity. The increased conductivity peak from ammonium sulfate is a convenient calibration for the end of the separation range. An *E. coli* ribosome standard elutes around 85 mL on the Sephacryl column. ACC activity is indicated by the "active" ranges.

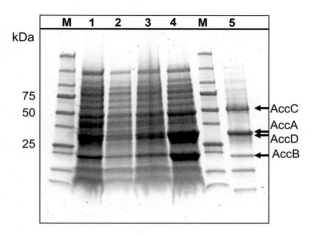

Fig. 3 SDS-PAGE of overexpressed ACC complex. M – molecular weight markers. Lane 1–clarified cell lysate. Lane 2–10 % ammonium sulfate pellet. Lane 3–20 % ammonium sulfate pellet. Lane 4–30 % ammonium sulfate pellet. Lane 5–active fraction from size exclusion chromatography. The SDS-PAGE gel was stained using Flamingo stain. AccA (33.3 kDa), AccD (35.2 kDa) and BirA (35.3 kDa) comigrate. AccB migrates at different rates depending upon sample preparation.

9. The concentration of the complex is determined using an extinction coefficient of 73,780 $M^{-1}cm^{-1}$ at 280 nm based on a 1:1:1:1 ratio of AccA:AccD:AccB:AccC (note 5).
10. The concentration of the complex is also determined using a colorimetric BCA assay for comparison.

4.4 Notes

1. Our reported protocol above stops ammonium sulfate fractionation at 30 %. However, there is quite a bit of ACC activity remaining in the ammonium sulfate supernatant. We expect another cut at 45 % ammonium sulfate would likely recover much of that activity. However, the yields were more than sufficient and the high specific activity at 30 % ammonium sulfate didn't justify going higher.
2. Typically, ammonium sulfate fractionation centrifugation steps are done at around 20,000g. The use of a swinging bucket centrifuge on the benchtop speeds up the processing steps and yields pellets that are less firm than at higher speeds. These pellets are easily resuspended into a slurry by pipetting with the small amount of liquid left after decanting the supernatant. Small portions (~10−20 μL) of the slurry are checked

for protein concentration, activity and purity (Fig. 3) to provide an estimate for how much volume to resuspend the pellets in for SEC.

3. Over an analytical Superose 6 column, the most active fractions correspond to the void volume. To further purify the active material, we switched to a HiPrep 16/60 Sephacryl S-500 HR column which can resolve much larger molecules. However, the S-500 HR column leads to protein with less activity than the Superose 6 column, likely due to the much longer run times leading to more dissociation of the complex. The addition of 5 % glycerol to the Superose 6 column increases activity of the fractions, however the runs take significantly longer due to increased back pressure. Nevertheless, we explored the effect of pH on the ammonium sulfate fractionation and SEC steps in the presence of glycerol and found no significant difference between pH 6.5 and 8.5.

4. We observe a 260/280 nm ratio that is somewhat higher than expected for our complex even after size exclusion chromatography, Fig. 2. *E. coli* ribosomes precipitate at 55 % ammonium sulfate (Kurland, 1971), suggesting we may have had ribosome contamination. We attempted to remove the suspected nucleic acid contamination through a few procedures. Ribosomes can be denatured by the addition of EDTA (Miall & Walker, 1969). Therefore, we included 5 or 50 mM EDTA and expected the inclusion of RNAse to help further degrade the denatured ribosomes. These preparations still had relatively high 260/280 nm ratios. In the original purification of *E. coli* ACC, a nucleic acid precipitation step with manganese chloride was used. We found the addition of 50 mM $MnCl_2$ to clarified lysate generated a pellet of material, that we removed prior to ammonium sulfate fractionation. Nevertheless, $MnCl_2$ failed to give a purified complex with a 260 nm value lower than 280. This suggests ATP/AMP or CoA are strongly bound to the proteins and carried through purification.

5. Due to the slightly high 260/280 ratio, the concentration of the complex was also estimated using a Bradford or BCA assay with bovine serum albumin as a calibration standard, which yielded similar results to the UV-Vis assay. One caveat with a Bradford assay is that the BCCP is known to bind Coomassie stain relatively poorly, however the small size of the BCCP compared to the other subunits decreases the protein concentration underestimation. Due to the poor staining of the BCCP by Coomassie stain on SDS-PAGE gels (Choi-Rhee & Cronan, 2003), we explored Flamingo stain which had less protein specific variability and is more reproducible than silver stain. Samples that are low in

protein concentration or that are in high salt concentration are precipitated with trichloroacetic acid and washed with cold acetone prior to SDS-PAGE analysis.

5. Activity assay by HPLC

There are many assays for ACC activity. The most convenient and popular method monitors ATP hydrolysis by the BC domain through coupled enzyme assay following NADH oxidation spectroscopically. However, this reaction is unsuitable for monitoring activity in crude lysates due to the presence of enzymes that would compete for NADH. Therefore, we chose to monitor activity using a discontinuous assay for the conversion of acetyl-CoA to malonyl-CoA via high-performance liquid chromatography (HPLC) detection, which is highly specific for ACC activity. While HPLC is time consuming, there are some benefits to the HPLC assay that may be overlooked. The ability to monitor for the hydrolysis of acetyl-CoA provides insight into the presence of contaminating enzymes, thus analyzing the ratio of hydrolyzed acetyl-CoA to malonyl-CoA ratio is an excellent measure of purity.

5.1 Equipment
- Microcentrifuge
- Agilent 1100 HPLC with UV/Vis detector
- Phenomenex Luna C18(2) 5 µm 100 Å, 250 × 4.6 mm column

5.2 Reagents
- 100 mM Acetyl-CoA
- 600 mM ATP
- 1 M sodium bicarbonate
- 1 M bis-tris propane pH 7
- 1 M KCl
- 25 % formic acid
- Acetonitrile
- 0.5 % trifluoroacetic acid

5.3 Procedure
1. Enzymatic reactions are set up in 1.5 mL tubes with 1 mL volumes of 250 µM acetyl-CoA, 4 mM ATP, 50 mM sodium bicarbonate, 10 mM $MgCl_2$ (note 1), 20 mM KCl and 100 mM bis-tris propane pH 7.0 and incubated at room temperature.

2. The enzymatic reactions are initiated by adding approximately 5 μM ACC, as cell lysate (note 2), resuspended ammonium sulfate pellet, or size exclusion chromatography fraction.
3. At various time points 100 μL of the reaction is withdrawn and quenched by the addition of 50 μL 25 % formic acid and the samples are centrifuged to remove precipitated protein.
4. 90 μL of the samples are applied to a reverse phase C18 column using an HPLC with UV/Vis monitoring at 260 nm. Analytes are resolved with an elution gradient of water with 0.5 % TFA to 20 % acetonitrile over 20 min.
5. The peaks for CoA (~9 min elution), malonyl–CoA (~10.5 min elution) and acetyl–CoA (~11 min elution) are integrated, summed and peak integrations used to determine percent conversion.
6. The HPLC analyses can be used to generate enzyme purification tables, Table 1, to evaluate the effect of changing variables in the general purification scheme outlined in this chapter (Burgess, 2009).

5.4 Notes

1. For preparations where we attempted nucleic acid precipitation with manganese chloride at 50 mM, we observed a dramatic decrease in the rate of hydrolysis in comparison to malonyl–CoA formation. In the early publications of *E. coli* ACC purification, manganese chloride was also used rather than magnesium chloride in subsequent activity assays (Alberts & Vagelos, 1968). Using our HPLC assay we have confirmed that 1 mM of magnesium or manganese chloride is sufficient for full catalytic activity, demonstrating manganese efficiently replaces magnesium. Thus, a switch from magnesium to manganese can increase the specificity of the reported HPLC assay to favor ACC.
2. One reason postulated for the absence of catalytic activity by *E. coli* ACC upon cell disruption is that dilution leads to disassembly of the ACC complex. However, based on our HPLC analysis, we notice that acetyl–CoA is quickly converted to CoA by cell lysate in the absence of overexpressed ACC. We conclude that enzymes such as citrate synthase, malate synthase, FabH and acetyltransferases are using acetyl–CoA rapidly and fatty acid biosynthetic enzymes use the malonyl–CoA product faster than it can build up at wild-type ACC levels. The original assays using ^{14}C bicarbonate labeling of acetyl–CoA to generate an acid-stable radioactive product ([3-^{14}C]malonyl–CoA) suffer from the same issues as our HPLC assay (Alberts & Vagelos, 1968).

Table 1 Representative *E. coli* ACC purification.

	Total protein (mg)[W]	Total activity[a]	Total hydrolysis[a]	Specific activity[b]	acetyl-CoA hydrolysis[b]	Yield %[Y]	Purity[P] (activity/ hydrolysis)
Clarified lysate[B]	2680	38,416	32347	14.3	12.1	100	1.2
10 % $(NH_4)_2SO_4$ pellet[B]	8	47	176	5.6	20.8	0.1	0.3
20 % $(NH_4)_2SO_4$ pellet[B]	7	35	176	4.9	24.5	0.1	0.2
30 % $(NH_4)_2SO_4$ pellet[B]	208	5425	1672	26.1	8.0	14.1	3.2
30 % supernatant[B]	1252	1339	19313	1.1	15.4	3.5	0.1
SEC[UV,A]	46	2837	448	62	9.8	7.4	6.3

[W]From 4 g of wet *E. coli* cell pellet from 1 L of bacterial culture.
[Y]Yield is based on total activity recovered at each step.
[P]Purity is rate of malonyl-CoA production divided by rate of CoA production.
[B]Protein concentration measured by Bradford assay.
[UV]Protein concentration measured by UV/Vis at 280 nm.
[A]Estimated from one run with ~5 % of the pellet, activity values from most active fractions.
[a]μmol product/min, activity is malonyl-CoA production rate, hydrolysis is CoA production rate
[b]μmol product/min/mg protein

6. Summary and conclusions

This chapter provides an update on the purification methods for bacterial heteromeric ACC complexes. The protocols here are inspired by the earliest reports of ACC purification but updated to reduce the amount of starting cell paste. For example, we can recover milligram quantities of very pure overexpressed ACC complex from a few grams of cell paste, rather than from kilograms. Importantly, we again demonstrate that the ammonium sulfate fraction method reliably preserves enzyme activity. In any purification a balance needs to be struck between purity, yield and activity. In our purification, we recover only a fraction of the starting activity. Since our goal is to solve a structure of the ACC complex we only needed a few milligrams of material. However, the very high purity and lack of acetyl-CoA hydrolysis after SEC is a promising starting point for further optimization.

With active complex in hand we were able to use methods outlined in the Methods in Enzymology Volume 579 (Crowther, 2016) to determine a cryo-EM structure of our isolated *E. coli* ACC complex (Xu et al., 2024). As outlined above, the cryo-EM structure has allowed us to rationalize where affinity tags should be avoided. Our cryo-EM structures also revealed that most of the protein-protein interactions holding the complex together are relatively small hydrophobic patches. Increasing salt concentrations stabilizes hydrophobic interactions. Thus, ammonium sulfate plays dual roles, salting out and increasing complex stability. Exploring different salts in the Hoffmeister series to precipitate the ACC complex and for subsequent SEC will likely yield more effective purifications.

The methods outlined in this chapter serve as a starting point to explore the differences in stability and activity of ACCs from different organisms. We have already purified active ACC complexes from *Acinetobacter baumanii*, *Staphylococcus aureus* and *Enterococcus faceium* using minioperons and ammonium sulfate fractionation. However, the ACC complexes from *A. baumanii* and *E. faceium* lack the coiled-coil domain on the N-termini of the AccA proteins and can't form stabilizing filaments. Despite being active after resuspension from ammonium sulfate pellets, they couldn't be easily purified by SEC. This observation reinforces the idea that stabilization of ACC complexes through mutagenesis might be a better means to increase catalytic activity, rather than raw overexpression for biofuel or polyketide production. Structural studies of the ACC complexes in the presence of inhibitors will also provide more information on how inhibition of individual domains affects the complexes, leading to the design of strain specific antibiotics.

Acknowledgments

This work is funded by the National Institute of Health grant R01 GM140290.

References

Abdel-Hamid, A. M., & Cronan, J. E. (2007). Coordinate expression of the acetyl coenzyme A carboxylase genes, accB and accC, is necessary for normal regulation of biotin synthesis in Escherichia coli. *Journal of Bacteriology, 189*(2), 369–376. https://doi.org/10.1128/JB.01373-06.

Alberts, A. W., & Vagelos, P. R. (1968). Acetyl CoA carboxylase. I. Requirement for two protein fractions. *Proceedings of the National Academy of Sciences of the United States of America, 59*(2), 561–568. https://doi.org/10.1073/pnas.59.2.561.

Athappilly, F. K., & Hendrickson, W. A. (1995). Structure of the biotinyl domain of acetyl-coenzyme A carboxylase determined by MAD phasing. *Structure (London, England: 1993), 3*(12), 1407–1419. https://doi.org/10.1016/s0969-2126(01)00277-5.

Berg, I. A., Kockelkorn, D., Buckel, W., & Fuchs, G. (2007). A 3-hydroxypropionate/4-hydroxybutyrate autotrophic carbon dioxide assimilation pathway in Archaea. *Science (New York, N. Y.), 318*(5857), 1782–1786. https://doi.org/10.1126/science.1149976.

Bilder, P., Lightle, S., Bainbridge, G., Ohren, J., Finzel, B., Sun, F., ... Waldrop, G. L. (2006). The structure of the carboxyltransferase component of acetyl-coA carboxylase reveals a zinc-binding motif unique to the bacterial enzyme. *Biochemistry, 45*(6), 1712–1722. https://doi.org/10.1021/bi0520479.

Blanchard, C. Z., Lee, Y. M., Frantom, P. A., & Waldrop, G. L. (1999). Mutations at four active site residues of biotin carboxylase abolish substrate-induced synergism by biotin. *Biochemistry, 38*(11), 3393–3400. https://doi.org/10.1021/bi982660a.

Broussard, T. C., Kobe, M. J., Pakhomova, S., Neau, D. B., Price, A. E., Champion, T. S., & Waldrop, G. L. (2013). The three-dimensional structure of the biotin carboxylase-biotin carboxyl carrier protein complex of E. coli acetyl-CoA carboxylase. *Structure (London, England: 1993), 21*(4), 650–657. https://doi.org/10.1016/j.str.2013.02.001.

Broussard, T. C., Price, A. E., Laborde, S. M., & Waldrop, G. L. (2013). Complex formation and regulation of Escherichia coli acetyl-CoA carboxylase. *Biochemistry, 52*(19), 3346–3357. https://doi.org/10.1021/bi4000707.

Burgess, R. R. (2009). Preparing a purification summary table. *Methods in Enzymology, 463*, 29–34. https://doi.org/10.1016/S0076-6879(09)63004-4.

Choi-Rhee, E., & Cronan, J. E. (2003). The biotin carboxylase-biotin carboxyl carrier protein complex of Escherichia coli acetyl-CoA carboxylase. *The Journal of Biological Chemistry, 278*(33), 30806–30812. https://doi.org/10.1074/jbc.M302507200.

Cronan, J. E. (2014). Biotin and lipoic acid: Synthesis, attachment, and regulation. *EcoSal Plus, 6*(1), https://doi.org/10.1128/ecosalplus.ESP-0001-2012.

Cronan, J. E. (2021). The classical, yet controversial, first enzyme of lipid synthesis: Escherichia coli acetyl-CoA carboxylase. *Microbiology and Molecular Biology Reviews: MMBR, 85*(3), e0003221. https://doi.org/10.1128/MMBR.00032-21.

Crowther, R. A. (Ed.). (2016). *The Resolution Revolution: Recent Advances in cryoEM Methods in Enzymology, 579.*

Davis, M. S., & Cronan, J. E., Jr. (2001). Inhibition of Escherichia coli acetyl coenzyme A carboxylase by acyl-acyl carrier protein. *Journal of Bacteriology, 183*(4), 1499–1503. https://doi.org/10.1128/JB.183.4.1499-1503.2001.

Davis, M. S., Solbiati, J., & Cronan, J. E., Jr. (2000). Overproduction of acetyl-CoA carboxylase activity increases the rate of fatty acid biosynthesis in Escherichia coli. *The Journal of Biological Chemistry, 275*(37), 28593–28598. https://doi.org/10.1074/jbc.M004756200.

De Queiroz, M. S., & Waldrop, G. L. (2007). Modeling and numerical simulation of biotin carboxylase kinetics: Implications for half-sites reactivity. *Journal of Theoretical Biology, 246*(1), 167–175. https://doi.org/10.1016/j.jtbi.2006.12.025.

Fall, R. R. (1976). Stabilization of an acetyl-coenzyme a carboxylase complex from Pseudomonas citronellolis. *Biochimica et Biophysica Acta, 450*(3), 475–480. https://doi.org/10.1016/0005-2760(76)90022-9.

Fall, R. R., Nervi, A. M., Alberts, A. W., & Vagelos, P. R. (1971). Acetyl CoA carboxylase: Isolation and characterization of native biotin carboxyl carrier protein. *Proceedings of the National Academy of Sciences of the United States of America, 68*(7), 1512–1515. https://doi.org/10.1073/pnas.68.7.1512.

Fall, R. R., & Vagelos, P. R. (1972). Acetyl coenzyme A carboxylase. Molecular forms and subunit composition of biotin carboxyl carrier protein. *The Journal of Biological Chemistry, 247*(24), 8005–8015. https://doi.org/10.1016/s0021-9258(20)81801-8.

Fall, R. R., & Vagelos, P. R. (1975). Biotin carboxyl carrier protein from Escherichia coli. *Methods in Enzymology, 35*, 17–25. https://doi.org/10.1016/0076-6879(75)35133-1.

Green, A. A., & Hughes, W. L. (1955). Protein fractionation on the basis of solubility in aqueous solutions of salts and organic solvents. *Methods in Enzymology, 1*, 67–90. https://doi.org/10.1016/0076-6879(55)01014-8.

Grossman, T. H., Kawasaki, E. S., Punreddy, S. R., & Osburne, M. S. (1998). Spontaneous cAMP-dependent derepression of gene expression in stationary phase plays a role in recombinant expression instability. *Gene, 209*(1-2), 95–103. https://doi.org/10.1016/s0378-1119(98)00020-1.

Guchhait, R. B., Polakis, S. E., Dimroth, P., Stoll, E., Moss, J., & Lane, M. D. (1974). Acetyl coenzyme A carboxylase system of Escherichia coli. *The Journal of Biological Chemistry, 249*(20), 6633–6645. https://doi.org/10.1016/s0021-9258(19)42203-5.

Guchhait, R. B., Polakis, S. E., & Lane, M. D. (1975a). Biotin carboxylase component of acetyl-CoA carboxylase from Escherichia coli. *Methods in Enzymology, 35*, 25–31. https://doi.org/10.1016/0076-6879(75)35134-3.

Guchhait, R. B., Polakis, S. E., & Lane, M. D. (1975b). Carboxyltransferase component of acetyl-CoA carboxylase from Escherichia coli. *Methods in Enzymology, 35*, 32–37. https://doi.org/10.1016/0076-6879(75)35135-5.

Janiyani, K., Bordelon, T., Waldrop, G. L., & Cronan, J. E., Jr. (2001). Function of Escherichia coli biotin carboxylase requires catalytic activity of both subunits of the homodimer. *The Journal of Biological Chemistry, 276*(32), 29864–29870. https://doi.org/10.1074/jbc.M104102200.

Kim, K. J., Kim, H. E., Lee, K. H., Han, W., Yi, M. J., Jeong, J., & Oh, B. H. (2004). Two-promoter vector is highly efficient for overproduction of protein complexes. *Protein Science: A Publication of the Protein Society, 13*(6), 1698–1703. https://doi.org/10.1110/ps.04644504.

Kurland, C. G. (1971). Purification of ribosomes from Escherichia coli. *Methods in Enzymology, 20*, 379–381. https://doi.org/10.1016/s0076-6879(71)20041-0.

Lane, M. D., Moss, J., & Polakis, S. E. (1974). Acetyl coenzyme A carboxylase. *Current Topics in Cellular Regulation, 8*, 139–195. https://doi.org/10.1016/b978-0-12-152808-9.50011-0.

McGoldrick, H. M., Roessner, C. A., Raux, E., Lawrence, A. D., McLean, K. J., Munro, A. W., ... Warren, M. J. (2005). Identification and characterization of a novel vitamin B12 (cobalamin) biosynthetic enzyme (CobZ) from Rhodobacter capsulatus, containing flavin, heme, and Fe-S cofactors. *The Journal of Biological Chemistry, 280*(2), 1086–1094. https://doi.org/10.1074/jbc.M411884200.

Miall, S. H., & Walker, I. O. (1969). Structural studies on ribosomes. II. Denaturation and sedimentation of ribosomal subunits unfolded in EDTA. *Biochimica et Biophysica Acta, 174*(2), 551–560. https://doi.org/10.1016/0005-2787(69)90284-6.

Mochalkin, I., Miller, J. R., Evdokimov, A., Lightle, S., Yan, C., Stover, C. K., & Waldrop, G. L. (2008). Structural evidence for substrate-induced synergism and half-sites reactivity in biotin carboxylase. *Protein Science: A Publication of the Protein Society, 17*(10), 1706–1718. https://doi.org/10.1110/ps.035584.108.

Nenortas, E., & Beckett, D. (1996). Purification and characterization of intact and truncated forms of the Escherichia coli biotin carboxyl carrier subunit of acetyl-CoA carboxylase. *The Journal of Biological Chemistry, 271*(13), 7559–7567. https://doi.org/10.1074/jbc.271.13.7559.

Radka, C. D., & Rock, C. O. (2022). Mining fatty acid biosynthesis for new antimicrobials. *Annual Review of Microbiology, 76*, 281–304. https://doi.org/10.1146/annurev-micro-041320-110408.

Salie, M. J., & Thelen, J. J. (2016). Regulation and structure of the heteromeric acetyl-CoA carboxylase. *Biochimica et Biophysica Acta, 1861*(9 Pt B), 1207–1213. https://doi.org/10.1016/j.bbalip.2016.04.004.

Shen, Y., Chou, C. Y., Chang, G. G., & Tong, L. (2006). Is dimerization required for the catalytic activity of bacterial biotin carboxylase? *Molecular Cell, 22*(6), 807–818. https://doi.org/10.1016/j.molcel.2006.04.026.

Smith, A. C., & Cronan, J. E. (2012). Dimerization of the bacterial biotin carboxylase subunit is required for acetyl coenzyme A carboxylase activity in vivo. *Journal of Bacteriology, 194*(1), 72–78. https://doi.org/10.1128/JB.06309-11.

Soriano, A., Radice, A. D., Herbitter, A. H., Langsdorf, E. F., Stafford, J. M., Chan, S., ... Black, T. A. (2006). Escherichia coli acetyl-coenzyme A carboxylase: Characterization and development of a high-throughput assay. *Analytical Biochemistry, 349*(2), 268–276. https://doi.org/10.1016/j.ab.2005.10.044.

Studier, F. W. (2005). Protein production by auto-induction in high density shaking cultures. *Protein Expression and Purification, 41*(1), 207–234. https://doi.org/10.1016/j.pep.2005.01.016.

Studier, F. W., & Moffatt, B. A. (1986). Use of bacteriophage T7 RNA polymerase to direct selective high-level expression of cloned genes. *Journal of Molecular Biology, 189*(1), 113–130. https://doi.org/10.1016/0022-2836(86)90385-2.

Tong, L. (2013). Structure and function of biotin-dependent carboxylases. *Cellular and Molecular Life Sciences: CMLS, 70*(5), 863–891. https://doi.org/10.1007/s00018-012-1096-0.

Wingfield, P. (1998). Protein precipitation using ammonium sulfate. *Current Protocols in Protein Science/Editorial Board, John E. Coligan ... [et al.], 13*(1), A.3F.1-A.3F.8. https://doi.org/10.1002/0471140864.psa03fs13.

Xiao, Q., Zhang, F., Nacev, B. A., Liu, J. O., & Pei, D. (2010). Protein N-terminal processing: substrate specificity of Escherichia coli and human methionine amino-peptidases. *Biochemistry, 49*(26), 5588–5599. https://doi.org/10.1021/bi1005464.

Xu, X., de Sousa, A. S., Boram, T. J., Jiang, W., & Lohman, J. R. (2024). Active E. coli heteromeric acetyl-CoA carboxylase forms polymorphic helical tubular filaments. *bioRxiv.* https://doi.org/10.1101/2024.05.28.596234.

Zarzycki, J., Brecht, V., Muller, M., & Fuchs, G. (2009). Identifying the missing steps of the autotrophic 3-hydroxypropionate CO_2 fixation cycle in Chloroflexus aurantiacus. *Proceedings of the National Academy of Sciences of the United States of America, 106*(50), 21317–21322. https://doi.org/10.1073/pnas.0908356106.

CHAPTER FOUR

Insights into the methodology of acetyl-CoA carboxylase inhibition

Mirela Tkalčić Čavužić ⓘ **, Brent A. Larson, and Grover L. Waldrop* ⓘ**
Department of Biological Sciences, Louisiana State University, Baton Rouge, LA, United States
*Corresponding author. e-mail address: gwaldro@lsu.edu

Contents

1.	Background	69
	1.1 Reaction	69
	1.2 Protein structure	69
2.	Inhibitors for analysis of ACC structure and function	70
	2.1 Biotin carboxylase	70
	2.2 Enzymatic assay	70
	2.3 Product inhibitors	73
	2.4 ATP analogs	74
	2.5 Carboxyphosphate analogs	75
	2.6 Substrate inhibition	76
	2.7 Inhibition by BCCP	77
	2.8 Inhibition by inactive mutants of biotin carboxylase	77
	2.9 Negative feedback inhibition by palmitoyl-acyl carrier protein	78
	2.10 Carboxyltransferase	79
	2.11 Enzymatic Assay	79
	2.12 Biotin analogs	80
	2.13 Inhibition by BCCP	81
	2.14 Inhibition by DNA	82
	2.15 Inhibition by RNA	83
3.	Inhibitors of ACC with antibacterial properties	85
4.	Biotin carboxylase	85
	4.1 ATP site	85
	4.2 Optimization of pyridopyrimidines	87
	4.3 Amino-oxazole derivatives	88
	4.4 Benzimidazole carboxamide derivatives	89
5.	Biotin binding site	89
	5.1 Sulfoamidobenzamide	89
6.	Unknown binding site	91
	6.1 Pyrrolocin C and equisetin	91
7.	Carboxyltransferase	91
	7.1 Pyrrolidinediones	91
	7.2 Cinnamon derivatives	94

Methods in Enzymology, Volume 708
ISSN 0076-6879, https://doi.org/10.1016/bs.mie.2024.10.017
Copyright © 2024 Elsevier Inc. All rights reserved, including those for text and data mining, AI training, and similar technologies.

7.3 Thailandamide A	96
7.4 Dual-ligand and heterobivalent inhibitors	96
8. Concluding remarks	98
Author note	98
References	98

Abstract

Acetyl-CoA carboxylase catalyzes the first committed and regulated step in fatty acid synthesis in all animals, plants and bacteria. In most Gram-positive and Gram-negative bacteria, the enzyme is composed of three proteins: biotin carboxylase, biotin carboxyl carrier protein and carboxyltransferase. The reaction consists of two half-reactions. The first half reaction is catalyzed by biotin carboxylase and involves the carboxylation of the vitamin biotin which is covalently attached to the biotin carboxyl carrier protein. The second half reaction catalyzed by carboxyltransferase involves the transfer of the carboxyl group from biotin to acetyl-CoA to form malonyl-CoA. This chapter will describe the inhibitors of both the biotin carboxylase and carboxyltransferase components of bacterial acetyl-CoA carboxylase. Inhibitors that were used in the elucidation of the structure and mechanism of the enzyme will be discussed first. The second half will focus on inhibitors that also possess antibacterial activity.

Acetyl–CoA carboxylase (ACC) catalyzes the first committed and regulated reaction in the synthesis of long–chain fatty acids in animals, plants and bacteria (Lane et al., 1974; Cronan & Waldrop, 2002; Waldrop et al., 2012; Cronan, 2021). Inhibition studies of ACC initially played a role in determining the structure and catalytic mechanism of the enzyme. However, about 20 years ago the bacterial form of ACC was found to be a novel target for antibacterial development (Freiberg et al., 2004). At which point inhibition of bacterial ACC took on a whole new purpose. In this chapter, the initial focus will be on inhibitors of bacterial ACC used for structure and function studies followed by inhibitors of bacterial ACC that have antibacterial properties. Inhibitors of human and eukaryotic ACC used in the treatment of cancer and nonalcoholic fatty liver disease (NAFLD) have been extensively reviewed elsewhere (Wu & Huang, 2020). A comprehensive analysis of ACC inhibition could not only lead to new insight into the mechanism of the enzyme but also lead to new approaches for developing inhibitors with antibacterial properties.

1. Background
1.1 Reaction

ACC is a multifunctional enzyme that catalyzes the carboxylation of acetyl-CoA to form malonyl-CoA. The sequence of reactions is shown in Fig. 1. The first reaction involves the carboxylation of the vitamin biotin by biotin carboxylase (BC). The source of CO_2 is bicarbonate which is activated with ATP to form a carboxyphosphate intermediate (Ogita & Knowles, 1988). The reaction requires two equivalents of Mg^{2+}, one of which is chelated to ATP such that the metal-nucleotide complex (Mg^{2+}-ATP) is the substrate (Waldrop et al., 2012). The vitamin biotin is covalently attached to a protein appropriately named biotin carboxyl carrier protein (BCCP). In the second reaction, carboxyltransferase (CT) transfers the carboxyl group on the BCCP-bound biotin to acetyl-CoA to form malonyl-CoA.

1.2 Protein structure

In most Gram-positive and Gram-negative bacteria, the biotin carboxylase, BCCP and carboxyltransferase are separate proteins that form a complex in vivo (Guchhait et al., 1974) but each protein can be isolated, and the biotin carboxylase and carboxyltransferase subunits retain their enzymatic activity and will use free biotin as a substrate (Guchhait et al., 1974).

The biotin carboxylase component is a homodimer where each monomer in the homodimer contains a complete active site (Waldrop et al., 1994a; Janiyani et al., 2001; de Queiroz & Waldrop, 2007). In contrast, the carboxyltransferase component is a heterotetramer where

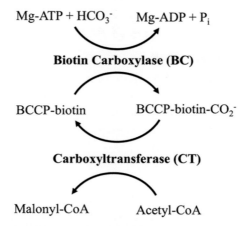

Fig. 1 Reaction catalyzed by acetyl-CoA carboxylase.

the active sites lie at the interface of an αβ dimer (Bilder et al., 2006; Silvers et al., 2016). The BCCP subunit is a monomer with two domains. The C-terminal domain (the last 87 residues) contains the lysine residue to which the biotin moiety is covalently attached (Athappilly & Hendrickson, 1995; Fall et al., 1976). The C-terminal domain folds independently and can be purified and the three-dimensional structure has been determined (Athappilly & Hendrickson, 1995; Chapman-Smith et al., 1994). Moreover, the C-terminal domain can act as a substrate for the both biotin carboxylase and carboxyltransferase (Blanchard et al., 1999a). On the other hand, the N-terminal domain (the first 69 residues) of BCCP has properties that suggest it is an intrinsically disordered protein. The first property is that the N-terminal domain does not migrate correctly on SDS-PAGE (Tompa, 2002; Blocquel et al., 2012), or on a SEC column (Uversky, 2002). On SDS-PAGE the N-terminal domain runs at about 16 kDa, while the calculated mass is 8 kDa based on amino acid sequence (Fig. 2).

Moreover, an intrinsically disordered N-terminal domain may explain why BCCP runs on SDS-PAGE at 22.5 kDa but has a calculated mass of 16.7 kDa (Li & Cronan, 1992a). Second, a crystal structure of the BC-BCCP complex showed electron density for the C-terminal domain but no electron density for the N-terminal domain (Broussard et al., 2013a). Third, intrinsically disordered proteins contain a higher-than-normal number of charged residues which are most often negative at neutral pH (Uversky, 2019). This is consistent with the N-terminal domain of BCCP having a theoretical and experimental pI of 4.5 (Li & Cronan, 1992a).

2. Inhibitors for analysis of ACC structure and function
2.1 Biotin carboxylase

The first studies on the inhibition of biotin carboxylase were used for analyzing the structure and function of ACC. Two fundamental techniques were utilized for the structure-function analyses: namely, x-ray crystallography and steady-state enzyme kinetics.

2.2 Enzymatic assay

For most of the kinetic inhibition studies described below, the activity of biotin carboxylase was determined using a common coupled enzyme assay for ADP shown in Fig. 3.

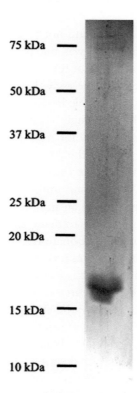

Fig. 2 Pure N-terminal domain (NTD) of biotin carboxyl carrier protein (BCCP) on a 15 % SDS-PAGE gel. The standards were run on the same gel and the positions of the known MW are depicted as lines.

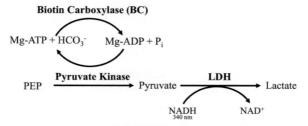

Fig. 3 Assay for biotin carboxylase. The activity of biotin carboxylase was determined using a coupled enzyme assay for ADP production using pyruvate kinase and lactate dehydrogenase (LDH) and the oxidation of NADH was followed at 340 nm.

The production of ADP was monitored using pyruvate kinase (30 U/ml) and lactate dehydrogenase (43 U/ml) and the oxidation of NADH was followed at 340 nm. Assays that measure the production of P_i (specifically the Enz check phosphate assay kit) were found not to be economically feasible. The activity of the coupling enzyme(s) was not particularly high. This required significantly increasing the amounts of the coupling enzyme(s) which in turn significantly increased the costs of the assay. Decreasing the amount of biotin carboxylase in the assay was not feasible because the signal to noise ratio of the phosphate assay was not high enough. Thus, the pyruvate kinase, lactate dehydrogenase coupled enzyme assay for ADP proved to be the most reliable assay for biotin carboxylase and for the ACC complex.

It is important to note that biotin carboxylase has three substrates: ATP, bicarbonate, and biotin. When doing inhibition studies, it is common to determine the inhibition pattern of an inhibitor versus each substrate. For biotin carboxylase it is only possible to determine inhibition patterns versus ATP and biotin. Variation of bicarbonate is difficult at best and inaccurate at worst. This is because the concentration of bicarbonate is determined by the pH of the reaction mixture. It is possible to vary bicarbonate, but it requires all the reactions been done in a stoppered cuvette and all solutions must be sparged with N_2 to remove the endogenous bicarbonate and CO_2 (Blanchard et al., 1999a). This was the method used to determine the K_m for bicarbonate of 0.37 ± 0.04 mM (Blanchard et al., 1999a). The inhibition assays of biotin carboxylase described below were run at pH 8.0 which gives a bicarbonate concentration of 0.5 mM (Asada, 1982) and is very close to the K_m value. In other words, no exogenous bicarbonate was added to the reaction because the concentration of bicarbonate at pH 8.0 is sufficient to measure initial velocities. Initial studies of biotin carboxylase added bicarbonate to the assay mixture (Blanchard et al., 1999a, 1999b; Levert et al., 2000), however, current assays of biotin carboxylase omit the bicarbonate from the assay mixture with no change in enzyme activity (Larson et al., 2020; Craft & Waldrop, 2022). Moreover, having a substrate whose concentration is pH dependent also precludes determining the pH dependence of inhibition (i.e. a pK_i vs. pH profile) for biotin carboxylase. All pH-dependent kinetic analyses of biotin carboxylase show a half-bell on the low side with a pK_a value of approximately 6.3 which is the pK_a value of carbonic acid (Asada, 1982) Adding a high concentration of bicarbonate (>20 mM) to the reaction mixture at any pH value is not feasible because if the reaction is open to the environment it will equilibrate with atmospheric CO_2 and the concentration of bicarbonate will decrease with decreasing pH (Asada, 1982).

2.3 Product inhibitors

One of the first inhibitors of biotin carboxylase was discovered using crystallography. One of the products of the reaction, phosphate (PO_4^-) was bound to the enzyme from *E. coli* when the three-dimensional structure was determined (Waldrop et al., 1994a, 1994b). The location of the phosphate group helped to identify the active site because this was the first three-dimensional structure of biotin dependent carboxylase. The finding was serendipitous and resulted from the crystallization solution being 10 mM potassium phosphate (pH 7.0). Interestingly, subsequent crystal structures of biotin carboxylase revealed that phosphate and the substrates bicarbonate and free biotin all bind in the same place in the active site (Chou et al., 2009).

The first crystal structural of biotin carboxylase with phosphate bound also revealed that the enzyme was in the ATP-grasp superfamily of proteins (Murzin, 1996; Galperin & Koonin, 1997). This family of enzymes catalyzes the coupling of a carboxyl group to either nitrogen or sulfur (Galperin & Koonin, 1997). ATP is used to activate the carboxyl group by forming an acylphosphate intermediate. The enzymes in the ATP-grasp superfamily have a flexible domain (more commonly called the B domain) that closes over the active site upon nucleotide binding (Waldrop et al., 2012; Thoden et al., 1999, 2000).

Crystallography was also useful in characterization of the product ADP and a structural analog of the product carboxybiotin (Fig. 4A), N1′-methoxycarbonyl biotin methyl ester (Fig. 4B) (Amspacher et al., 1999).

Unsurprisingly, ADP is bound in the ATP binding site while N1′-methoxycarbonyl biotin methyl ester is bound in the biotin binding site (Blanchard et al., 1999b). The crystal structure of biotin carboxylase bound to N1′-methoxycarbonyl biotin methyl ester was particularly informative because it showed that the carboxyl methyl ester moiety was orthogonal to

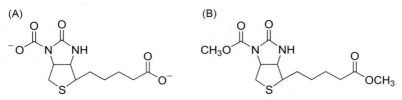

Fig. 4 Carboxybiotin Analog. (A) Carboxybiotin; (B) N1′-methoxycarbonyl biotin methyl ester.

$$H_2N\text{-}CO_2\text{-}PO_3{}^{2-} + Mg\text{-}ADP + d\text{-}biotin \xrightleftharpoons{BC} Mg\text{-}ATP + H_2N\text{-}CO_2{}^- \xrightarrow{spontaneous} NH_3 + CO_2$$

Fig. 5 Biotin carboxylase catalyzes the synthesis of ATP from carbamoyl phosphate and ADP.

the ureido ring of biotin when bound to the enzyme. In contrast, in the crystal structure of N1′-methoxycarbonyl biotin methyl ester the carboxyl methyl ester moiety is in plane with the ureido ring. The orthogonal conformation of carboxybiotin in the active site had been postulated because the resonance between the carboxyl group and the ureido ring is broken which lowers the energy barrier for carboxyl transfer to the acceptor molecule.

To determine the inhibition constants of ADP and Pi a slow alternate reaction catalyzed by biotin carboxylase was utilized. Biotin carboxylase catalyzes the synthesis of ATP from carbamoyl phosphate and ADP shown in Fig. 5 (Polakis et al., 1972).

The assay for this reaction is the standard coupled assay for ATP using hexokinase and glucose-6-phosphate dehydrogenase. Phosphate ($PO_4{}^-$) was found to be competitive with respect to carbamoyl phosphate with a K_i value of 2.2 $\pm$ 0.2 mM (Broussard et al., 2015). The K_m value for ADP, which in this case is a substrate, was 0.08 $\pm$ 0.01 mM (Blanchard et al., 1999a). The binding constant for the carboxybiotin analog N1′-methoxycarbonyl biotin methyl ester has not been determined.

2.4 ATP analogs

Two crystal structures of biotin carboxylase bound to the ATP analogs AMPPCP and AMPPNP (adenosine 5′-(β,γ-methylene)triphosphate and β,γ-imidoadenosine 5′-triphosphate, respectively) provided a structural understanding for why biotin carboxylase has a very slow ($k_{cat} = 0.07$ min^{-1}) bicarbonate-dependent ATPase activity (Blanchard et al., 1999a). The ATP analogs AMPPCP and AMPPNP have a methylene group and a nitrogen in place of the bridging oxygen between the β and γ phosphates, respectively The structure with AMPPCP and $HCO_3{}^-$ bound showed the closest oxygen on bicarbonate is 5.8 Å away from the γ-phosphate group of AMPPCP (Broussard et al., 2015). This is important because it prevents unproductive hydrolysis of ATP when biotin is not bound. The structure with AMPPNP provides another possible explanation for the very slow hydrolysis of ATP by biotin carboxylase in the absence of biotin (Mochalkin et al., 2008). The phosphate chain of AMPPNP is folded back on itself or, in other words, it is bound in an unproductive conformation. One or both structures could explain the slow bicarbonate-dependent ATPase activity of biotin carboxylase.

2.5 Carboxyphosphate analogs

Since the source of CO_2 for biotin dependent enzymes is bicarbonate the carboxyl group must be activated for nucleophilic attack by biotin. This is accomplished with ATP where bicarbonate reacts with the γ-phosphate group of ATP to form a carboxyphosphate intermediate (Ogita & Knowles, 1988). Carboxyphosphate (Fig. 6A) is a very unstable molecule with an estimated half-life of about 70 msecs (Ogita & Knowles, 1988).

In fact, carboxyphosphate has never been isolated from a biotin-dependent carboxylase. Therefore, stable analogs have been used to gain insight into the role of carboxyphosphate in the catalytic mechanism of biotin carboxylase. Phosphonoacetic acid is the closest analog to carboxyphosphate with the bridging oxygen replaced with a methylene group (Fig. 6B). Kinetic characterization revealed competitive inhibition versus ATP with a K_i value of 7.4 ± 1.8 mM (Blanchard et al., 1999b).

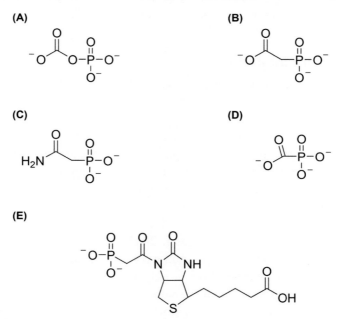

Fig. 6 Carboxyphosphate Analogs. (A) Carboxyphosphate; (B) Phosphonoacetic acid; (C) Phosphonoacetamide; (D) Phosphonoformate; (E) BP1 (N'-(phosphonoacetyl)biotin).

Crystallization of biotin carboxylase with phosphonoacetic acid has not been successful. However, phosphonoacetic acid was crystallized in the biotin carboxylase domain of pyruvate carboxylase (Lietzan et al., 2011). Phosphonoacetic acid bound in two possible orientations in the phosphate binding site observed in the first biotin carboxylase structure (Waldrop et al., 1994a). Moreover, the crystal structures of biotin carboxylase with the carboxyphosphate analogs, phosphonoacetamide (Fig. 6C) and phosphonoformate (Fig. 6D) were determined (Broussard et al., 2015). Both phosphonoacetamide and phosphonoformate also bound in the phosphate binding site observed in the first crystal structure of biotin carboxylase (Waldrop et al., 1994a).

To mimic the interaction between carboxyphosphate and biotin, the reaction intermediate N'-(phosphonoacetyl)biotin (referred to hereafter as BP1) shown in Fig. 6E was synthesized (Amspacher et al., 1999) and its inhibitory properties were characterized (Blanchard et al., 1999b). BP1 was competitive versus ATP with a K_i of 8.4 $\pm$ 1.1 mM and exhibited non-competitive inhibition versus biotin. This was the first biotin derived inhibitor of biotin carboxylase, and the inhibition patterns aided in determining the kinetic mechanism of the enzyme along with substrate inhibition studies described below.

2.6 Substrate inhibition

Biotin exhibits substrate inhibition when used with the isolated biotin carboxylase component of ACC (Blanchard et al., 1999b). This is most likely not observed in vivo because biotin is covalently attached to BCCP and never reaches the concentration needed to exhibit substrate inhibition. It is tempting to speculate that preventing substrate inhibition from occurring in vivo is why there is little to no free biotin in *E. coli* because it is covalently attached to the protein BCCP.

Nonetheless, substrate inhibition is usually considered a nuisance, however, it can be useful for determining the order of substrate binding and product release (Cook & Cleland, 2007). For biotin carboxylase, the substrate ATP was varied at non-inhibitory concentrations (see below) at different fixed levels of biotin both at non-inhibitory concentrations and at inhibitory concentrations. Then the slopes and y-intercepts of the double-reciprocal plot are graphed as a function of the biotin concentration. Both the slope and intercept graphs are non-linear indicating that biotin exhibits noncompetitive substrate inhibition. This means that biotin can bind when ATP concentrations are either very low or very high.

Combining the structural data with the kinetics of inhibition data led to a putative kinetic mechanism for substrate binding and product release in biotin carboxylase. Bicarbonate and ATP bind first in a random fashion followed by the binding of biotin. Carboxybiotin is the first product released followed by phosphate and then ADP.

It is important to note that substrate inhibition by ATP has also been reported (Chou & Tong, 2011). The type of substrate inhibition (e.g. competitive, uncompetitive, or noncompetitive) was not determined. However, a crystal structure of biotin carboxylase with two ADP molecules bound showed one ADP bound in the ATP binding site while the other ADP bound in the bicarbonate/biotin binding site. Again, while substrate inhibition by ATP may seem like a nuisance the crystal structure identifies the bicarbonate/biotin site as a potential target for inhibitors utilizing a nucleotide scaffold.

2.7 Inhibition by BCCP

A surprising type of substrate inhibition was observed when BCCP was tested as a substrate for biotin carboxylase. When biotin carboxylase activity was measured with full-length BCCP (i.e. containing both the N and C-terminal domains) in place of biotin there was no activity, that is, no hydrolysis of ATP to ADP and P_i (Broussard et al., 2013b). Only when carboxyltransferase and acetyl-CoA were added was activity observed. This makes sense physiologically because the cell would only want to hydrolyze ATP if carboxyltransferase and acetyl-CoA were present to accept the carboxyl group. That is, if biotin carboxylase was only bound to BCCP and not carboxyltransferase there would not be any hydrolysis of ATP. Since the C-terminal domain of BCCP which contains the biotin moiety was found to be an excellent substrate for biotin carboxylase (8000-fold increase in catalytic efficiency compared to biotin) (Blanchard et al., 1999a), the N-terminal domain must be the domain of BCCP that is causing the inhibition. The mechanism of inhibition by the N-terminal domain is not known.

2.8 Inhibition by inactive mutants of biotin carboxylase

Biotin carboxylase is a homodimer where each monomer contains a complete active site (Waldrop et al., 1994a). Therefore, does each monomer undergo catalysis independent of the other monomer or is there communication between the monomers? To answer this question heterodimers of biotin carboxylase were generated. In the heterodimer one subunit was wild type while the other subunit was from a mutant form of biotin carboxylase. The various mutant subunits had an active site mutation that decreased the activity 100-fold or more (Janiyani et al., 2001). The heterodimeric forms of biotin carboxylase were generated in vivo from one plasmid which contained the gene for the wild-type enzyme with an N-terminal FLAG tag and the gene for the mutant enzyme with an N-terminal His tag. The heterodimeric enzymes were purified by running a nickel affinity column first followed by an anti-FLAG column. Each of the heterodimeric enzymes had activity greater (0.35 to 11.5-fold) than the form of the enzyme where both subunits were mutant but significantly less (27.7 to 285.1-fold) than when both monomers were wild type. This indicated that the mutant form of the enzyme exerted a dominant-negative effect on the wild-type monomer. The interpretation of the data was an extreme form of "half the sites reactivity" where one active site is catalyzing a reaction while the other active site is releasing products. In other words, biotin carboxylase catalysis is oscillating, or the two monomers alternate their catalytic cycles. Modeling and numerical simulation of the data were consistent with this explanation of the results (de Queiroz & Waldrop, 2007). Moreover, a crystal structure of *E. coli* biotin carboxylase co-crystallized with the ATP analog $AMPPCF_2P$ showed binding to only a single subunit where the B-domain was in the closed conformation while the unliganded subunit was in the open conformation (Mochalkin et al., 2008). Thus, it is possible from a structural perspective for the two monomers of biotin carboxylase to be in two different conformations, one active and the other inactive.

2.9 Negative feedback inhibition by palmitoyl-acyl carrier protein

ACC exhibits classic negative feedback inhibition by the end-product of the fatty acid synthetic pathway, palmitoyl-acyl carrier protein (palmitoyl-ACP). The inhibition of acetyl-CoA carboxylase by palmitoyl-acyl carrier protein was first reported in 2001 but the mechanism of inhibition was not determined (Heath & Rock, 1996). When palmitoyl-ACP was tested as an inhibitor of the ACC complex which was formed in vivo and isolated, no inhibition was observed (Evans et al., 2017). However, when the three components of ACC, biotin carboxylase, BCCP and carboxyltransferase were added separately and incubated with palmitoyl-ACP for 5 min there was inhibition of enzymatic activity. The inhibition exhibited a pronounced hysteresis. This delay in the inhibition of ACC prevents inhibition by ACC by sudden spikes in the concentration of palmitoyl-ACP. In other words, ACC is only inhibited when the concentration of palmitoyl-ACP increases and remains elevated. No inhibition of isolated biotin carboxylase or carboxyltransferase was observed, suggesting that allosteric inhibition of the ACC complex was occurring. This hypothesis was supported by the fact that palmitoyl-ACP exhibited partial competitive inhibition with respect to acetyl-CoA which means the inhibitor and substrate bind at different sites. Kinetic studies indicated that the palmitoyl-ACP inhibited the binding of carboxyltransferase to the biotin carboxylase-BCCP complex. Structure activity studies of palmitoyl-ACP indicated that it was the pantothenic acid moiety that was involved in inhibiting ACC. In fact, pantothenic acid exhibited the same type of hysteretic inhibition that palmitoyl-ACP did. These results suggest that analogs of pantothenic acid could be exploited as possible antibacterial agents that target ACC.

2.10 Carboxyltransferase

In contrast to biotin carboxylase, the two main techniques used for characterization of inhibitors of the carboxyltransferase component of acetyl-CoA carboxylase were steady-state enzyme kinetics and electrophoretic mobility shift assays (EMSA).

2.11 Enzymatic Assay

The assay for the isolated carboxyltransferase component of ACC is shown in Fig. 7.

It involves running the reaction in reverse or the non-physiological direction where the carboxyl group of malonyl-CoA is transferred to biotin to form carboxybiotin and acetyl-CoA (Guchhait et al., 1975). The production of acetyl-CoA is coupled to citrate synthase and malate

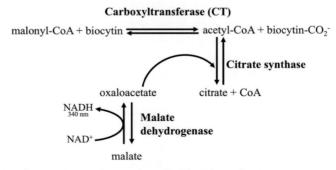

Fig. 7 Assay for measuring the activity of carboxyltransferase.

dehydrogenase. It is important to note that these two enzymes catalyze the first and last reactions of the Tricarboxylic Acid Cycle (TCA). The assay takes advantage of the fact that these two reactions play an important role in the regulation of the TCA cycle where the equilibrium for the reaction catalyzed by malate dehydrogenase lies far toward the substrate malate with only a small amount of oxaloacetate present at any given time. Malate is a very stable molecule while oxaloacetate being a β-keto acid is an unstable molecule. When acetyl-CoA is produced either by metabolism or in this case by carboxyltransferase it binds to citrate synthase along with oxaloacetate to produce citrate. This shifts the equilibrium of the malate dehydrogenase reaction towards the formation of oxaloacetate resulting in the oxidation of malate by NAD^+ to form NADH which absorbs at 340 nm. A typical reaction mixture contained the following: 0.5 mM NAD^+, 0.6 mg/ml BSA, 10 mM malic acid, 0.12 mg/ml malate dehydrogenase, 0.05 mg/ml citrate synthase, 100 mM HEPES-KOH pH 7.8. In the absence of carboxyltransferase, no background decarboxylation of oxaloacetate is observed and no lag in the time course is observed upon addition of carboxyltransferase.

Different biotin analogs were tested as substrates for the assay described above. Biotin methyl ester (Fig. 8A) and biocytin (Fig. 8B) were found to have V_{max} values almost 600-fold higher than biotin (Fig. 8C). Biocytin was used for all the inhibition studies described below because it was water soluble whereas biotin methyl ester must be dissolved in DMSO.

2.12 Biotin analogs

The first inhibition studies of carboxyltransferase examined the biotin analogs desthiobiotin (Fig. 8D) and 2-imidazolidone (Fig. 8E). Both compounds were poor inhibitors (K_i values of 54 mM for desthiobiotin and 1 M for 2-imidazolidone) of the enzyme (Blanchard & Waldrop, 1998). The one common structural feature of both compounds was neither of them contained a thiophene ring thereby demonstrating the importance of that moiety of biotin in binding to the enzyme. The other interesting finding with desthiobiotin and 2-imidazolidone is that both compounds exhibited parabolic noncompetitive inhibition which meant that two molecules of inhibitor bound to the enzyme. One inhibitor most likely binds in the biotin binding site but the binding site for the other inhibitor is not known.

Because desthiobiotin and 2-imidazolidone bound to carboxyltransferase with such weak affinity, a bisubstrate analog incorporating acetyl-CoA and biotin was synthesized and tested as an inhibitor (Fig. 8F)

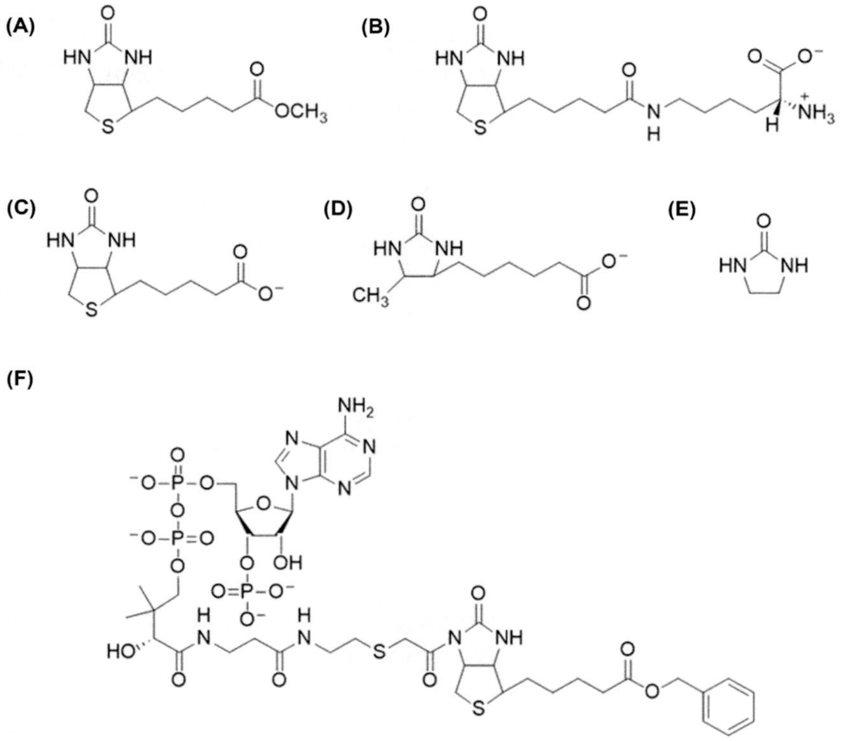

Fig. 8 Biotin analogs. (A) Biotin methyl ester; (B) Biocytin; (C) Biotin; (D) Desthiobiotin; (E) 2-imidazolidone; (F) Bisubstrate analog.

(Levert & Waldrop, 2002). The bisubstrate analog exhibited competitive inhibition versus malonyl-CoA and noncompetitive inhibition versus biocytin. These inhibition patterns indicate that substrate binding is ordered, with malonyl-CoA binding first. Moreover, the inhibition constant for the bisubstrate analog was 23 ± 2 μM. The bisubstrate analog had an affinity for carboxyltransferase that was 350 times greater than the substrate biotin. Unfortunately, the bisubstrate analog was not useful for structural studies of carboxyltransferase because NMR analysis showed that it underwent very slow decomposition (unpublished observations).

2.13 Inhibition by BCCP

The BCCP component inhibits or is not a substrate for isolated carboxyltransferase (Broussard et al., 2013b). However, as with biotin carboxylase, the C-terminal domain of BCCP which contains the biotin moiety was found to be an excellent substrate for carboxyltransferase in the reverse reaction (2000-fold increase in catalytic efficiency compared to biotin) (Blanchard et al., 1999a). Therefore, it is the N-terminal domain of BCCP that inhibits enzymatic activity, and the mechanism of inhibition is not known. Thus, the fact that the N-terminal domain of BCCP inhibits both biotin carboxylase and carboxyltransferase strongly suggest a regulatory role in the activity of ACC. To this end, the recent cryo-EM structure of ACC shows it is tubular with the N-terminal domain of BCCP facing inward and interacting with the N-terminal domains of other BCCP subunits to hold the tube together. Thus, the N-terminal domain does not inhibit biotin carboxylase and carboxyltransferase while all the subunits are incorporated into the active filamentous form. However, when biotin carboxylase and carboxyltransferase are not incorporated into the active filamentous form of ACC the N-terminal helps to make sure they do not have any enzymatic activity preventing wasteful use of ATP.

2.14 Inhibition by DNA

When the three-dimensional structure of the carboxyltransferase component of ACC was determined it was a surprise to find a zinc-finger domain on the amino terminus of the β subunit (Bilder et al., 2006). The zinc-finger domain was a Cys4 type and is only found in bacterial ACC. The function of the zinc-finger domain was not immediately apparent. Electrostatic surface potentials of carboxyltransferase showed the zinc-finger domain had a net-positive potential while the rest of protein had a net-negative potential. Therefore, because zinc-finger domains are commonly found in proteins that bind DNA, the ability of DNA to bind and inhibit carboxyltransferase was examined. In addition to enzyme kinetic analyses, DNA binding to carboxyltransferase was also analyzed by electrophoretic mobility shift assays (EMSA).

Carboxyltransferase was found to bind DNA non-specifically (Benson et al., 2008). A 4-nt ssDNA composed of all four nucleotides, as well as a 30-nt ssDNA and 30-nt dsDNA all inhibited carboxyltransferase activity. A thymidine dimer did not inhibit carboxyltransferase. Therefore, at least a 3-nt sequence is needed for binding to the enzyme. When binding of the 30-nt ssDNA to carboxyltransferase was characterized by EMSA, binding was determined to be cooperative with a half-maximal saturation of $0.48 \pm 0.02 \, \mu M$. Moreover, EMSA analysis of various dsDNA fragments ranging from 175 bp to 500 bp all showed cooperative binding with half-maximal saturation levels ranging from $0.67 \pm 0.04 \, \mu M$ to $0.95 \pm 0.03 \, \mu M$ and Hill coefficients of 1.5 ± 0.2 to 2.6 ± 0.2.

The 30-nt ssDNA was used as an inhibitor to determine the type of inhibition with respect to each substrate. Competitive inhibition was observed for both malonyl-CoA and biocytin with K_i values of $85 \pm 10\,\mu M$ versus malonyl-CoA and $34.2 \pm 4.0\,\mu M$ versus biocytin. The DNA mimic heparin also inhibited carboxyltransferase and exhibited competitive inhibition with respect to both substrates. However, the inhibition constants with heparin were $1.2 \pm 0.1\,\mu M$ versus malonyl-CoA and $2.2 \pm 0.3\,\mu M$ versus biocytin. The competitive inhibition patterns with DNA were confirmed by EMSA, which showed that malonyl-CoA and the bisubstrate analog described above (Fig. 8F) both inhibited formation of a carboxyltransferase-DNA complex.

While the kinetic and EMSA studies showed a linkage between catalysis and DNA binding, this raised the question of whether the zinc-finger domain was the structural feature that mediates this linkage. To answer this question, site-directed mutagenesis of the four cysteine residues that form the zinc-finger domain was performed (Meades et al., 2010a). Each cysteine residue was mutated to alanine individually and the double, triple, and quadruple mutants were constructed. Each mutant was tested for enzymatic activity and the ability to bind DNA using EMSA. The single mutants decreased enzymatic activity from 8 to 100-fold, while the double, triple, and quadruple mutants did not have any enzymatic activity. Analysis of DNA binding by EMSA showed the single mutants had diminished DNA binding activity, while all the multiple mutants did not bind DNA. Thus, the zinc-finger domain plays a significant role in both catalysis and nucleic acid binding.

The studies of inhibition by DNA and heparin showed that carboxyltransferase binds molecules with a net-negative charge, and that the enzymatic activity and nucleic acid binding properties are not separate functions. The two functions are intimately linked to each other: DNA binding interferes with catalysis, and vice versa. While DNA binding does not appear to be the physiological function of the zinc-finger domain, the detailed analysis of DNA binding could prove fruitful in the design of inhibitors of carboxyltransferase that target the zinc-finger domain and have antibacterial properties.

2.15 Inhibition by RNA

If carboxyltransferase binds DNA non-specifically then what is the role, if any, of the zinc-finger domain? The closest structural homolog of the zinc-finger domain was the zinc domain of the 50S ribosomal protein L37Ae

(Bilder et al., 2006; Meades et al., 2010a; Waldrop, 2011). Therefore, several different RNA molecules were tested for binding to carboxyltransferase using EMSA. The mRNA coding for the α and β subunits of carboxyltransferase were found to bind with the highest affinity with K_d values of 48 ± 8 nM and 83 ± 9 nM, respectively (Meades et al., 2010a). More importantly, the binding isotherm was hyperbolic and not cooperative as observed with DNA binding. This suggested the binding was specific for the mRNA coding for the α and β subunits of carboxyltransferase. Acetyl-CoA was found to inhibit binding of mRNA to carboxyltransferase. Moreover, the binding of carboxyltransferase inhibited translation of the mRNA coding for the α and β subunits, while acetyl-CoA relieved the inhibition. The inverse relationship between the substrate acetyl-CoA and mRNA coding for the α subunit of carboxyltransferase is illustrated in Fig. 9.

All of these findings for the binding of carboxyltransferase to the mRNA coding for the α and β subunits suggested a mechanism for the translational regulation of carboxyltransferase expression and activity (Meades et al., 2010a, 2011). The genes coding for the α and β subunits of carboxyltransferase are not in an operon but are located on opposite sides of the *E. coli* chromosome (Li & Cronan, 1992b). Yet, there must be stoichiometric expression of the genes to generate the $\alpha_2\beta_2$ heterotetramer of

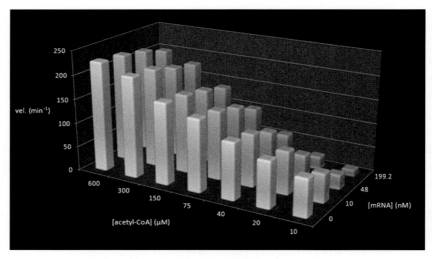

Fig. 9 Competition between Acetyl-CoA and mRNA.

carboxyltransferase. Moreover, ACC of which carboxyltransferase is a component is only needed during the log phase of *E. coli* growth. Therefore, a possible model for regulation of carboxyltransferase and ACC is that, during log phase of bacterial growth, acetyl-CoA levels significantly increase and compete with the mRNA for binding to carboxyltransferase which leads to an increase in enzymatic activity and a concomitant increase in the translation of carboxyltransferase. During stationary phase, acetyl-CoA levels decrease, and carboxyltransferase binds to the mRNA coding for the α and β subunits and inhibits translation.

3. Inhibitors of ACC with antibacterial properties

ACC catalyzes the first committed step in fatty acid biosynthesis in bacteria. Since fatty acids are primarily used for membrane biogenesis in bacteria, inhibition of ACC would be bactericidal. Therefore, small molecules that inhibit ACC could lead to the development of antibacterial agents. Given the increase in the number of antibiotic-resistant strains of pathogenic bacteria, inhibitors of ACC would be a novel class of antibacterial agents. Validation of ACC as a target for antibacterial agents comes from temperature sensitive mutations in the carboxyltransferase component and the observation that knockouts of the genes coding for any component of ACC could not be generated (Baba et al., 2006). Below are the compounds found to inhibit biotin carboxylase and carboxyltransferase and exhibit antibacterial activity.

4. Biotin carboxylase
4.1 ATP site
4.1.1 Pyridopyrimidines
The first compound to have antibacterial activity and inhibit biotin carboxylase was discovered by Pfizer (Miller et al., 2009). A whole-cell, high-throughput screen of 1.6 million compounds against *E. coli* yielded a class of molecules, the pyridopyrimidines, that inhibited growth. A representative example of a pyridopyrimidine is shown in Fig. 10A.

To determine the target of the pyridopyrimidines resistant strains of *E. coli*, *Haemophilus influenzae*, and *Moraxella catarrhalis* were selected using the compound in Fig. 10A and the mutations mapped to the gene coding

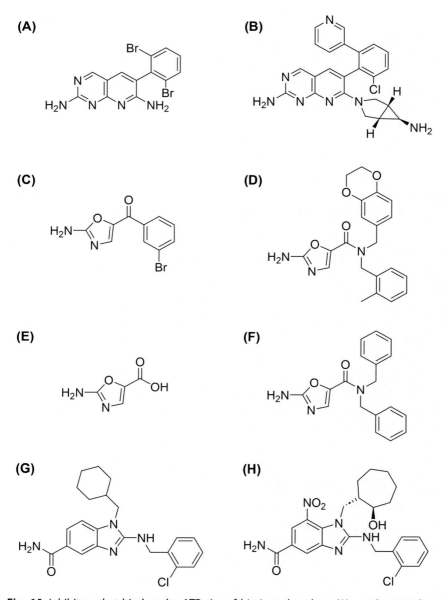

Fig. 10 Inhibitors that bind at the ATP site of biotin carboxylase. (A) pyridopyrimidine; (B) pyridopyrimidine derivative (7-(1R,5S)-6-Amino-3-azabicyclo[3.1.0]hexan-3-yl)-6-[2-chloro-6-(3-pyridyl)phenyl]pyrido[2,3-*d*]pyrimidin-2-amine; (C) amino-oxazole fragment; (D) amino-oxazole derivative (2-amino-N-(2,3-dihydro-1,4-benzodioxin-6-ylmethyl)-N-[(2-methylphenyl)methyl]-1.3-oxazole-5-carboxamide); (E) 2-aminooxazole-5-carboxylic acid; (F) amino-oxazole dibenzylamine; (G) benzimidazole carboxamide lead compound; (H) benzimidazole carboxamide optimized derivative (2-[(2-chlorophenyl)methylamino]-1-[[1R,2R)-2-hydroxycycloheptyl]methyl]-7-nitrosobenzimidazole-5-carboxamdie.

for biotin carboxylase. The resistant mutations localized in the ATP binding site of the enzyme. The molecule shown in Fig. 10A was found to have a K_d of 0.8 $\pm$ 0.2 nM using SPR. Crystal structures of *E. coli* biotin carboxylase and the pyridopyrimidines confirmed that the molecules do indeed bind in the ATP binding site. Interestingly, the library of 1.6 million molecules was assembled initially to target eukaryotic tyrosine kinases; screening the library against *E. coli* was an afterthought. Therefore, it's not surprising that this library would yield a molecule that binds in the ATP binding site of biotin carboxylase. The pyridopyrimidines were most effective against efflux compromised Gram-negative bacteria and exhibited little to no antibacterial activity versus Gram-positive bacteria. The reason for the difference in antibacterial activity was discovered in the resistant Gram-negative strains. One of the resistant mutations had threonine in place of an isoleucine at position 437 of biotin carboxylase (*E. coli* numbering). The position equivalent to 437 in Gram-positive bacteria is threonine. The crystal structure indicated that replacement of the isoleucine with threonine interfered with shape complementarity and contributed to the loss of hydrophobic interactions. While the pyridopyrimidines were not broad-spectrum antibacterial agents, their discovery firmly established that biotin carboxylase is a valid and novel target for antibacterial development.

4.2 Optimization of pyridopyrimidines

The former biotech company Achaogen in South San Francisco used the Pfizer pyridopyrimidine shown in Fig. 10A as a starting point for modifications to improve its antibacterial activity (Andrews et al., 2019). The Pfizer compound was a very effective antibacterial agent but only against efflux compromised strains of *E. coli* (Miller et al., 2009). The Achaogen group compared the physicochemical properties of the Pfizer compound to the physicochemical properties of other antibacterial agents that have cytoplasmic targets in Gram-negative bacteria. The antibacterial compounds with activity against Gram-negative bacteria were polar and charged whereas the Pfizer compound was non-polar and neutral. First, several derivatives at the 7-position of the pyridopyrimidine scaffold were synthesized and characterized for their inhibitory and antibacterial properties. These studies were followed by derivatization of the dihalo-aromatic ring of the Pfizer compound. Derivatives at these two vectors with the best biotin carboxylase inhibition and antibacterial properties were combined to generate a series of compounds, a representative of which is shown in

Fig. 10B. Compared to the original Pfizer pyridopyrimidine the Achaogen compound in Fig. 10B was more polar and the primary amine gave it a positive charge at neutral pH. The compound inhibited biotin carboxylase with an $IC_{50} \leq 15$ nM. Crystallographic analysis of the compound bound to biotin carboxylase from *Haemophilus influenzae* showed the primary amine interacted with two strictly conserved glutamic acid residues. Most importantly, the optimized physicochemical properties of the pyridopyrimidines led to antibacterial activity against wild-type *E. coli* and *Klebsiella pneumoniae*.

4.3 Amino-oxazole derivatives

After the discovery of the original pyridopyrimidine inhibitor validated biotin carboxylase as a target for antibacterial discovery, the Pfizer group turned to fragment-based drug design (FBDD) to find inhibitors of biotin carboxylase (Mochalkin et al., 2009). The premise behind FBDD is to screen molecules with molecular weights of less than 250 Da which increases the probability of finding a hit. Because the fragments are small, they are amenable to synthetic modification which is necessary because the binding affinities of the fragments are usually poor. A library of 5200 fragments were screened against biotin carboxylase using virtual screening and saturation transfer difference NMR (STD NMR). The hits derived from this initial screen were subjected to a secondary screen using a coupled enzyme assay for ACC that measures the production of phosphate. Six fragments were identified to have IC_{50} values less than 95 μM. The amino-oxazole fragment is shown in Fig. 10C and had an IC_{50} of 21.5 μM. Crystallographic analysis showed the fragment bound in the ATP binding site of the enzyme. Fragment growing was performed to generate the compound in Fig. 10D. This compound had an IC_{50} of 7 nM. Unfortunately, all the fragment-based compounds had little to no antibacterial activity against wild-type Gram-negative organisms because they were substrates for efflux pumps.

The fragment in Fig. 10C can be purchased in gram quantities as 2-aminooxazole-5-carboxylic acid (Fig. 10E). The carboxyl group serves as a convenient handle for the attachment of many different chemical moieties. Therefore, to significantly expand the number of amino-oxazole derived inhibitors, a computational approach was utilized (Mochalkin et al., 2009). A combinatorial library of 1.2×10^8 amino-oxazole derivatives was constructed and 9×10^6 of those derivatives underwent docking against biotin carboxylases from both Gram-positive and Gram-negative organisms. One of the most important conclusions was that attaching halogenated aromatic moieties to the amino-oxazole anchor significantly enhanced binding to the enzyme.

It is important to note that the inhibition patterns for an amino-oxazole inhibitor were different for the isolated biotin carboxylase component as compared to the ACC complex (Silvers et al., 2014). The amino-oxazole shown in Fig. 10F exhibited competitive inhibition versus ATP with the isolated biotin carboxylase component. This is consistent with the crystal structures showing that the amino-oxazole scaffold binds in the ATP binding site. However, the same amino-oxazole exhibited noncompetitive inhibition with the ACC complex. This is significant because the K_m for ATP in the ACC complex is about 1.7 μM (Broussard et al., 2013b). The intracellular concentration of ATP in log phase *E. coli* is 1–9 mM (Schneider & Gourse, 2004; Svenningsen et al., 2019). Thus, the ACC complex is always saturated with ATP. However, noncompetitive inhibition means the inhibitor can bind even when the substrate is at saturating levels. The molecular basis for the noncompetitive inhibition is not known but this finding will likely play a key role in determining the mechanism of catalysis by the ACC complex.

4.4 Benzimidazole carboxamide derivatives

While Pfizer was discovering the pyridopyrimidine inhibitors of biotin carboxylase, Schering-Plough discovered a completely different scaffold that inhibits biotin carboxylase (Cheng et al., 2009). Using affinity-selection mass spectrometry the Schering-Plough group screened a combinatorial library for ligands that bound to biotin carboxylase. One of the lead compounds is shown in Fig. 10G. Crystallographic studies showed the compound bound in the ATP binding site of biotin carboxylase. The three lead compounds underwent extensive optimization to generate 76 derivatives which were tested for their inhibition properties and antibacterial activity. The derivative shown in Fig. 10H had an IC_{50} of 0.01 μM and the lowest MIC value of 0.2 μg/ml against *E. coli*.

5. Biotin binding site
5.1 Sulfoamidobenzamide

So far, all the inhibitors of biotin carboxylase with antibacterial activity bind in the ATP site. However, there is one compound that binds in the biotin binding site. The company Microbiotix developed an in vivo screen for inhibitors of fatty acid synthesis in *E. coli* and *Pseudomonas aeruginosa* (Wallace et al., 2015). The compound shown in Fig. 11 was the most potent hit from that screen.

The molecule, 4-[[2-chloro-5-(phenylcarbamoyl)phenyl]sulphonylamino]benzoate, contains a sulfoamidobenzamide core and will be referred to as SABA hereafter. SABA exhibited antibacterial activity against efflux compromised strains of *E. coli* and *P. aeruginosa* and resistant mutations mapped to the gene coding for biotin carboxylase. Competitive inhibition patterns were observed when either ATP or biotin were varied (Craft & Waldrop, 2022). Since the inhibition patterns were inconclusive as to the binding site for SABA, multiple inhibition analyses were used. Multiple inhibition studies define the topological relationship between two inhibitors (e.g. do they bind to the same site or different sites) (Yonetani & Theorell, 1964; Northrop & Cleland, 1974). In multiple inhibition studies, initial velocities are measured at varying concentrations of one inhibitor while the other inhibitor is held constant. The first inhibitor is varied again but at a higher fixed concentration of the second inhibitor. The substrates are held at subsaturating levels. The data are plotted as 1/velocity versus the first inhibitor. To determine if SABA binds in the ATP site, the amino-oxazole shown in Fig. 10F was used, while desthiobiotin (Fig. 8D) was used to evaluate whether SABA bound to the biotin site. The multiple inhibition analysis showed that the binding of SABA and desthiobiotin were mutually exclusive, whereas SABA and the amino-oxazole could bind simultaneously. Thus, SABA binds in the biotin site. The tetrahedral sulfonamide moiety in SABA was observed by x-ray crystallography to bind in the phosphate binding site of biotin carboxylase (unpublished observations).

The explanation for the competitive inhibition pattern of SABA versus ATP was determined to be due to the observation that SABA could bind to the enzyme-ADP complex (Craft & Waldrop, 2022). The first two

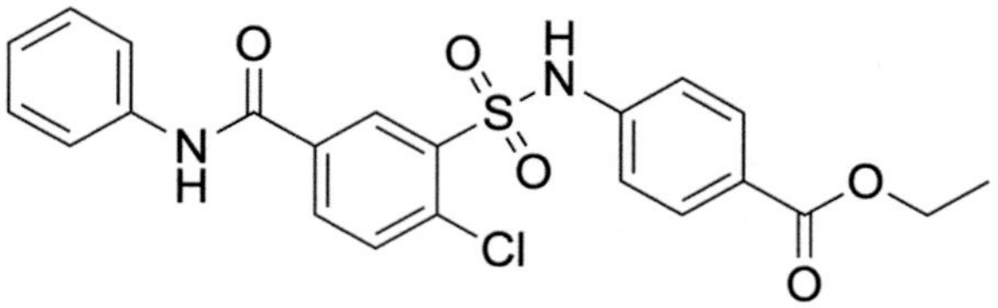

Fig. 11 SABA (4-[[2-chloro-5-(phenylcarbamoyl)phenyl]sulphonylamino]benzoate). Inhibitor that binds at the biotin site of biotin carboxylase.

products released after catalysis are carboxybiotin and phosphate leaving an enzyme–ADP complex. SABA binds and prevents the release of ADP (Fig. 12). The mechanism of inhibition by SABA is very similar to the mechanism of the common antibacterial agent triclosan on enoyl–ACP reductase (Ward et al., 1999).

6. Unknown binding site
6.1 Pyrrolocin C and equisetin

Pyrrolocin C and Equisetin (Fig. 13A and B, respectively) are fungal natural products with antibacterial activity against Gram-positive organisms such as *Staphylococcus aureus* and *Streptococcus pneumonia* (Larson et al., 2020; Patridge et al., 2015; Vesonder et al., 1979; Jadulco et al., 2014; Kakule et al., 2015). Metabolism studies showed a decrease in malonyl-CoA in the presence of Pyrrolocin C or Equisetin suggesting the target was acetyl-CoA carboxylase. Both Pyrrolocin C and Equisetin were found to inhibit the biotin carboxylase component of ACC (Larson et al., 2020), however, the binding site or mechanism of inhibition is unknown.

7. Carboxyltransferase
7.1 Pyrrolidinediones

In 2004 Freiberg et al. from Bayer discovered that the antibacterial agent moiramide B (Fig. 14A) inhibited the carboxyltransferase component of

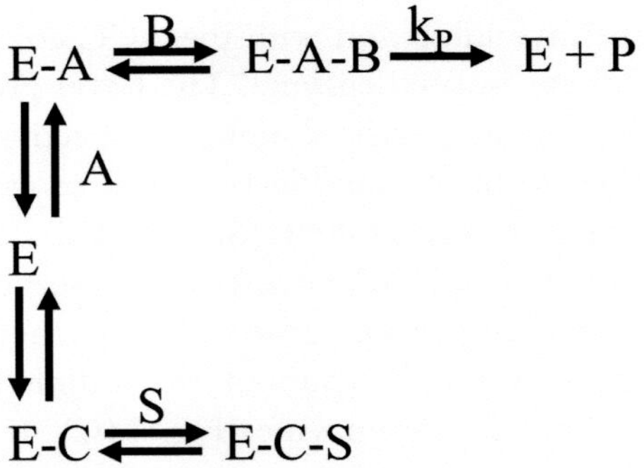

$$E\text{-}A \xrightleftharpoons{B} E\text{-}A\text{-}B \xrightarrow{k_P} E + P$$

$$\Updownarrow A$$

$$E$$

$$\Updownarrow$$

$$E\text{-}C \xrightleftharpoons{S} E\text{-}C\text{-}S$$

Fig. 12 Mechanism of inhibition of biotin carboxylase by SABA. In Fig. 12: E is biotin carboxylase, A is ATP, B is Biotin, C is ADP and S is SABA.

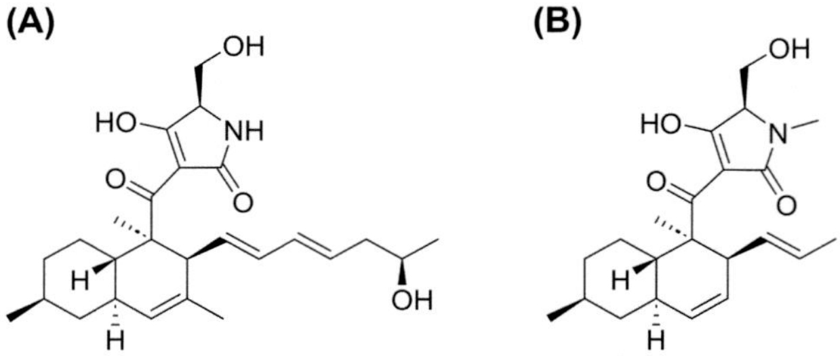

Fig. 13 (A) Pyrrolocin; (B) Equisetin.

ACC (Freiberg et al., 2004). This was the first report of a compound with antibacterial activity inhibiting ACC.

Moiramide B is a natural product that was isolated from a strain of *Pseudomonas fluorescens* found in the Moira Sound in Alaska (Vesonder et al., 1979). It is in the class of molecules called pyrrolidinediones which also includes andrimid (Fig. 14B). Andrimid was discovered in 1987 from a broth culture of a species of *Enterobacter* (Fredenhagen et al., 1987), while the Bayer group found that andrimid also inhibits carboxyltransferase (Freiberg et al., 2004). Moiramide B was found to exhibit competitive inhibition versus malonyl-CoA and noncompetitive inhibition versus biocytin. The K_i value versus malonyl-CoA was 5 nM. These patterns are consistent with the ordered addition of substrates with malonyl-CoA binding first and suggested that moiramide B bound to the active sites for both substrates. It is important to note that moiramide B also exhibited competitive inhibition versus acetyl-CoA in the ACC complex (Freiberg et al., 2004). This contrasts with inhibitors of biotin carboxylase, which exhibited noncompetitive inhibition with the ACC complex but competitive inhibition with the isolated enzyme. The Bayer group reported that moiramide B showed broad spectrum antibacterial activity.

The crystal structure of moiramide B bound to carboxyltransferase from *S. aureus* was determined in 2016 (Silvers et al., 2016). Consistent with the inhibition patterns, moiramide B was observed in the binding sites for both substrates. Surprisingly, the carbon in the succinimide ring that connects to the carbonyl group of the valine fragment is sp^2-hybridized when bound to the enzyme (Fig. 14C). It is not known if this is the enol tautomer or enolate anion. The corresponding C-H bond in andrimid has a pK_a of 6.8, thus, it is not clear if the molecule binds to the enzyme as the enol/enolate or the keto tautomer.

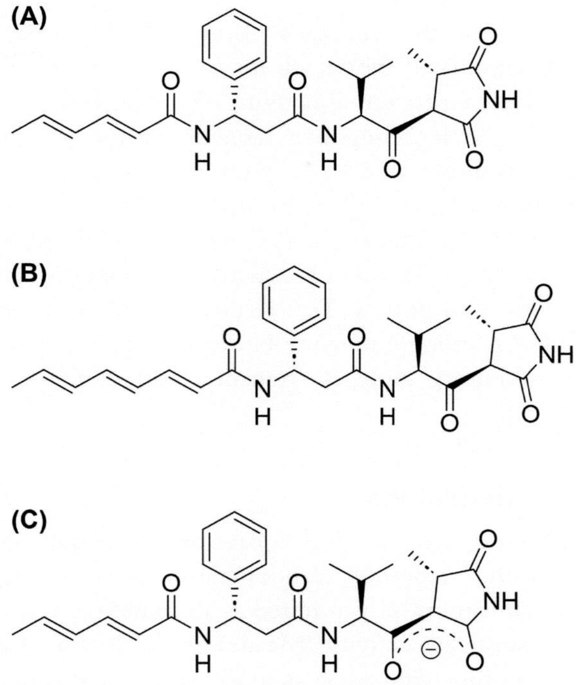

Fig. 14 (A) Moiramide B; (B) Andrimid; (C) enolate of Moiramide B.

Nevertheless, this tautomerization provides the succinimidyl amides of the pyrrolidinedione head group stronger electron density so the oxygen atoms can interact with the two oxyanion holes in the enzyme. The structural data was consistent with the findings of Pohlmann et al. where derivatives of the pyrrolidinedione head group abolished binding to the enzyme (Pohlmann et al., 2005). Structure activity studies of moiramide B found that every part of the molecule except the unsaturated fatty acid fragment was involved in binding to the enzyme (Silvers et al., 2016). This is consistent with the structural studies which showed the unsaturated fatty acid fragment did not interact with any part of the enzyme. In contrast, antibacterial testing of the fragments revealed that none of the fragments exhibited any antibacterial activity against efflux compromised *E. coli*. Only the complete moiramide B molecule had antibacterial activity. These results demonstrated that the role of the unsaturated fatty acid tail was to allow moiramide B entry into the cytoplasm.

Lastly, the discovery that moiramide B targeted the carboxyltransferase subunit of ACC stimulated the search for other inhibitors of the enzyme that might possess antibacterial activity. This interest prompted the development of a high-throughput screening assay for the carboxyltransferase which is shown in Fig. 15 (Santoro et al., 2006).

The production of acetyl-CoA by carboxyltransferase is coupled to citrate synthase which produces CoASH which reacts with 5,5′-dithiobis (2-nitrobenzoic acid) aka Ellman's Reagent which absorbs light at 412 nm after reaction with thiol groups. This results in a high-throughput screen where reactions containing a non-inhibitory molecule turn yellow, while those with an inhibitory molecule remain colorless in the wells of the microtiter plate.

7.2 Cinnamon derivatives

Cinnamon oil has long been known to have antibacterial properties. Since moiramide B firmly established carboxyltransferase as a target for antibacterial agents, cinnamon oil was tested as an inhibitor, and it did indeed inhibit carboxyltransferase activity (Meades et al., 2010b). Cinnamon oil did not inhibit biotin carboxylase activity. Since trans-cinnamaldehyde (Fig. 16A) makes up 70 % of cinnamon oil, it was tested for inhibition of carboxyltransferase . Trans-cinnamaldehyde was a competitive inhibitor versus biocytin with a K_i value of 3.8 ± 0.6 mM and an uncompetitive inhibitor versus malonyl-CoA. This combination of inhibition patterns indicates that trans-cinnamaldehyde binds in the biotin binding site after the binding of malonyl-CoA. The trans-cinnamaldehyde analog

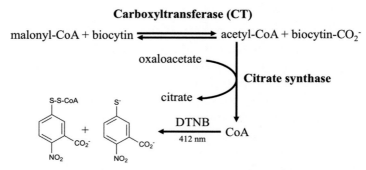

Fig. 15 High-throughput screening assay for carboxyltransferase. DTNB is 5,5′-dithiobis(2-nitrobenzoic acid.

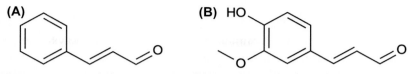

Fig. 16 Cinnamon derivatives: (A) Trans-cinnamaldehyde; (B) 4-hydroxy-3-methoxy-trans-cinnamaldehyde.

4-hydroxy-3-methoxy-trans-cinnamaldehyde (Fig. 16B) was also found to be an inhibitor of carboxyltransferase. Cinnamon oil, trans-cinnamaldehyde and 4-hydroxy-3-methoxy-trans-cinnamaldehyde all showed modest antibacterial activity against several strains of S. aureus.

Because 4-hydroxy-3-methoxy-trans-cinnamaldehyde absorbs light at 340 nm, it interfered with the coupled enzyme assay for carboxyltransferase. As a result, another assay was developed to measure the inhibition of carboxyltransferase. Capillary electrophoresis (CE) was used to measure the activity and, by inference, inhibition of carboxyltransferase. The substrate malonyl-CoA and the product acetyl-CoA differ by a single negative charge. CE can easily separate those two molecules, and they can be detected at 260 nm since both molecules contain a nucleotide. The enzyme is incubated with the substrates with and without inhibitor and aliquots are removed as a function of time and subjected to CE which quenches the reaction. The peaks corresponding to the substrate and product can be integrated to quantitate their levels. In the absence of inhibitor, the substrate peak, malonyl-CoA, decreases with time while the product peak, acetyl-CoA, increases with time. In the presence of 4-hydroxy-3-methoxy-trans-cinnamaldehyde there was a very slow decrease in malonyl-CoA and a very slow increase in acetyl-CoA, indicating inhibition. The advantages of CE for measuring inhibition of enzyme activity are that it is a direct measurement (i.e. there are no coupling enzymes that could interact with the inhibitor). Moreover, the volume of each injection is about 50 nanoliters, thereby conserving valuable reagents and enzymes. The disadvantage of CE is that initial velocities cannot be measured because the length of a single CE run is longer than the time course for the initial velocity. This means that the type or mechanism of inhibition (e.g. competitive) cannot be determined. CE assays of the bacterial ACC complex (Bryant et al., 2013) and human ACC (Nguyen et al., 2019) have also been developed.

7.3 Thailandamide A

The natural product thailandamide A is a linear polyene polyketide produced by *Burkholderia thailandensis* (Fig. 17) (Wu & Seyedsayamdost, 2018). This compound exhibited antibacterial activity against Gram-positive organisms. The target for Thailandamide A is carboxyltransferase (Wu & Seyedsayamdost, 2018; Wozniak et al., 2018). This was determined by mapping of resistant mutations. The mechanism by which thailandamide A inhibits carboxyltransferase has not been determined.

7.4 Dual-ligand and heterobivalent inhibitors

Treatment of bacterial infections with antibacterial agents inevitably leads to resistance. Resistance is particularly likely if the antibacterial agent interacts with a single target. One possible approach to circumvent bacterial resistance is to combine two antibacterial agents into a single molecule. The hypothesis is that the antibacterial agent will kill the organism before it can acquire the necessary number of resistant mutations. This was the approach taken by Silvers et al (Silvers et al., 2014). to inhibit the ACC complex. The inhibitor was a dual-ligand that contained the biotin carboxylase inhibitor amino-oxazole and the carboxyltransferase inhibitor moiramide B (Fig. 18A).

The two inhibitors were linked with a 15 (20 Å) and 7 (9 Å) carbon hydrocarbon chain. The dual-ligand with a 15-carbon linker did not exhibit any antibacterial activity because it was not transported into the cytoplasm. In contrast, the dual-ligand with the 7-carbon linker did exhibit antibacterial activity against wild-type Gram-positive organisms. Thus, the size of the linker was critical for entering the cytoplasm. Most importantly, the mutant selection window (MSW) for the dual-ligand was lower than the MSW for either of the parent compounds (Silvers et al., 2014). The smaller the MSW for a compound the lower the probability of developing resistance (Zhao & Drlica, 2002).

The major shortcoming of the dual-ligand studied by Silvers et al (Silvers et al., 2014). was that the amino-oxazole inhibitor of the biotin carboxylase component was about 50-fold less potent than the moiramide

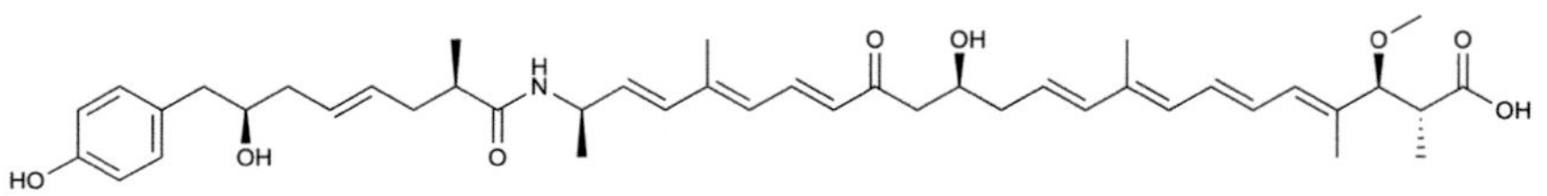

Fig. 17 Thailandamide A.

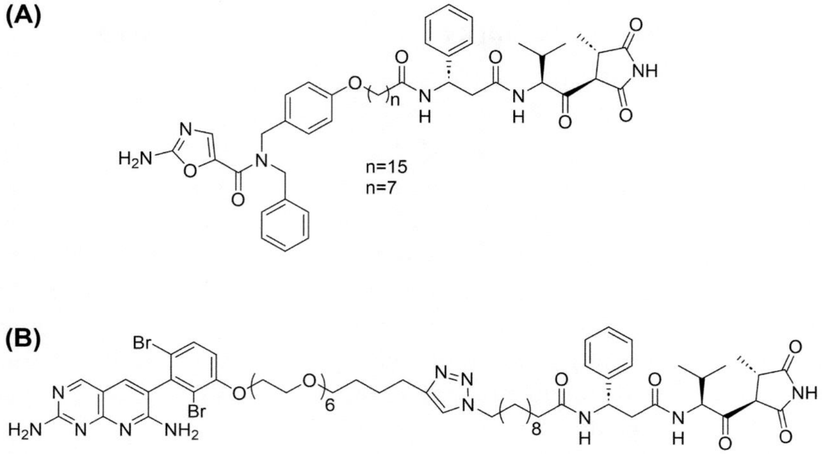

Fig. 18 (A) Dual-ligand inhibitor (2-amino-N-benzyl-N-[[4-[8-[[(1S)-3-[[(2S)-3-methyl-1-[(3R,4S)-4-methyl-2,5-dioxopyrrolidin-3-yl]-1-oxobutan-2-yl]amino]-3-oxo-1-phenylpropyl]amino]-8-oxooctoxy]phenyl]methyl]-1,3-oxazole-5-carboxamide; (B) Heterobivalent inhibitor (11-(4-(22-(2,4-Dibromo-3-(diaminopyrido[2,3-*d*]pyrimidin-6-yl)phenoxy)-5,8,11,14,17,20-hexaoxadocos-1-yl)-1H-1,2,3-triazol-1-yl)-N-((1S)-3-methyl-1-((3R,4S)-4-methyl-2,5-dioxopyrrolidin-3-yl)-1-oxobutan-2-yl)amino)-3-oxo-1-phenylpropyl)-undecanamide.

B inhibitor of the carboxyltransferase component. To address this discrepancy in affinities between the two parent compounds, Tonge and coworkers synthesized heterobivalent compounds where the biotin carboxylase inhibitor was replaced with the pyridopyrimidine shown in Fig. 8A (Cifone et al., 2022). The carboxyltransferase inhibitor remained moiramide B. The pyridopyrimidine and moiramide B parent compounds were linked with polyethylene glycol coupled to a triazole with an eight-carbon saturated hydrocarbon. Linkers of various lengths were tested for inhibition and antibacterial activity. The heterobivalent compound with a 50 Å linker (Fig. 18B) was found to bind with the highest affinity (0.2 nM) to the ACC complex and exhibited a post–antibiotic effect of 1–4 h which was significantly longer than either of the two parent compounds alone. Thus, both studies using dual–ligand or heterobivalent inhibitors (i.e. one compound with two different targets) strongly support the notion that antibacterial agents can be created that have a therapeutic effect but do not give rise to bacterial resistance during treatment.

8. Concluding remarks

The initial inhibition studies of ACC played a pivotal role in understanding the three-dimensional structure and chemical and kinetic properties of the enzyme. With the discovery that ACC was a potential target for antibacterial compounds, new inhibitors of synthetic and natural origin were discovered and continue to be discovered. It will be interesting to see if the use of AI programs such as AlphaFold will aid in the discovery of new chemical matter for the inhibition of ACC as well as identify binding sites for inhibitors for which there is no structural information. Given the never-ending problem of antibiotic-resistant bacteria, inhibition studies of ACC will continue to play a central role in the discovery of new antibacterial therapeutics.

Author note

We have no known conflict of interest to disclose.

References

Amspacher, D. R., Blanchard, C. Z., Fronczek, F. R., Saraiva, M. C., Waldrop, G. L., & Strongin, R. M. (1999). Synthesis of a reaction intermediate analogue of biotin-dependent carboxylases via a selective derivatization of biotin. *Organic Letters, 1*(1), 99–102.

Andrews, L. D., Kane, T. R., Dozzo, P., Haglund, C. M., Hilderbrandt, D. J., Linsell, M. S., ... Wlasichuk, K. B. (2019). Optimization and mechanistic characterization of pyridopyrimidine inhibitors of bacterial biotin carboxylase. *Journal of Medicinal Chemistry, 62*(16), 7489–7505.

Asada, K. (1982). Biological carboxylations. In S. Inoue, & N. Yamazaki (Eds.). *Organic and bio-organic chemistry of carbon dioxide*. New York: John Wiley & Sons.

Athappilly, F. K., & Hendrickson, W. A. (1995). Structure of the biotinyl domain of acetyl-coenzyme A carboxylase determined by MAD phasing. *Structure (London, England: 1993), 3*(12), 1407–1419.

Baba, T., Ara, T., Hasegawa, M., Takai, Y., Okumura, Y., Baba, M., ... Mori, H. (2006). Construction of Escherichia coli K-12 in-frame, single-gene knockout mutants: The Keio collection. *Molecular Systems Biology, 2* 2006 0008.

Benson, B. K., Meades, G., Jr., Grove, A., & Waldrop, G. L. (2008). DNA inhibits catalysis by the carboxyltransferase subunit of acetyl-CoA carboxylase: Implications for active site communication. *Protein Science: A Publication of the Protein Society, 17*(1), 34–42.

Bilder, P., Lightle, S., Bainbridge, G., Ohren, J., Finzel, B., Sun, F., ... Melnick, M. (2006). The structure of the carboxyltransferase component of acetyl-coA carboxylase reveals a zinc-binding motif unique to the bacterial enzyme. *Biochemistry, 45*(6), 1712–1722.

Blanchard, C. Z., Amspacher, D., Strongin, R., & Waldrop, G. L. (1999b). Inhibition of biotin carboxylase by a reaction intermediate analog: Implications for the kinetic mechanism. *Biochemical and Biophysical Research Communications, 266*(2), 466–471.

Blanchard, C. Z., Chapman-Smith, A., Wallace, J. C., & Waldrop, G. L. (1999a). The biotin domain peptide from the biotin carboxyl carrier protein of Escherichia coli acetyl-CoA carboxylase causes a marked increase in the catalytic efficiency of biotin carboxylase and carboxyltransferase relative to free biotin. *The Journal of Biological Chemistry, 274*(45), 31767–31769.

Blanchard, C. Z., Lee, Y. M., Frantom, P. A., & Waldrop, G. L. (1999a). Mutations at four active site residues of biotin carboxylase abolish substrate-induced synergism by biotin. *Biochemistry, 38*(11), 3393–3400.

Blanchard, C. Z., & Waldrop, G. L. (1998). Overexpression and kinetic characterization of the carboxyltransferase component of acetyl-CoA carboxylase. *The Journal of Biological Chemistry, 273*(30), 19140–19145.

Blocquel, D., Habchi, J., Gruet, A., Blangy, S., & Longhi, S. (2012). Compaction and binding properties of the intrinsically disordered C-terminal domain of Henipavirus nucleoprotein as unveiled by deletion studies. *Molecular Biosystems, 8*(1), 392–410.

Broussard, T. C., Kobe, M. J., Pakhomova, S., Neau, D. B., Price, A. E., Champion, T. S., & Waldrop, G. L. (2013a). The three-dimensional structure of the biotin carboxylase-biotin carboxyl carrier protein complex of E. coli acetyl-CoA carboxylase. *Structure (London, England: 1993), 21*(4), 650–657.

Broussard, T. C., Pakhomova, S., Neau, D. B., Bonnot, R., & Waldrop, G. L. (2015). Structural analysis of substrate, reaction intermediate, and product binding in haemophilus influenzae biotin carboxylase. *Biochemistry, 54*(24), 3860–3870.

Broussard, T. C., Price, A. E., Laborde, S. M., & Waldrop, G. L. (2013b). Complex formation and regulation of Escherichia coli acetyl-CoA carboxylase. *Biochemistry, 52*(19), 3346–3357.

Bryant, S. K., Waldrop, G. L., & Gilman, S. D. (2013). A capillary electrophoretic assay for acetyl coenzyme A carboxylase. *Analytical Biochemistry, 437*(1), 32–38.

Chapman-Smith, A., Turner, D. L., Cronan, J. E., Jr., Morris, T. W., & Wallace, J. C. (1994). Expression, biotinylation and purification of a biotin-domain peptide from the biotin carboxy carrier protein of Escherichia coli acetyl-CoA carboxylase. *The Biochemical Journal, 302*(Pt 3), 881–887.

Cheng, C. C., Shipps, G. W., Jr., Yang, Z., Sun, B., Kawahata, N., Soucy, K. A., ... Mann, P. (2009). Discovery and optimization of antibacterial AccC inhibitors. *Bioorganic & Medicinal Chemistry Letters, 19*(23), 6507–6514.

Chou, C. Y., & Tong, L. (2011). Structural and biochemical studies on the regulation of biotin carboxylase by substrate inhibition and dimerization. *The Journal of Biological Chemistry, 286*(27), 24417–24425.

Chou, C. Y., Yu, L. P., & Tong, L. (2009). Crystal structure of biotin carboxylase in complex with substrates and implications for its catalytic mechanism. *The Journal of Biological Chemistry, 284*(17), 11690–11697.

Cifone, M. T., He, Y., Basu, R., Wang, N., Davoodi, S., Spagnuolo, L. A., ... Walker, S. G. (2022). Heterobivalent inhibitors of acetyl-CoA carboxylase: Drug target residence time and time-dependent antibacterial activity. *Journal of Medicinal Chemistry, 65*(24), 16510–16525.

Cook, P. F., & Cleland, W. W. (2007). *Enzyme kinetics and mechanism* (1st ed.). Garland Science Publishing.

Craft, M. K., & Waldrop, G. L. (2022). Mechanism of biotin carboxylase inhibition by ethyl 4-[[2-chloro-5-(phenylcarbamoyl)phenyl]sulphonylamino]benzoate. *Journal of Enzyme Inhibition and Medicinal Chemistry, 37*(1), 100–108.

Cronan, J. E., Jr., & Waldrop, G. L. (2002). Multi-subunit acetyl-CoA carboxylases. *Progress in Lipid Research, 41*(5), 407–435.

Cronan, J. E. (2021). The classical, yet controversial, first enzyme of lipid synthesis: Escherichia coli acetyl-CoA carboxylase. *Microbiology and Molecular Biology Reviews: MMBR, 85*(3), e0003221.

de Queiroz, M. S., & Waldrop, G. L. (2007). Modeling and numerical simulation of biotin carboxylase kinetics: Implications for half-sites reactivity. *Journal of Theoretical Biology, 246*(1), 167–175.

Evans, A., Ribble, W., Schexnaydre, E., & Waldrop, G. L. (2017). Acetyl-CoA carboxylase from Escherichia coli exhibits a pronounced hysteresis when inhibited by palmitoyl-acyl carrier protein. *Archives of Biochemistry and Biophysics, 636*, 100–109.

Fall, R. R., Glaser, M., & Vagelos, P. R. (1976). Acetyl coenzyme A carbosylase. Circular dichroism studies of Escherichia coli biotin carboxyl carrier protein. *The Journal of Biological Chemistry, 251*(7), 2063–2069.

Fredenhagen, A., Tamura, S. Y., Kenny, P. T. M., Komura, H., Naya, Y., Nakanishi, K., ... Kita, H. (1987). Andrimid, a new peptide antibiotic produced by an intracellular bacterial symbiont isolated from a brown planthopper. *Journal of the American Chemical Society, 109*(14), 4409–4411.

Freiberg, C., Brunner, N. A., Schiffer, G., Lampe, T., Pohlmann, J., Brands, M., ... Ziegelbauer, K. (2004). Identification and characterization of the first class of potent bacterial acetyl-CoA carboxylase inhibitors with antibacterial activity. *The Journal of Biological Chemistry, 279*(25), 26066–26073.

Galperin, M. Y., & Koonin, E. V. (1997). A diverse superfamily of enzymes with ATP-dependent carboxylate-amine/thiol ligase activity. *Protein Science: A Publication of the Protein Society, 6*(12), 2639–2643.

Guchhait, R. B., Polakis, S. E., Dimroth, P., Stoll, E., Moss, J., & Lane, M. D. (1974). Acetyl coenzyme A carboxylase system of Escherichia coli. Purification and properties of the biotin carboxylase, carboxyltransferase, and carboxyl carrier protein components. *The Journal of Biological Chemistry, 249*(20), 6633–6645.

Guchhait, R. B., Polakis, S. E., & Lane, M. D. (1975). Carboxyltransferase component of acetyl-CoA carboxylase from Escherichia coli. *Methods in Enzymology, 35*, 32–37.

Heath, R. J., & Rock, C. O. (1996). Regulation of fatty acid elongation and initiation by acyl-acyl carrier protein in Escherichia coli. *The Journal of Biological Chemistry, 271*(4), 1833–1836.

Jadulco, R. C., Koch, M., Kakule, T. B., Schmidt, E. W., Orendt, A., He, H., ... Pond, C. (2014). Isolation of pyrrolocins A–C: cis- and trans-decalin tetramic acid antibiotics from an endophytic fungal-derived pathway. *Journal of Natural Products, 77*(11), 2537–2544.

Janiyani, K., Bordelon, T., Waldrop, G. L., & Cronan, J. E., Jr. (2001). Function of Escherichia coli biotin carboxylase requires catalytic activity of both subunits of the homodimer. *The Journal of Biological Chemistry, 276*(32), 29864–29870.

Kakule, T. B., Jadulco, R. C., Koch, M., Janso, J. E., Barrows, L. R., & Schmidt, E. W. (2015). Native promoter strategy for high-yielding synthesis and engineering of fungal secondary metabolites. *ACS Synthetic Biology, 4*(5), 625–633.

Lane, M. D., Moss, J., & Polakis, S. E. (1974). Acetyl coenzyme A carboxylase. *Current Topics in Cellular Regulation, 8*(0), 139–195.

Larson, E. C., Lim, A. L., Pond, C. D., Craft, M., Cavuzic, M., Waldrop, G. L., ... Barrows, L. R. (2020). Pyrrolocin C and equisetin inhibit bacterial acetyl-CoA carboxylase. *PLoS One, 15*(5), e0233485.

Levert, K. L., Lloyd, R. B., & Waldrop, G. L. (2000). Do cysteine 230 and lysine 238 of biotin carboxylase play a role in the activation of biotin? *Biochemistry, 39*(14), 4122–4128.

Levert, K. L., & Waldrop, G. L. (2002). A bisubstrate analog inhibitor of the carboxyl-transferase component of acetyl-CoA carboxylase. *Biochemical and Biophysical Research Communications, 291*(5), 1213–1217.

Li, S. J., & Cronan, J. E., Jr (1992a). The gene encoding the biotin carboxylase subunit of Escherichia coli acetyl-CoA carboxylase. *The Journal of Biological Chemistry, 267*(2), 855–863.

Li, S. J., & Cronan, J. E., Jr (1992b). The genes encoding the two carboxyltransferase subunits of Escherichia coli acetyl-CoA carboxylase. *The Journal of Biological Chemistry, 267*(24), 16841–16847.

Lietzan, A. D., Menefee, A. L., Zeczycki, T. N., Kumar, S., Attwood, P. V., Wallace, J. C., ... Maurice, S. (2011). M: Interaction between the biotin carboxyl carrier domain and the biotin carboxylase domain in pyruvate carboxylase from Rhizobium etli. *Biochemistry, 50*(45), 9708–9723.

Meades, G., Cai, X., Thalji, N. K., Waldrop, G. L., & De Queiroz, M. (2011). Mathematical modelling of negative feedback regulation by carboxyltransferase. *IET Systems Biology, 5*(3), 220–228.

Meades, G., Jr., Benson, B. K., Grove, A., & Waldrop, G. L. (2010a). A tale of two functions: Enzymatic activity and translational repression by carboxyltransferase. *Nucleic Acids Research, 38*(4), 1217–1227.

Meades, G., Jr., Henken, R. L., Waldrop, G. L., Rahman, M. M., Gilman, S. D., Kamatou, G. P., ... Gibbons, S. (2010b). Constituents of cinnamon inhibit bacterial acetyl CoA carboxylase. *Planta Medica, 76*(14), 1570–1575.

Miller, J. R., Dunham, S., Mochalkin, I., Banotai, C., Bowman, M., Buist, S., ... Huband, M. D. (2009). A class of selective antibacterials derived from a protein kinase inhibitor pharmacophore. *Proceedings of the National Academy of Sciences of the United States of America, 106*(6), 1737–1742.

Mochalkin, I., Miller, J. R., Evdokimov, A., Lightle, S., Yan, C., Stover, C. K., & Waldrop, G. L. (2008). Structural evidence for substrate-induced synergism and half-sites reactivity in biotin carboxylase. *Protein Science: A Publication of the Protein Society, 17*(10), 1706–1718.

Mochalkin, I., Miller, J. R., Narasimhan, L., Thanabal, V., Erdman, P., Cox, P. B., ... Stover, C. K. (2009). Discovery of antibacterial biotin carboxylase inhibitors by virtual screening and fragment-based approaches. *ACS Chemical Biology, 4*(6), 473–483.

Murzin, A. G. (1996). Structural classification of proteins: New superfamilies. *Current Opinion in Structural Biology, 6*(3), 386–394.

Nguyen, T. H., Waldrop, G. L., & Gilman, S. D. (2019). Capillary electrophoretic assay of human acetyl-coenzyme A carboxylase 2. *Electrophoresis, 40*(11), 1558–1564.

Northrop, D. B., & Cleland, W. W. (1974). The kinetics of pig heart triphosphopyridine nucleotide-isocitrate dehydrogenase. II. Dead-end and multiple inhibition studies. *The Journal of Biological Chemistry, 249*(9), 2928–2931.

Ogita, T., & Knowles, J. R. (1988). On the intermediacy of carboxyphosphate in biotin-dependent carboxylations. *Biochemistry, 27*(21), 8028–8033.

Patridge, E. V., Darnell, A., Kucera, K., Phillips, G. M., Bokesch, H. R., Gustafson, K. R., ... Plummer, M. (2015). Pyrrolocin A, a 3-decalinoyltetramic acid with selective biological activity, isolated from amazonian cultures of the novel endophyte diaporthales sp. E6927E. *Natural Product Communications, 10*(10), 1649–1654.

Pohlmann, J., Lampe, T., Shimada, M., Nell, P. G., Pernerstorfer, J., Svenstrup, N., ... Freiberg, C. (2005). Pyrrolidinedione derivatives as antibacterial agents with a novel mode of action. *Bioorganic & Medicinal Chemistry Letters, 15*(4), 1189–1192.

Polakis, S. E., Guchhait, R. B., & Lane, M. D. (1972). On the possible involvement of a carbonyl phosphate intermediate in the adenosine triphosphate-dependent carboxylation of biotin. *The Journal of Biological Chemistry, 247*(4), 1335–1337.

Santoro, N., Brtva, T., Roest, S. V., Siegel, K., & Waldrop, G. L. (2006). A high-throughput screening assay for the carboxyltransferase subunit of acetyl-CoA carboxylase. *Analytical Biochemistry, 354*(1), 70–77.

Schneider, D. A., & Gourse, R. L. (2004). Relationship between growth rate and ATP concentration in Escherichia coli: A bioassay for available cellular ATP. *The Journal of Biological Chemistry, 279*(9), 8262–8268.

Silvers, M. A., Pakhomova, S., Neau, D. B., Silvers, W. C., Anzalone, N., Taylor, C. M., & Waldrop, G. L. (2016). Crystal structure of carboxyltransferase from Staphylococcus aureus bound to the antibacterial agent moiramide B. *Biochemistry, 55*(33), 4666–4674.

Silvers, M. A., Robertson, G. T., Taylor, C. M., & Waldrop, G. L. (2014). Design, synthesis, and antibacterial properties of dual-ligand inhibitors of acetyl-CoA carboxylase. *Journal of Medicinal Chemistry, 57*(21), 8947–8959.

Svenningsen, M. S., Veress, A., Harms, A., Mitarai, N., & Semsey, S. (2019). Birth and resuscitation of (p)ppGpp induced antibiotic tolerant persister cells. *Scientific Reports, 9*(1), 6056.

Thoden, J. B., Blanchard, C. Z., Holden, H. M., & Waldrop, G. L. (2000). Movement of the biotin carboxylase B-domain as a result of ATP binding. *The Journal of Biological Chemistry, 275*(21), 16183–16190.

Thoden, J. B., Kappock, T. J., Stubbe, J., & Holden, H. M. (1999). Three-dimensional structure of N5-carboxyaminoimidazole ribonucleotide synthetase: A member of the ATP grasp protein superfamily. *Biochemistry, 38*(47), 15480–15492.

Tompa, P. (2002). Intrinsically unstructured proteins. *Trends in Biochemical Sciences, 27*(10), 527–533.

Uversky, V. N. (2019). Intrinsically disordered proteins and their "mysterious" (meta) physics. *Frontiers in Physics, 7*.

Uversky, V. N. (2002). What does it mean to be natively unfolded? *European Journal of Biochemistry / FEBS, 269*(1), 2–12.

Vesonder, R. F., Tjarks, L. W., Rohwedder, W. K., Burmeister, H. R., & Laugal, J. A. (1979). Equisetin, an antibiotic from Fusarium equiseti NRRL 5537, identified as a derivative of N-methyl-2,4-pyrollidone. *The Journal of Antibiotics, 32*(7), 759–761.

Waldrop, G., Holden, H. M., & Rayment, I. (1994b). Preliminary X-ray crystallographic analysis of biotin carboxylase isolated from Escherichia coli. *Journal of Molecular Biology, 235*(1), 367–369.

Waldrop, G. L., Holden, H. M., & ST Maurice, M. (2012). The enzymes of biotin dependent CO(2) metabolism: What structures reveal about their reaction mechanisms. *Protein Science: A Publication of the Protein Society, 21*(11), 1597–1619.

Waldrop, G. L., Rayment, I., & Holden, H. M. (1994a). Three-dimensional structure of the biotin carboxylase subunit of acetyl-CoA carboxylase. *Biochemistry, 33*(34), 10249–10256.

Waldrop, G. L. (2011). The role of symmetry in the regulation of bacterial carboxyltransferase. *Biomolecular Concepts, 2*(1–2), 47–52.

Wallace, J., Bowlin, N. O., Mills, D. M., Saenkham, P., Kwasny, S. M., Opperman, T. J., ... Moir, D. T. (2015). Discovery of bacterial fatty acid synthase type II inhibitors using a novel cellular bioluminescent reporter assay. *Antimicrobial Agents and Chemotherapy, 59*(9), 5775–5787.

Ward, W. H., Holdgate, G. A., Rowsell, S., McLean, E. G., Pauptit, R. A., Clayton, E., ... Jude, D. A. (1999). Kinetic and structural characteristics of the inhibition of enoyl (acyl carrier protein) reductase by triclosan. *Biochemistry, 38*(38), 12514–12525.

Wozniak, C. E., Lin, Z., Schmidt, E. W., Hughes, K. T., & Liou, T. G. (2018). Thailandamide, a fatty acid synthesis antibiotic that is coexpressed with a resistant target gene. *Antimicrobial Agents and Chemotherapy, 62*(9).

Wu, X., & Huang, T. (2020). Recent development in acetyl-CoA carboxylase inhibitors and their potential as novel drugs. *Future Medicinal Chemistry, 12*(6), 533–561.

Wu, Y., & Seyedsayamdost, M. R. (2018). The polyene natural product thailandamide A inhibits fatty acid biosynthesis in gram-positive and gram-negative bacteria. *Biochemistry, 57*(29), 4247–4251.

Yonetani, T., & Theorell, H. (1964). Studies on liver alcohol hydrogenase complexes. 3. Multiple inhibition kinetics in the presence of two competitive inhibitors. *Archives of Biochemistry and Biophysics, 106*, 243–251.

Zhao, X., & Drlica, K. (2002). Restricting the selection of antibiotic–resistant mutant bacteria: Measurement and potential use of the mutant selection window. *The Journal of Infectious Diseases, 185*(4), 561–565.

CHAPTER FIVE

Characterization of guanidine carboxylases

M. Sinn[a,*], J. Techel[a,b], A. Joachimi[a], and J.S. Hartig[a,b,*]

[a]Department of Chemistry, University of Konstanz, Germany
[b]Konstanz Research School Chemical Biology (KoRS–CB), University of Konstanz, Germany
*Corresponding authors. e-mail address: malte.sinn@uni-konstanz.de; joerg.hartig@uni-konstanz.de

Contents

1. Introduction	106
2. Before you begin	108
3. Key resources table	109
4. Materials and equipment	111
5. Step-by-step method details	112
5.1 Isolation of guanidine assimilating bacteria	112
5.2 Sequencing of the guanidine assimilating bacteria	113
5.3 Cloning of guanidine and urea carboxylase expression vectors	114
5.4 Expression and purification of carboxylases	116
5.5 Note	118
5.6 Characterization of carboxylases	118
5.7 Optional: Comparative proteome analysis	120
6. Limitations	121
7. Optimization and troubleshooting	121
7.1 No colonies observed during isolation	121
8. Safety considerations and standards	121
9. Conclusion	122
References	122

Abstract

Guanidine metabolism has been an overlooked aspect of the global nitrogen cycle until RNA sensors (riboswitches) were discovered in bacteria that bind the nitrogen-rich compound. The associated genes were initially proposed to detoxify guanidine from the cells. We were intrigued by a genetic organization where the guanidine riboswitch is located upstream of an operon comprising a carboxylase, two putative hydrolases, and an assigned allophanate hydrolase. An ABC transporter is located on the same operon with a periplasmic binding domain that is indicative of an importer. Therefore, we hypothesized that certain bacteria actively import guanidine and assimilate the nitrogen. To test this hypothesis, we searched for bacteria that were able to assimilate guanidine. We isolated three enterobacteria (*Raoultella terrigena* str. JH01, *Erwinia rhapontici* str. JH02 and *Klebsiella michiganensis* str. JH07) that utilize

Methods in Enzymology, Volume 708
ISSN 0076-6879, https://doi.org/10.1016/bs.mie.2024.10.013
Copyright © 2024 Elsevier Inc. All rights are reserved, including those for text and data mining, AI training, and similar technologies.

guanidine efficiently as a nitrogen source. Proteome analyses demonstrate that the expression of the guanidine riboswitch-associated carboxylase, in conjunction with associated hydrolases and transport genes, is markedly elevated in the presence of guanidine. Subsequent analysis of the carboxylases that are homologous to urea carboxylase confirmed the substrate preference of guanidine over urea. This chapter outlines a procedure for the isolation of guanidine-assimilating bacteria and the analysis of their proteome to identify enzymes responsible for guanidine degradation. Finally, an assay for the characterization of the endogenous guanidine carboxylases in comparison with the endogenous urea carboxylase from *E. rhapontici* is described.

1. Introduction

Guanidine is a nitrogen-rich compound that is widely used in explosives and propellants, as a chaotropic agent for protein unfolding and as a slow-release nitrogen fertilizer in agriculture (Güthner, Mertschenk, & Schulz, 2014). Guanidine was discovered already 150 years ago as a thermal degradation product of guanine. Despite early reports of fungi that converted guanidine into urea (Iwanoff & Awetissowa, 1931) and reports about guanidine occurrence in plants and environmental samples (Kato, Yamagata, & Tsukahara, 1986; Schulze, 1893; Wishart et al., 2018), the role of guanidine in biological systems remained unknown. The discovery of four classes of guanidine riboswitches revived a new interest in the biology of guanidine (Lenkeit, Eckert, Hartig, & Weinberg, 2020; Nelson, Atilho, Sherlock, Stockbridge, & Breaker, 2017; Salvail, Balaji, Yu, Roth, & Breaker, 2020; Sherlock & Breaker, 2017; Sherlock, Malkowski, & Breaker, 2017). Guanidine riboswitches are structured RNAs that specifically bind guanidine and regulate the gene expression of downstream genes. They are widespread in bacteria and regulate the expression of guanidine exporters (Kermani, Macdonald, Gundepudi, & Stockbridge, 2018) and enzymes that degrade guanidine (Funck et al., 2022; Schneider et al., 2020; Sinn, Hauth, Lenkeit, Weinberg, & Hartig, 2021; Wang et al., 2021). First thought to be involved in the detoxification of guanidine, we got intrigued by the presence of ABC transporters regulated by guanidine riboswitches that comprised a periplasmic binding domain suggesting a putative role in the active import of guanidine. Thus we hypothesized that guanidine could be used as a nitrogen source by bacteria. In recent years, it was shown that bacteria evolved two systems that enable them to assimilate guanidine either by direct hydrolysis to urea (Funck et al., 2022; Wang et al., 2021) or by carboxylation and stepwise degradation to ammonia and CO_2 (Schneider et al., 2020; Sinn et al., 2021).

We isolated three different enterobacteria (*Raoultella terrigena*, *Klebsiella michiganensis*, and *Erwinia rhapontici*) that were able to utilize guanidine as nitrogen source. We sequenced their genomes and performed a comparative proteome analysis. An operon controlled by a guanidine riboswitch was found to be upregulated if the cells grew on guanidine (Sinn et al., 2021). The operon comprised genes encoding an ABC-transporter and genes annotated as urea carboxylase, urea carboxylase-associated genes 1 and 2 and allophanate hydrolase (Fig. 1). Carboxylases that are associated with guanidine riboswitches have been shown to prefer guanidine over urea as a substrate (Nelson et al., 2017), so we will refer to them as guanidine carboxylases (Gca). Schneider et al. demonstrated that the urea carboxylase associated genes 1 and 2 encode for a dimeric carboxyguanidine deiminase (Schneider et al., 2020) that catalyzes the hydrolysis of carboxyguanidine to allophanate and ammonia (Fig. 1). *E. rhapontici* harbored a guanidine carboxylase as well as another gene annotated as urea carboxylase at a different locus. We cloned, heterologously expressed and purified both proteins and evaluated their substrate preferences. As anticipated, the guanidine riboswitch-associated enzyme showed a substrate preference for guanidine with a k_{cat}/K_M of 100 ± 4 s^{-1}mM^{-1} compared to 1.1 ± 0.04 s^{-1}mM^{-1} for urea. In contrast, the urea carboxylase from the other locus preferred urea as a substrate with a k_{cat}/K_M of 3.0 ± 0.2 s^{-1}mM^{-1} compared to 0.008 ± 0.002 s^{-1}mM^{-1} for guanidine. Investigation of the phylogeny of urea/guanidine carboxylases revealed that they clustered in distinct clades (Sinn et al., 2021).

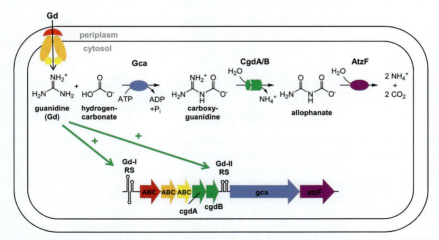

Fig. 1 Scheme of the guanidine carboxylase pathway and its genetic organization. Guanidine induces gene expression (green arrows) by binding to the guanidine riboswitches (Gd-I and Gd-II). Note that in some cases *atzF* is encoded at a different genetic locus.

In this chapter, we describe a procedure for isolating bacteria that are able to grow on guanidine as the sole nitrogen source, their identification, and genome sequencing. Comparative proteome analysis is presented as a method to identify enzymes potentially involved in the degradation of guanidine. We describe the cloning of two carboxylases found in *E. rhapontici* with an N-terminal 6x histidine tag for nickel affinity chromatography, their heterologous expression and purification. Finally, an assay for the characterization of the carboxylases is outlined. The strategy used to find bacteria that grow on guanidine as a nitrogen source and to determine the underlying molecular mechanism (Fig. 2) can be easily transferred to other (nitrogen-) compounds.

2. Before you begin

Timing: 1 d.
1. Bacterial cultivation
   a. 5x M8 salts (42.5 g/L Na$_2$HPO$_4$·2H$_2$O, 15 g/L KH$_2$PO$_4$, 2.5 g/L NaCl), autoclaved
   b. 1 M MgCl$_2$, 1 M CaCl$_2$, autoclaved
   c. 100x trace element solution (10 mM EDTA, 3 mM FeCl$_3$, 0.62 mM ZnCl$_2$, 76 μM CuCl$_2$, 42 μM CoCl$_2$, 162 μM H$_3$BO$_3$; 8 μM MnCl$_2$), sterile filtered
   d. 1000x vitamin solution (0.1 g/L cyanocobalamin, 0.08 g/L 4-aminobenzoic acid, 0.02 g/L D-(+)-biotin, 0.2 g/L niacin, 0.1 g/L Ca-D-

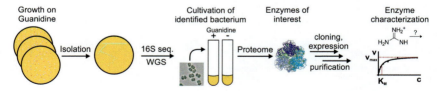

Fig. 2 Strategy for the isolation of guanidine-assimilating bacteria and the determination of the underlying molecular mechanism. Bacteria are grown on solid minimal medium supplemented with guanidine as the sole nitrogen source. Colonies are picked and transferred to fresh solid minimal medium until homogenous colonies are obtained. 16S rRNA sequencing is used to identify the isolated species and the genome of the isolated strains is sequenced. For comparative proteome analysis, the isolated species are grown in liquid minimal medium containing either ammonia or guanidine as nitrogen source. The whole proteome is extracted and subjected to LC-MS analysis. Proteins that are upregulated if the bacteria grew on guanidine are examined and evaluated bioinformatically to determine candidates that might be involved in the degradation of guanidine. Enzymes of interest are cloned, expressed and purified. Finally, the reaction of the enzymes is characterized.

(+)-pantothenic acid, 0.3 g/L pyridoxamine hydrochloride, 0.2 g/L thiamine hydrochoride), sterile filtered
   e. 50% glycerol, autoclaved (or alternative carbon source)
   f. 100 mM guanidine, sterile filtered
   g. 10x PBS (80.06 g/L NaCl, 2.01 g/L KCl, 14.24 g/L Na$_2$HPO$_4$ * 2 H$_2$O, 2.72 g/L KH$_2$PO$_4$, adjust pH to 7.4)
2. Cloning of guanidine carboxylases
   a. 0.5x TBE buffer (5.4 g/L TRIS, 2.75 g/L Boric acid, 2 mL of 0.5 M EDTA (pH 8.0))
   b. 6x DNA loading dye (Mix 0.2 mL 1 M Tris-HCl (pH 7,6), 12 mL Glycerol, 2.4 mL 0.5 M EDTA (pH 8,0), 6 mg Bromophenol blue, 6 mg Xylencyanol and 5.4 mL H$_2$O for a total volume of 20 mL)
   c. SOC medium (20 g/L tryptone, 5 g/L yeast extract, 0.5 g/L NaCl, autoclaved); add 10 mL filter-sterilized MgCl$_2$ (1 M), 10 mL filter-sterilized MgSO$_4$ (1 M) and 20 mL of filter-sterilized glucose (20% (w/v))
3. Purification of guanidine carboxylases
   a. Lysis Buffer (20 mM Tris (pH 8), 200 mM NaCl, 20 mM imidazole, EDTA free protease inhibitor (cOmplete™), 0.02 mg/ml lysozyme)
   b. Wash buffer (20 mM Tris (pH 8), 200 mM NaCl, 20 mM imidazole)
   c. Elution buffer (20 mM Tris (pH 8), 200 mM NaCl, 500 mM imidazole)
   d. Tris buffer (20 mM Tris (pH 8), 200 mM NaCl)
   e. 6x SDS loading buffer (Mix 7 mL 1 M Tris-HCl (pH 6,8), 3 mL Glycerol, 1 g SDS, 0.98 g DTT and 6 mg Bromophenol blue for a total volume of 10 mL)

3. Key resources table

Key resource table is below.

Table 1 List of key resources.

Reagent or resource	Source	Identifier
Experimental Models: Organisms/Strains		
Escherichia coli XL10 gold		
Escherichia coli BL21 (DE3) gold		

(continued)

Table 1 List of key resources. (*cont'd*)

Reagent or resource	Source	Identifier
Oligonucleotides		
FH186	E.rhapontici _gibson_pet28a_fwd GCTAGCATGACTGGTGG ACAGCAAATGGG	Sigma-Aldrich
FH187	E.rhapontici _gibson_pet28a_rev ATGGCTGCCGCGCGGCAC	Sigma-Aldrich
FH188	E.rhapontici_gibson_uca_fwd tggtgccgcgcggcagccat ATGTTTAATACCGTACTC ATCGCTAAC	Sigma-Aldrich
FH189	E.rhapontici _gibson_uca_rev tgtccaccagtcatgctagc TTATTCCAGCCATAGCAGGG	Sigma-Aldrich
FH190	E.rhapontici _gibson_gca_fwd tggtgccgcgcggcagccat ATGTTTACAAAACTGCT GATTGCTAATC	Sigma-Aldrich
FH191	E.rhapontici _gibson_gca_rev tgtccaccagtcatgctagc TTAAACGCCCATCACCACC	Sigma-Aldrich
	16S rRNA_fwd AGAGTTTGATCMTGGCTCAG	Sigma-Aldrich
	16S rRNA_rev CGGTTACCTTGTTACGACTT	Sigma-Aldrich
	Sequencing primer forward TCCCCATCGGTGATGTC	Sigma-Aldrich
	Sequencing primer reverse CTAGTTATTGCTCAGCGG	Sigma-Aldrich
Recombinant DNA		
pET28a(+)		Novagen

4. Materials and equipment

- Agar-Agar, BioScience, granulated (Carl Roth, 1347)
- Agarose Gel electrophoresis apparatus (Bio-Rad, Sub-Cell GT Agarose Gel Electrophoresis Systems)
- Agarose Gel Imager (Vilber Lourmat, E-Box CX5)
- Agarose GTQ (Carl Roth, 6352)
- Amicon® centrifugal filter 3 kDa cut-off (Merck)
- ATP (Fisher Scientific, 10304340)
- cOmplete™, EDTA-free Protease Inhibitor Cocktail (Roche, 11873580001)
- DNA-Oligos (Sigma Aldrich) Primers; diluted to 5 µM
- DNeasy Blood & Tissue Kit (Qiagen, 69504)
- dNTPs (Fisher Scientific, 10520651)
- *Dpn*I (NEB, R0176 L)
- DTT (Carl Roth, 6908)
- Electroporation Cuvettes 100 µL, 1 mm Gap width (Roth, PP38.1)
- Eppendorf Electroporator 2510
- Filter papers (Machery-Nagel, 531 018)
- GeneRuler 1 kb DNA Ladder, ready to use (Fisher Scientific, 11812124)
- Gibson Assembly® Master Mix (NEB, E2611 L)
- Guanidine hydrochloride (Carl Roth, 0037)
- Imidazole (Carl Roth, X998)
- IPTG (Carl Roth, CN08)
- Kanamycin (Carl Roth, T832)
- LB Broth (Lennox) (Carl Roth, X964)
- Microtitration plates ROTILABO® F-profile (Carl Roth, 9293.1)
- Midori Green Advance (Biozym, 617004)
- Mini-PROTEAN® Tetra Cell (Bio-Rad, 1658001)
- Multimode Microplate Reader (Tecan, SPARK)
- NADH-Dinatriumsalz (Carl Roth, AE12.1)
- Ni-NTA-Agarose (Qiagen, 30210)
- PageRuler™ Plus Prestained Protein Ladder (Fisher Scientific, 26619)
- PD-10 desalting column (3.5 mL Sephadex G25) (Cytiva)
- pET28a (+) (Sigma Aldrich, 69864-3)
- Phospho(enol)pyruvic acid monosodium salt hydrate (Sigma Aldrich, P0564)
- Phusion HF DNA Polymerase, Hot Start II (Fisher Scientific, 10628439)
- Pierce™ BCA Protein Assay Kit (Thermo Scientific, 23227)

- Polypropylene column, 1 mL (Qiagen, 34924)
- Pyruvate Kinase/Lactic Dehydrogenase enzymes from rabbit muscle (Sigma Aldrich, P0294)
- Rotiphorese 10xSDS-PAGE (Carl Roth, 3060.2)
- Rotiphorese Blau R (Carl Roth, 3074)
- ROTIPHORESE®Blue R (Carl Roth, 3074)
- ROTIPHORESE®Gel 40 (37.5:1) (Carl Roth, T802)
- Shaking Incubator (Infors HT, Ecotron)
- Sonifier (Branson, S-250D)
- Steril petri dishes (Sarstedt, 82.1473.001)
- TRIS (Carl Roth, 4855)
- Urea (Carl Roth, 3941)
- Zymoclean Gel DNA Recovery Kit (Zymo Research, D4001)
- Zyppy ® Plasmid Miniprep Kit (Zymo Research, D4037)

5. Step-by-step method details
5.1 Isolation of guanidine assimilating bacteria

Timing: 1–2 weeks.
1. Prepare minimal medium plates that contain 5 mM guanidine as nitrogen source: Dissolve 10 g agar-agar in 300 mL deionized water in a 500 mL flask with a magnetic stir bar.
2. Pre-warm 5x M8 salts to 40 °C and add 100 ml to the freshly autoclaved agar solution while still hot on a magnetic stirrer. Wait until solution cooled down to app. 50–60 °C.
3. Add 1 mL of 1 M $MgCl_2$ solution and 50 μl of 1 M $CaCl_2$.
4. Add 5 mL of 100x trace element solution and 0.5 mL of 1000x vitamin solution.
5. Add 10 mL of 50 % (v/v) glycerol and 25 mL of 100 mM guanidine.
6. Fill up to 500 mL with sterile deionized water.
7. Immediately cast app. 25 mL of the medium into sterile petri dishes.
8. Let medium solidify for app. 3 h at RT.
9. Collect environmental sample. In our case, the sample was taken from the lake shore sediment of the Lake of Constance (47°41′44.2′N, 9°11′35.1′E).
10. Prepare a suspension of app. 5 g sediment with 10 ml lake water.
11. Filter your sample through a sterile paper filter and collect the flow through in a sterile vessel.

Characterization of guanidine carboxylases | 113

12. Rinse with an additional 10 mL of lake water.
13. Prepare a serial (100, 1k, 10k, 100k fold) dilution of your sample in sterile 1x PBS.
14. Streak 100 µl of each dilution onto separate guanidine containing minimal medium plates.
15. Incubate at RT for 96 h or until colonies are visible.
16. Pick single colonies and streak onto fresh plates.
17. Repeat step 15 and 16 until homogenous colonies are obtained.

5.2 Sequencing of the guanidine assimilating bacteria

Timing: 1 week.

18. Pick a single colony and inoculate 10 mL M8 minimal medium that contains 1 % glycerol and 5 mM guanidine as nitrogen source.
19. Incubate the liquid culture at 30 °C and 200 rpm until bacterial growth is visible.
20. Extract genomic DNA with DNeasy Blood & Tissue Kit, using the protocol "Pretreatment for Gram-Negative Bacteria" followed by the protocol "Purification of Total DNA from Animal Tissues (Spin-Column Protocol)" outlined in the kit handbook.
21. Measure the DNA concentration with Tecan Nanoquant plate in the Tecan platereader.
22. Set up a PCR with the following composition: 17.5 µL ultrapure water, 10 µL 5x Phusion HF buffer, 5 µL dNTP mixture (2 mM each), 6 µL 16S rRNA_fwd primer (5 µM), 6 µL 16S rRNA_rev primer (5 µM), 1 µL template DNA (10 ng/µL), 4 µl DMSO, 0.5 µL Phusion HF polymerase.
23. Run PCR following the thermocycling conditions in Table 2.
24. Dissolve 0.32 g agarose in 40 ml 0.5x TBE add 1 µl of Midori Green Advance DNA stain and cast a 6 × 10 cm 0.8 % agarose gel.

Table 2 Thermocycling conditions for Phusion PCR with 16S rRNA primers.

Step	Temp.	Time	Cycles
Initial–Denaturation	98 °C	30 sec	1
Denaturation	98 °C	10 sec	
Annealing	60 °C	30 sec	25
Elongation	72 °C	40 sec	
Final–Elongantion	72 °C	7 min	1

25. Mix 5 μL of the PCR reaction with 1 μl of 6x agarose loading dye and load on agarose gel and add 3 μl of GeneRuler 1 kb DNA Ladder next to the sample to assess the correct size of the PCR product.
26. Perform agarose gel electrophoresis for 45 min at 120 V.
27. Check for a PCR product at app. 1500 bp under UV illumination.
28. Excise DNA product from agarose gel on a UV illuminated stage.
29. Clean up the DNA from agarose gel with Zymoclean Gel DNA Recovery Kit according to the manufacturer's protocol.
30. Measure the DNA concentration with Tecan Nanoquant plate in the Tecan platereader.
31. Send the DNA with 16S rRNA_fwd primer and 16S rRNA_rev primer for sequencing
32. For a whole genome sequencing send the genomic DNA for whole genome sequencing e.g. Novogene, UK.

5.3 Cloning of guanidine and urea carboxylase expression vectors

Timing: 1–2d.

In *Erwinia rhapontici* str. JH02 two putative urea carboxylases (uca) are found. One is associated with a guanidine riboswitch and we hypothesized that it is a guanidine carboxylase (gca). We cloned expression vectors for both genes and examined the substrate specificity of both enzymes. Here the cloning for 6x His tagged variants of Uca and Gca for nickel affinity chromatography purification by Gibson Assembly is described.

33. Set up a PCR with primers FH186/FH187 and pET28a(+) as template for backbone generation:
    21.5 μL ultrapure water, 10 μL 5x Phusion HF buffer, 5 μL dNTP mixture (2 mM each), 6 μL primer FH186 (5 μM), 6 μL primer FH187 (5 μM), 1 μL pET28a(+) (10 ng/μL), 0.5 μL Phusion HF polymerase.
34. Run PCR following the thermocycling conditions in Table 3.
35. Add 1 μL *Dpn*I to PCR reaction and incubate for 20 min to remove plasmid template
36. Dissolve 0.32 g agarose in 40 ml 0.5x TBE add 1 μl of Midori Green Advance DNA stain and cast a 6 × 10 gel. Immediately insert a comb with large pockets suitable to hold 60 μl of sample aside a smaller pocket for the DNA ladder.
37. Add 10 μl 6x agarose loading dye to the PCR.

Characterization of guanidine carboxylases

Table 3 Thermocycling conditions for backbone generation.

Step	Temp.	Time	Cycles
Initial-Denaturation	98 °C	30 sec	1
Denaturation	98 °C	10 sec	
Annealing	68 °C	30 sec	25
Elongation	72 °C	150 sec	
Final-Elongantion	72 °C	7 min	1

38. Load the 60 μl of PCR on the agarose gel and add 3 μl of GeneRuler 1 kb DNA Ladder next to the sample to assess the correct size of the PCR product.
39. Perform agarose gel electrophoresis for 45 min at 120 V.
40. Excise DNA product from agarose gel on a UV illuminated stage. Check product size (6000 bp)
41. Clean up the DNA from agarose gel with Zymoclean Gel DNA Recovery Kit according to the manufacturer's protocol.
42. Determine DNA concentration of the backbone as described.
43. Set up PCR with FH188/FH189 or FH190/191 with genomic DNA from step 20 for generating DNA fragments containing the uca or gca gene respectively:

    21.5 μL ultrapure water, 10 μL 5x Phusion HF buffer, 5 μL dNTP mixture (2 mM each), 6 μL primer FH188 or FH190 (5 μM), 6 μL primer FH189 or FH191 (5 μM), 1 μL genomic DNA (25 ng/μL), 0.5 μL Phusion HF polymerase.
44. Run PCR following the thermocycling conditions in Table 4.
45. Cast agarose gel as described before, but dissolve 0.6 g agarose in 40 ml 0.5x TBE.
46. Repeat DNA purification for the gene fragments as described before (steps 37–42). Estimated product size is 3600 bp for both fragments.
47. Set up the Gibson Assembly reaction:

    10 μl Gibson Assembly mastermix, 50 ng of backbone, 25 ng of uca or gca fragment respectively. Fill up to 20 μL with ultrapure water and incubate at 50 °C for 20 min.
48. Add 40 μL ultrapure water to the Gibson assembly reaction.
49. Add 1.5 μl of the diluted reaction to 80 μL of electrocompetent E. coli XL10 gold.

Table 4 Thermocycling conditions for uca/gca DNA fragment generation.

Step	Temp.	Time	Cycles
Initial–Denaturation	98 °C	30 sec	1
Denaturation	98 °C	10 sec	
Annealing	62 °C	30 sec	25
Elongation	72 °C	90 sec	
Final-Elongantion	72 °C	7 min	1

50. Transfer into a chilled electroporation cuvette on ice.
51. Perform electroporation at 1800 V in an Eppendorf Electroporator 2510.
52. Immediately transfer into 900 µl pre-warmed SOC medium and incubate for 30 min at 37 °C.
53. Streak 50 µl of transformed cells on a LB agar plate that contains 50 µg/mL kanamycin. Incubate over night at 37 °C.
54. Inoculate 5 mL LB supplemented with 50 µg/mL kanamycin with one colony of the transformed cells.
55. Incubate for app. 6 h at 37 °C and 210 rpm.
56. Use 2–4 mL for plasmid preparation using the Zyppy ® Plasmid Miniprep Kit according to the manufacturer's protocol. Elute in 30 µl ultrapure water.
57. Determine the DNA concentration as described above.
58. Combine 5 µl of plasmid (30–100 ng/µL) with 5 µl of forward or reverse sequencing primer (5 µM) and send for sequencing e.g. AZENTA Life Science.
59. Check plasmid integrity by sequence alignment of in-situ generated expression vector and sequencing results.
60. Transform E. coli BL21 (DE3) gold with expression constructs (10 ng/µL) (see steps 49–52), streak only 10 µl of a 1:100 dilution of transformed cells on the agar plate.

5.4 Expression and purification of carboxylases

Timing: 1–2 d.

61. Prepare a starter culture by inoculating 5 mL of LB broth containing 50 µg/mL kanamycin with a single colony of E. coli BL21 (DE3) harboring the pET28a-His6-uca or gca plasmid.

Characterization of guanidine carboxylases 117

62. Incubate the starter culture overnight at 37 °C, shaking at 200 rpm.

63. Pre-heat 50 mL of LB broth containing 50 µg/mL kanamycin in a 250 mL Erlenmeyer flask to 37 °C.

64. Inoculate the tempered LB broth with 0.5 mL of starter culture.

65. Cultivate at 37 °C, 200 rpm to an OD600 of approximately 0.5.

66. Induce protein expression by adding Isopropyl β-d-1-thiogalactopyranoside (IPTG) to a final concentration of 0.5 mM.

67. Incubate the induced cultures at 18 °C for approximately 16 h.

68. Harvest cells by centrifugation at 4 °C, 4000 rpm for 20 min.

69. Discard the supernatant and store the pellet at −20 °C until use.

70. Resuspend the frozen pellet in 2 mL lysis buffer and incubate on ice for 30 min.

71. Use a sonifier to lyse the cells on ice (30 % amplitude, 1.5 s ON, 1.5 s OFF, 3 min duty cycle). Repeat the sonication cycle as needed to achieve lysis.

72. Centrifuge the lysate in two 1 mL fractions at 4 °C, 14,000 rpm for 10 min to remove cell debris and insoluble material.

73. Filter the supernatant through a 0.2 µm filter and take a 20 µL sample for later analysis by SDS-PAGE.

74. Resuspend one pellet in 1.2 mL Tris buffer, take a 20 µL sample for later analysis by SDS-PAGE, then discard the pellets.

75. Prepare a 1 mL polypropylene column with 0.5 mL Ni-NTA Agarose beads.

76. Let the beads settle by letting the buffer flow through completely.

77. Load the filtered supernatant onto the Ni-NTA column and incubate at 4 °C on a rotary shaker for at least 60 min.

78. Collect 20 µL of the Flow through for SDS-PAGE analysis and discard the rest.

79. Wash the column with 3 × 1 mL of wash buffer and collect 20 µL SDS-PAGE (sodium dodecyl sulfate polyacrylamide gel electrophoresis) samples of the first wash step, otherwise discard the flow through.

80. Elute the protein with 1 mL elution buffer.

81. For buffer exchange, prepare a PD-10 desalting column by washing it with 5 × 5 mL of Tris buffer.

82. Apply the eluted protein and discard flow through until the protein has fully entered the column material.

83. Add 1.5 mL of Tris buffer to the column to elute the protein from the column, collect the eluted protein (approx. 1.5 mL) and set aside a 20 µL SDS-PAGE sample.

84. Apply the protein solution to an Amicon centrifugal filter (Merck) with a 3 kDa cut off and increase protein concentration by centrifugation at 4 °C, 14,000 rpm in intervals until the desired concentration is reached (~2-3 mg/mL).

85. Snap freeze the protein by pipetting drops of approx. 20–60 μL directly into liquid nitrogen, collect all droplets and store at −80 °C until use.

86. To monitor enzyme purification, run SDS-PAGE.

    a. Bio-Rad Mini-PROTEAN® Tetra Cell gel electrophoresis set up was used according to manufacturer's instructions.

    b. Prepare samples for SDS-PAGE by adding 4 μl 6x SDS loading buffer to each collected 20 μl fractions and heat to 96 °C for 5 min.

    c. Use a separating gel containing 14 % acrylamide (ROTIPHOR-ESE®Gel 40 (37.5:1)).

    d. Prepare electrophoresis chamber with 1x SDS-PAGE running buffer (prepared from ROTIPHORESE®10x SDS-PAGE) and add the gel.

    e. Load 7 μl of denatured protein samples from step b and marker (PageRuler™ Plus Prestained Protein Ladder) and run gel for 1 h at 30 mA.

    f. Cover the gel with 50 % ethanol, 10 % acetic acid in a box with lid. Heat up for 20–30 s in a microwave (700 W). Incubate under agitation for 10 min. Discard the solution.

    g. Cover the gel with 5 % ethanol, 7.5 % acetic acid add 700 μl of Rotiphorese®Blue R under agitation. Heat up in a microwave (700 W) for 20–30 s. Incubate for 10–15 min until bands are visible. Discard solution. Incubation in ultrapure water for 1 h or overnight might increase the contrast of bands to background

5.5 Note

Step 63: The LB broth should take up at most 1/5 of the volume of the Erlenmeyer flask.

Step 71: After two cycles of sonication the lysate should be distinctly clearer than before Fig. 3.

5.6 Characterization of carboxylases

Timing: 1.5 days.

Characterization of the Uca/Gca activity was performed as described by Kanamori et al. (Kanamori, Kanou, Atomi, & Imanaka, 2004) by

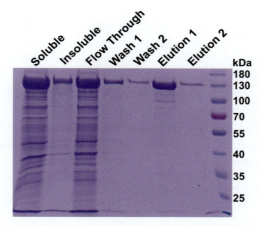

Fig. 3 SDS-PAGE analysis of Gca purification. Coomassie stained SDS-PA gel after electrophoresis with different fractions from the purification scheme loaded: the soluble and insoluble fraction after lysis, the flow through after loading the soluble lysate onto the Ni-NTA agarose column, and the wash and elution steps.

coupling the ATP consumption of the carboxylation reaction with the oxidation of NADH via pyruvate kinase and lactate dehydrogenase. The initial carboxylation reaction consumes ATP and produces ADP, which is used by the pyruvate kinase to convert phosphoenolpyruvate (PEP) to pyruvate. Pyruvate is converted to lactate by lactate dehydrogenase, oxidizing NADH to NAD^+. The rate of NADH oxidation (decrease in absorbance at 340 nm) is measured, which reflects the rate of the carboxylation reaction.

87. Create a reaction master mix containing: 20 mM Tris-HCl (pH 8.0), 200 mM NaCl, 2 mM DTT, containing Uca or Gca (1 μM), 8 mM $MgCl_2$, 8 mM $NaHCO_3$, 1 mM PEP, 0.8 mM NADH, and 5 U/ml pyruvate kinase/lactate dehydrogenase enzyme mix.
88. Add 5 μl of different concentrations of guanidine or urea to a 96 well plate (0–200 mM). Perform each reaction in triplicates. Always include a non-substrate control to account for background ATP hydrolysis. If a background rate of ATP hydrolysis is observed in the absence of carboxyltransferase domain substrate (here guanidine or urea), it should be subtracted.
89. Add 45 μl master mix to the substrate and incubate for 2–5 min at R.T.
90. Start the reaction by adding 4 μl 12.5 mM ATP, to a final concentration of 1 mM, to each well.

91. Use a plate reader to monitor the decrease in NADH absorbance at 340 nm, 25 °C

92. Use the slope of absorbance decrease over time to evaluate the carboxylation rate.

93. Determine velocity by defining 1 U of activity equals 1 μmol NAD + formation per minute.

94. Use GraphPad Prism 6 or equivalent software to plot velocity against substrate concentration, assuming Michaelis-Menten kinetics, to determine the kinetic parameters

5.7 Optional: Comparative proteome analysis

Timing: 1–2 d.

95. Inoculate 5 mL LB with Raoultella terrigena str. JH01, Erwinia rhapontici str. JH02 or Klebsiella michiganensis str. JH07 in a glass reaction tube

96. Grow overnight at 30 °C and 210 rpm

97. Dilute bacteria in 1x M8 medium without nitrogen source to an OD of 0.05

98. Inoculate 50 mL 1x M8 minimal media supplemented with 5 mM guanidine or 15 mM ammonium and incubate at 30 °C and 210 rpm.

99. Measure OD600 of the cultures and harvest the same amount of cells as calculated from OD600 * Volume after approximately 10 hr.

100. Resuspend cells in 3 mL 1x PBS and lyse by sonication with a Branson Sonifier (30 % amplitude, 0.5 s ON, 1 s OFF, 2 min duty cycle)

101. Determine the total protein amount with e.g. the PierceTM BCA Protein Assay Kit from Thermo Scientific according to the manufacturer's protocol.

102. Load 50 μg total protein for each sample on a 14 % SDS PA-gel. Load a protein marker next to your samples.

103. Run electrophoresis at 30 mA per gel until the sample has completely entered the resolving gel.

104. Cut out the proteins. Use the ladder to estimate the molecular weight of your sample.

105. Transfer the gel piece into a 1.5 mL reaction tube for subsequent in-gel digestion and LC-MS analysis.

106. LC-MS should be conducted in a semi-quantitative manner, thus individual proteins are comparable between the samples.

107. Search for proteins that are highly upregulated when cells are grown with guanidine as the nitrogen source compared to cells grown with ammonia. The annotation of proteins can also be used as a guide to search for proteins whose proposed activity matches the reactions they are supposed to catalyze.

6. Limitations

If you want to search for bacteria able to degrade your compound of interest it might be important to screen different minimal medium formula, e.g. in regard to the carbon source. Carbon source preference can vary largely between species. Other factors to consider are temperature, salt composition, cofactor requirements and oxygen concentration. A good approximation can be concluded from the environment, where the sample was taken, especially in regard to temperature and oxygen availability.

7. Optimization and troubleshooting
7.1 No colonies observed during isolation

- test different carbon sources like glucose, glycerol, carbonic acids
- change formulation of the minimal medium: We use a derivative of M9 minimal medium that is high in phosphate and salt.
- Supplement the medium with trace elements/vitamins
- Test different incubation temperatures

8. Safety considerations and standards

Bear in mind that isolated bacteria might be infectious. Protective clothes like lab coat, goggles and gloves should be worn whenever possible. Gloves should be disinfected with 80 % ethanol before being discarded after handling bacterial samples. Before and after working with bacterial cultures, the working area should be thoroughly disinfected with 80 % ethanol. Using a sterile hood can significantly reduce the risk of infections and contaminations, especially before the isolated bacteria are characterized.

9. Conclusion

Here, we present a straightforward strategy that enabled us to isolate bacteria capable of using guanidine as their sole nitrogen source and to elucidate the underlying molecular mechanism. We found that some bacteria possess a guanidine carboxylase homologous to urea carboxylase, but the enzymes showed orthogonal substrate specificity, preferring guanidine over urea. In contrast to urea carboxylases, the guanidine carboxylases are associated with carboxyguanidine deiminase genes (Schneider et al., 2020). The presented approach was also successfully applied in our lab to isolate *Pseudomonas canavaninivorans* that utilizes the plant arginine antimetabolite canavanine as the sole energy, carbon and nitrogen source (Hauth, Buck, & Hartig, 2022). Following that, we were able to elucidate the enzymes necessary for the degradation of canavanine in *P. canavaninivorans* (Hauth, Buck, Stanoppi, & Hartig, 2022). The comparative proteome data also revealed a standalone canavanyl-tRNA^{Arg} deacylase, that edits mischarged tRNA^{Arg} (Hauth, Funck, & Hartig, 2023). With the presented protocol, the degradation of other nitrogenous compounds can be targeted. In principle, guanidine can be replaced by any compound of interest making this approach also useful for the discovery of novel degradative enzyme systems.

References

Funck, D., Sinn, M., Fleming, J. R., Stanoppi, M., Dietrich, J., Lopez-Igual, R., ... Hartig, J. S. (2022). Discovery of a Ni^{2+}-dependent guanidine hydrolase in bacteria. *Nature, 603*(7901), 515–521. https://doi.org/10.1038/s41586-022-04490-x.

Güthner, T., Mertschenk, B., & Schulz, B. (2014). Guanidine and derivatives. In V. C. H. Wiley (Vol. Ed.), *Ullmann's fine chemicals: 2,* (pp. 657–672). Weinheim, Germany: Wiley-VCH Verlag GmbH & Co. KGaA.

Hauth, F., Buck, H., & Hartig, J. S. (2022). *Pseudomonas canavaninivorans* sp. nov., isolated from bean rhizosphere. *International Journal of Systematic and Evolutionary Microbiology, 72*(1), https://doi.org/10.1099/ijsem.0.005203.

Hauth, F., Buck, H., Stanoppi, M., & Hartig, J. S. (2022). Canavanine utilization via homoserine and hydroxyguanidine by a PLP-dependent gamma-lyase in *Pseudomonadaceae* and *Rhizobiales. RSC Chemical Biology, 3*(10), 1240–1250. https://doi.org/10.1039/d2cb00128d.

Hauth, F., Funck, D., & Hartig, J. S. (2023). A standalone editing protein deacylates mischarged canavanyl-tRNAArg to prevent canavanine incorporation into proteins. *Nucleic Acids Research, 51*(5), 2001–2010. https://doi.org/10.1093/nar/gkac1197.

Iwanoff, N. N., & Awetissowa, A. N. (1931). The fermentative conversion of guanidine in urea. *Biochemische Zeitschrift, 231,* 67–78 Retrieved from ://WOS:000200827000010.

Kanamori, T., Kanou, N., Atomi, H., & Imanaka, T. (2004). Enzymatic characterization of a prokaryotic urea carboxylase. *Journal of Bacteriology, 186*(9), 2532–2539. https://doi.org/10.1128/jb.186.9.2532-2539.2004.

Kato, T., Yamagata, M., & Tsukahara, S. (1986). Guanidine compounds in fruit-trees and their seasonal-variations in citrus (Citrus-Unshiu Marc). *Journal of the Japanese Society for Horticultural Science, 55*(2), 169–173. https://doi.org/10.2503/jjshs.55.169.

Kermani, A. A., Macdonald, C. B., Gundepudi, R., & Stockbridge, R. B. (2018). Guanidinium export is the primal function of SMR family transporters. *Proceedings of the National Academy of Sciences of the United States of America, 115*(12), 3060–3065. https://doi.org/10.1073/pnas.1719187115.

Lenkeit, F., Eckert, I., Hartig, J. S., & Weinberg, Z. (2020). Discovery and characterization of a fourth class of guanidine riboswitches. *Nucleic Acids Research, 48*(22), 12889–12899. https://doi.org/10.1093/nar/gkaa1102.

Nelson, J. W., Atilho, R. M., Sherlock, M. E., Stockbridge, R. B., & Breaker, R. R. (2017). Metabolism of free guanidine in bacteria is regulated by a widespread riboswitch class. *Molecular Cell, 65*(2), 220–230. https://doi.org/10.1016/j.molcel.2016.11.019.

Salvail, H., Balaji, A., Yu, D., Roth, A., & Breaker, R. R. (2020). Biochemical validation of a fourth guanidine riboswitch class in bacteria. *Biochemistry, 59*(49), 4654–4662. https://doi.org/10.1021/acs.biochem.0c00793.

Schneider, N. O., Tassoulas, L. J., Zeng, D., Laseke, A. J., Reiter, N. J., Wackett, L. P., & Maurice, M. S. (2020). Solving the conundrum: Widespread proteins annotated for urea metabolism in bacteria are carboxyguanidine deiminases mediating nitrogen assimilation from guanidine. *Biochemistry, 59*(35), 3258–3270. https://doi.org/10.1021/acs.biochem.0c00537.

Schulze, E. (1893). Ueber einige stickstoffhaltige Bestandtheile der Keimlinge von *Vicia sativa*. *Zeitschrift für Physiologische Chemie, 17*, 193–216. https://vlp.mpiwg-berlin.mpg.de/library/data/lit16884?.

Sherlock, M. E., & Breaker, R. R. (2017). Biochemical validation of a third guanidine riboswitch class in bacteria. *Biochemistry, 56*(2), 359–363. https://doi.org/10.1021/acs.biochem.6b01271.

Sherlock, M. E., Malkowski, S. N., & Breaker, R. R. (2017). Biochemical validation of a second guanidine riboswitch class in bacteria. *Biochemistry, 56*(2), 352–358. https://doi.org/10.1021/acs.biochem.6b01270.

Sinn, M., Hauth, F., Lenkeit, F., Weinberg, Z., & Hartig, J. S. (2021). Widespread bacterial utilization of guanidine as nitrogen source. *Molecular Microbiology, 116*(1), 200–210. https://doi.org/10.1111/mmi.14702.

Wang, B., Xu, Y., Wang, X., Yuan, J. S., Johnson, C. H., Young, J. D., & Yu, J. (2021). A guanidine-degrading enzyme controls genomic stability of ethylene-producing cyanobacteria. *Nature Communications, 12*(1), 5150. https://doi.org/10.1038/s41467-021-25369-x.

Wishart, D. S., Feunang, Y. D., Marcu, A., Guo, A. C., Liang, K., Vazquez-Fresno, R., ... Scalbert, A. (2018). HMDB 4.0: The human metabolome database for 2018. *Nucleic Acids Research, 46*(D1), D608–D617. https://doi.org/10.1093/nar/gkx1089.

CHAPTER SIX

Methods to study prFMN-UbiD mediated (de)carboxylation

Dominic R. Whittall and David Leys*
Manchester Institute of Biotechnology, University of Manchester, Manchester, United Kingdom
*Corresponding author. e-mail address: david.leys@manchester.ac.uk

Contents

1. Introduction	126
2. In vivo production of *holo*-UbiD/*apo*-UbiD	129
2.1 Equipment	130
2.2 Buffers and reagents	130
2.3 Molecular biology	131
2.4 Heterologous production	131
2.5 Purification of *holo*-UbiD	132
3. In vitro reconstitution of *apo*-UbiD using prFMN^{reduced}	132
3.1 Equipment	133
3.2 Buffers and reagents	133
3.3 Production of prFMN^{reduced}	134
3.4 Reconstitution of *apo*-UbiD with prFMN^{reduced}	134
3.5 Notes	135
4. Synthesis of DMAP	135
4.1 Equipment	135
4.2 Reagents	135
4.3 Procedure	136
4.4 Preparing bis-triethylammonium phosphate (TEAP) solution	136
4.5 Preparing dimethylallyl monophosphate (DMAP)	136
4.6 Notes	136
5. In vitro oxidative maturation of prFMN containing UbiD	137
5.1 Equipment	137
5.2 Buffers and reagents	137
5.3 Procedure	138
6. Characterisation of prFMN species via ESI-MS	138
6.1 Equipment	138
6.2 Buffers and reagents	139
6.3 Procedure	139
7. ^{1}H NMR monitored enzyme catalysed deuterium exchange	139
7.1 Equipment	140
7.2 Buffers and reagents	140
7.3 Procedure	140

Methods in Enzymology, Volume 708
ISSN 0076-6879, https://doi.org/10.1016/bs.mie.2024.10.020
Copyright © 2024 Elsevier Inc. All rights are reserved, including those for text and data mining, AI training, and similar technologies.

8. (De)carboxylation assays monitored by HPLC/LC-MS	140
8.1 Equipment	141
8.2 Buffers and reagents	141
8.3 Procedure	141
9. Preparation of whole-cell UbiD-biocatalysts	142
9.1 Equipment	143
9.2 Buffers and reagents	143
9.3 Procedure	143
10. UbiD-substrate screening through monitoring flavin spectral shift	144
10.1 Equipment	144
10.2 Buffers and reagents	144
10.3 Reconstitution of UbiD with FMN	144
10.4 Spectral shift assay	144
11. Experimental strategies to drive carboxylation	145
12. Conclusion	146
References	147

Abstract

The microbial UbiX-UbiD system facilitates the reversible (de)carboxylation of alpha, beta-unsaturated carboxylic acids, including aromatic compounds. The direct C-H carboxylation presents an attractive method for functionalisation and carbon capture but is difficult to achieve under mild conditions. Hence, UbiD-mediated $Csp2$-H activation can serve as a versatile tool for developing new biocatalytic routes to transform aryl or alkene compounds and carbon dioxide into valuable commodity chemicals. UbiD activity is dependent on the prenylated flavin cofactor, prFMN, produced by UbiX. Oxidative maturation of the prFMN^{reduced} UbiX product into the active prFMN^{iminium} is a critical prerequisite for UbiD activity. However, efficiency of prFMN incorporation and oxidative maturation can vary considerably batch to batch and between distinct UbiD enzymes. Herein, we present detailed protocols for the production, reconstitution and characterisation of active UbiD enzymes, including UbiD-mediated (de)carboxylation.

1. Introduction

Members of the widespread microbial UbiD enzyme family facilitate a diverse range of (de)carboxylative chemistry, participating in numerous biochemical processes (Bloor et al., 2023; Roberts & Leys, 2022). Typical substrates of UbiD enzymes are alpha, beta–unsaturated carboxylic acids, including acrylic and (hetero)aromatic acids. Both the natural and evolved functionalities of UbiD enzymes, such as in bio–synthesis of plastic precursor FDCA by HmfF (Payne et al., 2019), or isobutene production by evolved Fdc (Saaret et al. 2021), hold tremendous potential for use within a variety of industrial processes, including pharmaceutical synthesis, biofuel production, and the generation of high-value secondary metabolites.

UbiD enzymes are dependent on a heavily modified flavin cofactor, prenylated FMN (prFMN) for catalysis. The recent discovery and identification of prFMN has prompted a renewed interest in the UbiD enzyme family (Leys, 2018). The flavin prenyltransferase UbiX catalyses the formation of prFMN[reduced] from reduced FMNH$_2$ by utilising the isoprenoid precursor dimethylallyl monophosphate (DMAP) as a prenylating agent (White et al., 2015). The *apo*-UbiD enzyme then binds prFMN[reduced], which undergoes oxidative maturation within the UbiD to form the catalytically relevant species, prFMN[iminium] (Fig. 1). The prFMN[iminium] cofactor has azomethine ylide characteristics and performs covalent catalysis based on 1, 3 dipolar cycloaddition with the substrate, yielding a substrate-

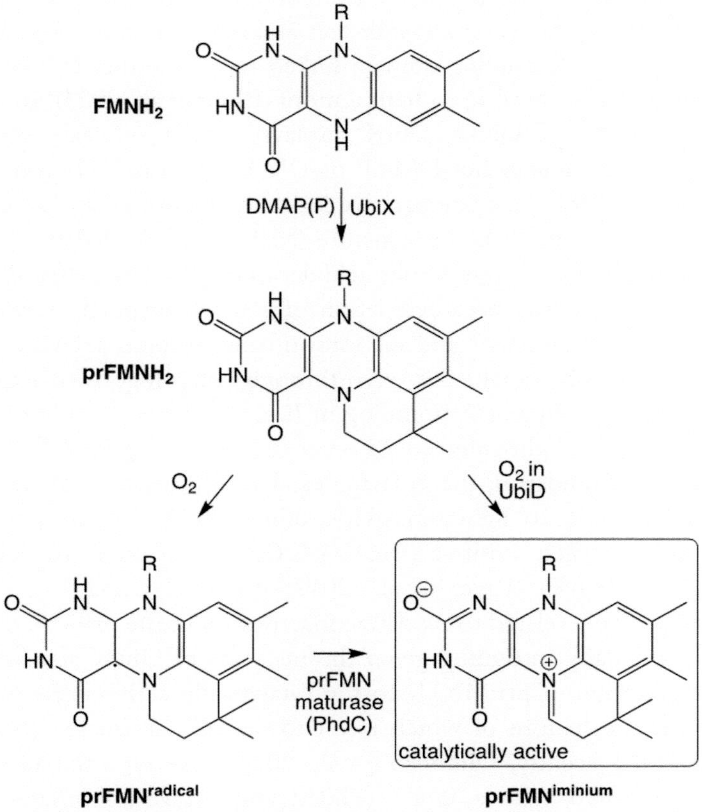

Fig. 1 Pathway of prFMN biosynthesis and oxidative maturation.

prFMN^{iminium} adduct. This is followed by reversible (de)carboxylation and cycloelimination, resulting in product formation (Payne et al., 2015). While the key elements of prFMN maturation and catalysis have been elucidated for *A. niger* Fdc as the model UbiD enzyme (Payne et al., 2015; Marshall et al., 2017; Bailey et al., 2018; Balaikaite et al., 2020), considerable variability has emerged in terms of oligomeric structure, oxidative maturation and mode of covalent catalysis employed (reviewed in Bloor et al., 2023; Roberts & Leys, 2022).

Ubiquitously present in microbes, *ubiD* genes are associated with a variety of metabolic pathways (Meganathan, 2001). Consequently, UbiD acid substrates exhibit exceptional diversity, though the presence of an alpha-beta unsaturated carboxylic acid group remains consistent in all. This versatility is in part explained by the highly diverse UbiD quaternary structure and accompanying active site configuration. Comparison of the existing UbiD crystal structures indicates a common fold incorporating an N-terminal prFMN binding domain linked to a C-terminal oligomerisation domain via a central alpha helical motif (Bloor et al., 2023). In contrast to the versatility of UbiD, UbiX displays a strict substrate specificity towards FMNH$_2$ and either DMAP or DMAP(P) as prenyl group donor (Wang et al., 2018). This comparative lack of chemical versatility is linked to the highly conserved UbiX structure (Marshall et al., 2019).

The model UbiD system ferulic acid decarboxylase from *Aspergillus niger* (AnFdc1) catalyses the decarboxylation of a range of phenylacrylic acids and is dependent on Mn^{2+} and K$^+$ ions in order to bind prFMN. Active, *holo*-AnFdc1 can be obtained in vivo through heterologous co-expression of the *fdc1* gene with a *ubiX* homolog in *E. coli*. However, AnFdc1 is light sensitive, with light-induced isomerisation of prFMN^{iminium} to prFMN^{ketamine} rendering the cofactor catalytically inactive (Bailey et al., 2018). In addition to light sensitivity, other UbiD enzymes have been shown to be oxygen sensitive (AroY/VdcCD), or sensitive to both light and oxygen (HmfF) (Payer et al., 2017; Payne et al., 2019). Oxygen sensitivity is likely related to oxidative inactivation of the prFMN cofactor. This is particularly intriguing given the necessity of UbiD-mediated oxidative maturation of prFMNH$_2$ to the catalytically active prFMN^{iminium}, the specific mechanism of which is yet to be fully elucidated (Balaikaite et al., 2020). Recently, DiRocco (et al., 2024) have reported an enzyme (PhdC) associated with a UbiD (PhdA) and prFMN synthase (PhdB) capable of facilitating the one-electron oxidation of prFMN^{radical} to the active, prFMN^{iminium} form (Fig. 1). Co-expression of PhdC with HmfF, resulted in the production of active, *holo*-HmfF – suggesting the general function of PhdC as a prFMN maturase. AnFdc1 is a dimer, with the active site being situated at the junction between the two domains. The relative position of the prFMN binding domain to the oligomerisation domain shows considerable variation across different crystal structures – highly suggestive of catalytically-relevant domain motion (Marshall et al., 2021), although the extend of which this applies across the wider family is unclear.

Stability of the *apo*-AnFdc1 dimer is enhanced upon binding of the prFMN cofactor, which promotes a more compact conformation (Beveridge et al., 2016). Consequently, the crystal structures obtained for AnFdc1 represent a closed form of the enzyme, which is associated with substrate binding. In contrast, crystal structures of other UbiD enzymes in the closed, substrate-bound conformation appear more challenging to obtain (Roberts & Leys, 2022), with the majority of structures representing an open state. To what extend prFMN binding is linked to domain motion is not firmly established, although UbiD enzymes that crystallise readily in the open form tend to show weaker cofactor binding affinity. On account of this, heterologous expression and purification of many UbiD enzymes often results in the production of *apo*-UbiD protein. Consequently, in vitro cofactor reconstitution is an essential step in the production of *holo*-UbiD enzymes.

Here, we detail the heterologous expression and purification of both *holo*- and *apo*-UbiD, along with various techniques for in vitro reconstitution and oxidative maturation. Methods to study prFMN species by ESI-MS are also included. In addition, we also present methods detailing the study of UbiD catalytic activity through application of NMR, HPLC and LC-MS analytical techniques. Finally, we describe methods to study UbiD-mediated compound (de)carboxylation using purified enzymes and whole-cell biocatalysis.

2. *In vivo* production of *holo*-UbiD/*apo*-UbiD

The native *E. coli* UbiX activity levels are too low to sustain the production of high levels of *holo*-UbiD enzymes. Therefore, co-expression with a highly active UbiX is necessary in order to improve production levels of *holo*-UbiD enzymes in vivo. While our lab has attempted to co-express UbiX homologues cognate to the target UbiD enzyme, we have found that the largest quantities of prFMN incorporated UbiD enzymes have been obtained when co-expressing with UbiX enzymes derived from

E. coli or *Pseudomonas aeruginosa* (Payne et al., 2015). Supplementation of *E. coli* cultures overexpressing UbiX with prenol has also been shown to boost production of prFMN (Wang et al., 2018). In addition, supplementing growth media and purification buffers with K^+ and Mn^{2+}, two ions essential for prFMN binding in many UbiD enzymes, improves retention levels of prFMN and hence production levels of *holo*-UbiD (Fig. 2). More recently, Zhu et al. (2023) reported the construction of an in vivo prFMN synthesis pathway within *E. coli* that demonstrated a 3.8-fold enhancement of prFMN production relative to a UbiX overexpression strain.

2.1 Equipment
- Orbital shaking incubator
- Centrifuge and rotor
- Cell disruptor
- Ni-NTA agarose resin
- 10-DG desalting column
- Anaerobic glove box

2.2 Buffers and reagents
- 1 L of autoclaved LB medium
- 50 mg/ml ampicillin (filter sterilised through a 0.2 μM syringe filter)
- 50 mg/ml kanamycin (filter sterilised through a 0.2 μM syringe filter)
- 0.25 M IPTG (filter sterilised through a 0.2 μM syringe filter)
- 1 M $MnCl_2$ (filter sterilised through a 0.2 μM syringe filter)

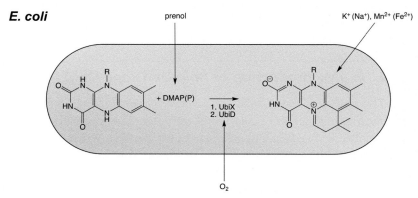

Fig. 2 Overview of the in vivo production strategy for *holo*-UbiD.

- 1 L Ni-Equilibration buffer: 50 mM Tris-HCl, 200 mM NaCl, 1 mM $MnCl_2$, pH 7.5, prepared in Milli-Q water
- 1 L Ni-Elution buffer: 50 mM Tris-HCl, 200 mM NaCl, 1 mM $MnCl_2$, 500 mM imidazole, pH 7.5, prepared in Milli-Q water
- 1 L 10-DG buffer: 20 mM Tris-HCl, 100 mM NaCl, pH 7.5, prepared in Milli-Q water
- Store buffers at 4 °C prior to use

2.3 Molecular biology

A. niger Fdc1 was cloned into pET30a (Kan resistant) at the *NdeI/XhoI* restriction sites, yielding a protein containing a C-terminal His-tag. The location of the His-tag may influence prFMN binding or enzyme activity, depending on the UbiD homologue. *E. coli* UbiX was cloned into the *NdeI/XhoI* sites of pET21b (Amp resistant), producing an untagged protein.

Both constructs were co-transformed into *E. coli* BL21 (DE3). Successful co-transformants demonstrated resistance to both ampicillin and kanamycin.

Co-expression of *A. niger* Fdc1 and *E. coli* UbiX can also be achieved by cloning both genes into a single plasmid containing two multiple cloning sites; or by expressing Fdc1 and UbiX on two vectors with different origins of replication.

2.4 Heterologous production

Pre-culture a single co-transformant colony in 50 ml sterile LB containing 50 µg/ml ampicillin and 50 µg/ml kanamycin. Incubate overnight within a shaking incubator at 37 °C. The following day, inoculate 1 L of autoclaved LB (supplemented with 50 µg/ml ampicillin and 50 µg/ml kanamycin) with 5 ml of overnight pre-culture. Grow at 37 °C and 180 rpm within a shaking incubator, until an OD_{600} of 0.6–1.0 is reached. At this point, induce the culture by adding 0.25 mM IPTG (final concentration), and supplement with 1 mM $MnCl_2$ (final concentration). Mn^{2+} ions are essential for prFMN binding in Fdc1, in addition to AroY, VdcCD and other UbiD homologues. It has also been suggested that UbiD enzyme PCD utilises Fe^{2+} for oxygen-independent electron-transfer during oxidative maturation of prFMN (Mergelsberg et al., 2017). Reduce the temperature to 17 °C and grow the cultures overnight. The following day, harvest the culture via centrifugation, storing the resulting pellets at -20 °C until required.

2.5 Purification of *holo*-UbiD

Some members of the UbiD family are sensitive to light and/or to oxygen (Bailey et al., 2018; Payer et al., 2017). Hence, taking precautions against exposure to light/oxygen during purification is highly recommended. To minimise light exposure, wrap columns and bottles used during purification in aluminium foil, and make use of black Eppendorf tubes during elution and fractionation steps. If the target UbiD enzyme is highly sensitive to oxygen, perform all protein purification steps within an anaerobic glove box.

As a general rule, protein purification should be performed at 4 °C to minimise degradation of the target protein. However, if purifying anaerobically (i.e. within a glove box or chamber), purification at low temperatures may prove impractical. The decision to purify either aerobically (at 4 °C) or anaerobically (at an ambient temperature) should depend on the thermostability and oxygen sensitivity of the UbiD in question.

Resuspend cell pellets in Ni-Equilibration buffer and supplement with RNAse, DNAse, lysozyme and EDTA-free protease inhibitor tablets. Stir in a cold room until fully homogenous, before lysing via cell disruption at 20,000 psi. Centrifuge the lysate at $100,000 \times g$ (4 °C) for 60 min. Filter the lysate through a $0.45\,\mu M$ syringe filter, before applying to 5 ml Ni-NTA agarose resin equilibrated with Ni-Equilibration buffer, retaining the flow through. Wash the resin with 4 column volumes of Ni-Equilibration buffer containing 10 mM imidazole, collecting the flow through. Elute the bound protein from the resin by adding 2 column volumes of Ni-Elution buffer containing 250 mM imidazole, collecting 1 ml fractions in individual tubes. Subject all collected fractions to SDS-PAGE analysis to locate fractions containing UbiD protein. Pool fractions containing pure protein and if required, reduce the volume of the pooled fractions using a centrifugal concentrator. Apply to a 10-DG column (equilibrated with 10-DG buffer) to remove imidazole, eluting with 10-DG buffer into 0.5-1 ml fractions. In some instances, the protein is coloured (prFMN^{radical} is purple, prFMN^{iminium} a faint yellow) − this enables simple visual or spectroscopic detection of fractions containing protein. Alternatively, Bradford reagent can be used. Combine fractions containing purified protein, aliquot and snap freeze in liquid nitrogen.

3. *In vitro* reconstitution of *apo*-UbiD using prFMN^{reduced}

Despite co-expression with UbiX, many UbiD enzymes purify predominantly in the *apo*-form as verified by a lack of absorption above

Methods to study prFMN-UbiD mediated (de)carboxylation

280 nm and no or little activity. In addition, levels of active prFMN^{iminium} incorporation can vary between batches of purified UbiD. Consequently, in vitro reconstitution of *apo*-UbiD enzymes with prFMN is often required to produce high levels of active *holo*-UbiD. This necessitates production of a stock of prFMN^{reduced}, which can be made by the UbiX enzyme under anaerobic conditions. The UbiX enzyme from *Pseudomonas aeruginosa* provides an efficient means of producing prFMN from $FMNH_2$ and dimethylallyl monophosphate (DMAP, White et al., 2015). While other UbiX homologues can utilise the di-phosphorylated dimethylallyl pyrophosphate (DMAP(P)) in the production of prFMN, the use of PaUbiX is preferable due to its stability in solution and the relatively straightforward organic synthesis procedure of DMAP (Fig. 3) (White et al., 2015; Arunrattanamook & Marsh, 2018).

3.1 Equipment
- Anaerobic glove box
- Microcentrifuge
- UV–Vis spectrophotometer housed in anaerobic glove box (optional)

3.2 Buffers and reagents
- Anaerobic buffer (i.e. purged with N_2)
- FMN
- DMAP
- Purified PaUbiX
- Purified Fre reductase and NADH or sodium dithionite (to serve as reductant)

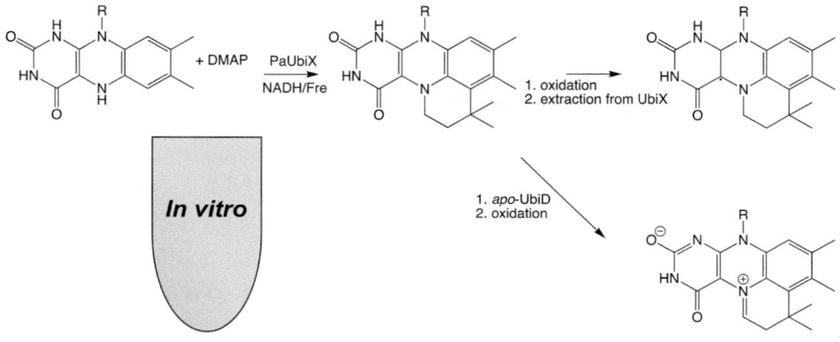

Fig. 3 Overview of the in vitro production strategy for *holo*-UbiD.

3.3 Production of prFMN^{reduced}

All buffers must be rendered anaerobic through purging with nitrogen gas. All reagents must be dissolved in anaerobic buffer (specific to the UbiD being reconstituted) and left in the anaerobic glove box to equilibrate. To minimise exposure to light, we recommend that the reaction be performed inside a black Eppendorf tube. A standard reaction should contain 50 μM of UbiX and 1 mM of FMN. FMN is then reduced by the addition of 10 μM Fre reductase and 2 mM NADH; following addition of NADH, the solution should appear colourless. If using dithionite, titrate from a 100 mM stock until the solution is colourless. The reaction should then be left for a period of 15 min to ensure all FMN is reduced. Post-reduction, DMAP is added to a final concentration of 5 mM. The reaction is then left to proceed at ambient temperature within the glove box for a minimum of 60 min. The reaction can be left to proceed overnight – if doing so, the concentration of NADH should be increased to 10 mM in order to avoid background oxidation depleting NADH levels. Lastly, the prFMN^{reduced} reaction mixture is filtered using a centrifugal concentrator to remove UbiX and Fre. Exposure of the filter membrane to oxygen provides a useful means of checking for successful production of prFMN^{reduced}. Development of a purple colour on the filter corresponds to a prFMN^{radical} species, and hence confirms the presence of prFMN^{reduced} within the filtrate. The filtered stock of prFMN^{reduced} can then be transferred to a second black Eppendorf within the anaerobic glove box as a means of limiting exposure to light. Quantification of prFMN^{reduced} is non-trivial, given its oxygen sensitivity. Under the conditions used, (near) total conversion of FMN (which can be readily quantified through absorption at 450 nm in the oxidised state) occurs, allowing sufficiently accurate estimation of the prFMN^{reduced} product concentration.

3.4 Reconstitution of *apo*-UbiD with prFMN^{reduced}

Reconstitution must be performed within an anaerobic glove box. Add the prFMN reduced filtrate to the UbiD, along with MnCl$_2$ and KCl, to a molar ratio of 2:2:2:1 (prFMN reduced:MnCl$_2$:KCl:UbiD). Incubate the mixture for a minimum of 10 min, before adding to a desalting column equilibrated with anaerobic buffer – this will remove excess prFMN^{reduced} and NAD$^+$. Elute the reconstituted UbiD protein according to the manufacturer's protocol.

3.5 Notes

Using dithionite as a reductant risks the formation of prFMN–sulphite adducts – thus, we developed a protocol that makes use of the natural FMN reductase, Fre (Marshall et al., 2017; Shepherd et al., 2015). Following removal of Fre by centrifugal filtration, prFMN^{reduced} will be readily oxidised to the purple prFMN^{radical} species. Oxygen content of the anaerobic glove box must therefore be kept to a minimum at all times. The desalting column should be washed with anaerobic buffer containing dithionite to remove all remaining oxygen from the resin. Dithionite should then be removed from the column via thorough washing (>4 column volumes) with anaerobic buffer.

4. Synthesis of DMAP

Despite the availability of DMAP and DMAP(P) from commercial suppliers, issues with high costs and variable purity often arise. Consequently, we opt to synthesise DMAP in-house, employing the protocol described below.

4.1 Equipment
- Round-bottomed and Erlenmeyer flasks (both 25 ml)
- Glass chromatography column
- Measuring cylinder, magnetic stirrer bar, rubber septa, needles, stirring plates, laboratory balance, rotary evaporator, vacuum pump, oven.
- Pasteur pipettes and bulb droppers
- Disposable syringes (3 and 6 ml)
- TLC plates (TLC silica gel 60, Merck, cat. no. 105553)
- NMR tubes and NMR spectrophotometer

4.2 Reagents
- Concentrated phosphoric acid
- Triethylamine
- Acetonitrile
- 3-Methylbut-2-en-1-ol
- Trichloroacetonitrile
- Isopropanol
- NH_4OH (38 % aqueous)
- H_2O
- D_2O
- Silica gel (pore size 60 Å, particle size 240-400 mesh)

4.3 Procedure

Protocol was adapted from literature (Keller & Thompson, 1993).

4.4 Preparing bis-triethylammonium phosphate (TEAP) solution

First, prepare solution A by adding 2.5 ml of concentrated phosphoric acid to 9.4 ml of acetonitrile. Then prepare solution B by adding 11.0 ml of triethylamine to 10.0 ml acetonitrile. Slowly mix 9.4 ml of solution A with 10.0 ml solution B whilst stirring to form TEAP solution.

4.5 Preparing dimethylallyl monophosphate (DMAP)

First, prepare 100 ml of quenching solution - isopropanol:NH_4OH:H_2O at a volume ratio of 6:2:0.5. Add 2 nmol of 3-methyl-2-buten-1-ol to 10.0 ml of trichloroacetonitrile. Stir in a round-bottomed flask at 25 °C and 600 rpm. Add 10.0 ml of TEAP solution in a dropwise manner to the one-pot reaction until a pale yellow solution forms. Following complete addition of the 10.0 ml TEAP solution, quench the reaction by adding 10.0 ml of quenching solution and allow to cool at an ambient temperature (Fig. 4).

Purify the crude product by dry column vacuum chromatography using silica gel equilibrated with quenching solution. Monitor the purification using TLC plates, utilising the remaining quenching solution as a mobile phase. Soak the TLC plates in potassium permanganate solution and leave to air dry. Yellow spots should be visible against a dark purple background. Following removal of the remaining solvent in vacuo, the product will appear as a pale yellow solid. Confirm product identity using ^{1}H and ^{31}P NMR.

4.6 Notes

A notable colour change occurs during the DMAP synthesis reaction. Following addition of the TEAP solution to the alcohol/trichloroacetonitrile solution, the solution should begin to turn yellow. If the reaction mix starts to turn orange, stop adding TEAP solution and quench the reaction. Orange or brown solutions tend to indicate the formation of DMAP(P) and DMAP (P)(P), rather than the desirable yellow colour that indicates DMAP

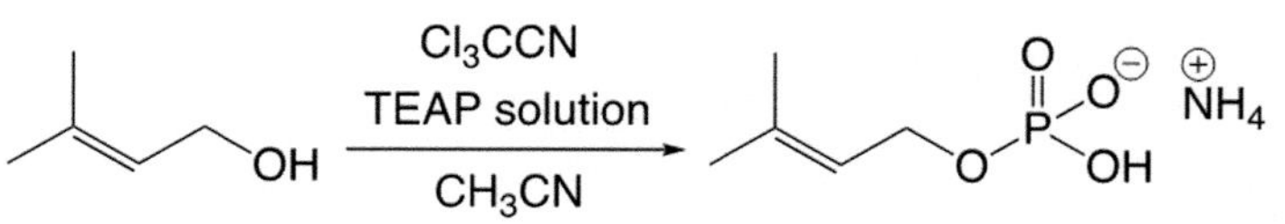

Fig. 4 Summary of dimethylallyl monophosphate (DMAP) production.

formation. NMR data of the purified compound should be consistent with values reported in the literature (Lira et al., 2013).

^{1}H NMR (400 MHz, D$_2$O): δ ppm 5.48–5.41 (m, ^{1}H, (CH$_3$)$_2$C=CH), 4.47 (t, J=7.3 Hz, ^{1}H, ^{1}H from (CH$_3$)$_2$C=CHCH$_2$), 4.36 (t, J=6.8 Hz, ^{1}H, ^{1}H from (CH$_3$)$_2$C=CHCH$_2$), 1.77 (d, J=3.1 Hz, ^{3}H, CH$_3$), 1.72 (d, J=5.8 Hz, ^{3}H, CH$_3$). ^{13}C NMR (126 MHz, D$_2$O): δ ppm 140.1 (C), 119.5 (d, J=7.6 Hz, (CH$_3$)$_2$C=CH), 62.1 (d, J=5.0 Hz, (CH$_3$)$_2$C=CHCH$_2$), 24.9 (CH$_3$), 17.1 (CH$_3$). MS (ESI$^-$) m/z (%): 165.0 (M-NH$_4^+$).

5. *In vitro* oxidative maturation of prFMN containing UbiD

Exposing prFMN^{reduced}:UbiD to oxygen can lead to its transformation into either the active prFMN^{iminium} species or the inactive prFMN^{radical} species. Both the outcome of this transformation and the degree of activation exhibit considerable variability, likely depending on the specific UbiD or UbiD variant used. While the exact mechanism of oxygen-mediated oxidative maturation remains unclear, our research has confirmed that UbiD enzymes play an essential role in forming active prFMN^{iminium} species capable of supporting enzymatic activity (Balaikaite et al., 2020). In addition to molecular oxygen, a number of alternative oxidants can facilitate the oxidative maturation of prFMN^{reduced}:UbiD or prFMN^{radical}:UbiD, with the latter (inactive) species being stable in the presence of oxygen. Notably, potassium ferricyanide can effectively replace molecular oxygen to oxidise the reduced prFMN species to form the active iminium species (Roberts et al., 2023). Additionally, proteins transcribed from genes located on the same operon as *ubiD* and *ubiX* may assist with the oxidative maturation of prFMN^{reduced} to prFMN^{iminium} via the prFMN^{radical} species, as demonstrated by the phenazine-1-carboxylate decarboxylase operon in *Mycobacterium fortuitum* (DiRocco et al., 2024).

5.1 Equipment
- Anaerobic glovebox
- UV–Vis spectrophotometer in anaerobic glovebox (optional)

5.2 Buffers and reagents
- Anaerobic buffer (i.e. purged with N$_2$)
- Potassium ferricyanide
- UbiD substrate (for example, cinnamic acid for Fdc1)

5.3 Procedure

Within an anaerobic glove box, oxidise 25 μM of reconstituted (i.e. *holo-*) UbiD enzyme with 1–4 molar equivalents of potassium ferricyanide, and incubate at an ambient temperature for 10–30 min. This should be done within a sealed vial and using buffers that have been thoroughly purged with nitrogen gas.

Potassium ferricyanide should be flushed with nitrogen and dissolved in anaerobic buffer within the glovebox. A concentrated stock should be prepared, before being diluted to working concentrations using anaerobic buffer.

The enzyme maturation process can then be tracked by determining the initial rate of substrate decarboxylation, which is achieved by following the consumption of substrate by UV–visible spectroscopy. For Fdc1, this can be done by measuring the rate of cinnamic acid consumption at 270 nm using the extinction coefficient of $\varepsilon_{270} = 20,000 \, M^{-1} \, cm^{-1}$. A typical 500 μl assay contains 1 μM of enzyme and 80 μM of cinnamic acid in 50 mM KCl, 50 mM NaPi (pH 6) within a 10 mm path length cuvette at 25 °C.

Initial rates of reaction versus varying concentration of cinnamic acid can be used to determine apparent k_{cat} values. Due to incorporation levels of $prFMN^{iminium}$ varying between preparations, k_{cat} values (in addition to all other Michalis–Menten parameters) should be reported as apparent. Full oxidative maturation will be obtained at a two-equivalent excess of oxidant relative to the enzyme. Excess oxidant can result in a reduction in activity levels, although the mechanism of this is yet to be fully elucidated. In the case of *A. niger* Fdc1, highly active enzyme has an apparent k_{cat} approaching $10 \, s^{-1}$ at room temperature.

6. Characterisation of prFMN species via ESI-MS

As previously discussed, oxidative maturation of $prFMN^{reduced}$ to $prFMN^{iminium}$ is required for (de)carboxylase catalytic activity. The study of oxidative maturation of prFMN is key to assessing potential (de)carboxylase activity of members of the UbiD family and their variants. Electrospray ionisation mass spectrometry (ESI-MS) provides an additional means of observing prFMN species following UbiD-mediated oxidative maturation.

6.1 Equipment
- Anaerobic glovebox
- UV-Vis spectrophotometer in anaerobic glovebox (optional)

- PD-10 desalting column
- Crimp-sealable amber mass spectrometry vials
- LC-MS instrument (running in positive ion mode) and corresponding analysis software

6.2 Buffers and reagents
- 100 mM ammonium acetate pH 7.0 (anaerobic, purged with N_2)
- LC-MS grade solvents (i.e. water and acetonitrile with an appropriate ion pairing agent such as trifluoracetic acid)

6.3 Procedure
Within the anaerobic glove box, desalt *holo*-UbiD enzymes into 100 mM ammonium acetate pH 7.0 (rendered anaerobic via thorough purging with nitrogen) using a PD-10 column equilibrated as described elsewhere in the chapter. Transfer into amber, crimp-sealable mass spectrometry vials to minimise cofactor hydrolysis through light and oxygen exposure. Samples should contain at least 10 μM of *holo*-UbiD protein. Using an appropriate LC instrument, inject 5 μl of sample into 5 % acetonitrile (containing 0.1 % trifluoroacetic acid) and desalt inline in order to release the cofactor from the enzyme complex. Elute over a period of 1 min using 95 % acetonitrile (0.1 % TFA). Analyse the resulting ions using a mass spectrometer run in positive mode and deconvolute using the corresponding analysis software.

MH^+ ion masses of prFMN cofactors extracted from Fdc1 and variants correspond to the following species: 525.16 Da, prFMN^{iminium}; 526.16 Da, prFMN^{radical}; 527.16 Da, prFMN^{reduced}; 541.17 Da, prFMN$^{N5\ amide}$; 543.19 Da, prFMN^{hydroxylated}; 559.18 Da, prFMN^{hydroperoxide} (Bailey et al., 2018; Balaikaite et al., 2020).

7. ^{1}H NMR monitored enzyme catalysed deuterium exchange

The recent determination of the 1,3-dipolar cycloaddition mechanism of Fdc1 has uncovered an additional means of testing UbiD enzymes for (de)carboxylase activity against putative substrates (Bailey et al., 2019). Specifically, during the carboxylation direction of the reaction, Glu282 (as numbered in *A. niger* Fdc1) removes a proton from the styrene cycloadduct. If levels of CO_2 are insufficient, this intermediate is re-protonated, leading to the release of styrene. Notably, when this reaction is conducted in deuterium oxide (D_2O), the trans-vinyl proton in styrene is specifically replaced with a deuteron (Bailey et al., 2018). The ability to catalyse this proton exchange reaction is a key mechanistic feature of UbiD enzymes, and can serve as a means to identify potential substrates (Mondal et al., 2024). This change can be monitored using proton nuclear magnetic resonance (^{1}H NMR) (Ferguson et al., 2016).

7.1 Equipment
- Anaerobic glovebox
- UV–Vis spectrophotometer in anaerobic glovebox (optional)
- PD-10 desalting column
- NMR sample tubes
- NMR spectrometer (Operating at a frequency of 400 MHz or higher)

7.2 Buffers and reagents
- Anaerobic buffer prepared in deuterium oxide (D_2O)
- Putative UbiD substrate

7.3 Procedure
Typical assays should be prepared in 500 µl of anaerobic 100 mM NaPi buffer (pD 6.0, roughly equivalent to pH 5.6) in deuterium oxide. Stocks of *holo*-UbiD should be exchanged into anaerobic D_2O buffer using a PD-10 desalting column equilibrated with anaerobic D_2O buffer and prepared as described elsewhere in this chapter. Incubate 5 µM of the resulting *holo*-UbiD stock with 10 mM of putative substrate within a sealable vial. Ensure suitable control samples (i.e. enzyme and substrate free samples) are also prepared. Incubate overnight at 30 °C (this may vary depending on the UbiD under investigation). Transfer samples to NMR tubes and acquire data. Check for depletion of proton signal(s) in samples containing both enzyme and substrate relative to the appropriate control samples to identify which (if any) protons are being exchanged. Levels of deuterium incorporation (mol%) can be calculated by comparing peak heights of the corresponding proton signals in UbiD-containing samples relative to those within enzyme free control samples.

8. (De)carboxylation assays monitored by HPLC/LC-MS

High-Performance Liquid Chromatography (HPLC) enables the rapid separation, identification, and quantification of both polar and non-polar compounds, rendering it ideal for studying the products of both carboxylation and decarboxylation assays (Payne et al., 2019; Payne et al., 2021; Gahloth et al., 2022; Roberts et al., 2023). Compound separation depends on the interaction of each component of the sample, dissolved in a solvent (the mobile phase) with the packing material of the column (the stationary phase). Compounds can then be identified based on their column retention times, and quantified according to the intensity and area of their detected peaks. This approach relies on the use of standards, i.e. running samples containing reference compounds that correspond to the expected products. If such reference compounds are unavailable (or unknown), or product peaks do not separate sufficiently, use of Liquid Chromatography Mass Spectrometry may be more desirable. LC-MS feeds compounds separated using the same principles of HPLC into a mass spectrometer, enabling the mass of each ion to be accurately determined. This data can then be used to interpret compound molecular weight, structure and overall identity.

8.1 Equipment
- Anaerobic glovebox
- UV-Vis spectrophotometer in anaerobic glovebox
- PD-10 desalting column
- Crimp-sealable amber mass spectrometry vials
- HPLC instrument equipped with UV detector with corresponding analysis software
- Suitable HPLC column (i.e. C18)
- LC-MS instrument and corresponding analysis software
- Suitable LC-MS column (i.e. C18)

8.2 Buffers and reagents
- Anaerobic buffer (purged with N_2)
- UbiD substrate
- $KHCO_3$ (carboxylation reactions only)
- HPLC/LC-MS grade solvents (i.e. water and acetonitrile with an appropriate ion pairing agent such as trifluoracetic acid)

8.3 Procedure
All reactions should be set up within an anaerobic glove box or suitable anaerobic chamber and using crimp sealable mass spectrometry vials. Typical decarboxylation assays should contain 500 μl of 10 mM substrate in 50 mM NaPi, 50 mM KCl, 1 mM $MnCl_2$, pH 6. For carboxylation reactions, assays should contain 500 μl of 50 mM substrate in 100 mM KPi,

50 mM KCl, 1 mM $MnCl_2$, pH 6 and between 0.05 − 1 M $KHCO_3$ (final pH will increase up to 7.5, depending on amount of carbonate added). Carbonate stocks should be diluted with anaerobic buffer immediately prior to addition to each reaction mix.

Transfer *holo*-UbiD into anaerobic buffer using a pre-equilibrated PD-10 desalting column, as detailed earlier in this chapter. Both (de)carboxylase assays should be prepared both with and without 20 μM of UbiD enzyme. After sealing, vials should be removed from the anaerobic box and incubated overnight at 50 °C. Post-incubation, 50 μl of the reaction should be added to 450 μl of 50 % acetonitrile/water (to form a 1:10 dilution). Samples should then be centrifuged at $16100 \times g$ in order to precipitate any remaining protein − this helps to prevent blockages within the HPLC or LC-MS equipment. Following transfer into suitable vials, samples are ready for HPLC and/or LC-MS analysis.

For HPLC, selection of both mobile and stationary phases will vary depending on the compounds under analysis. For our work involving (de) carboxylation of cinnamic acid to styrene, we have found that a 5 μM C18 100 A (250 × 4.6 mm) column and acetonitrile/water in a 50/50 mix (containing 0.1 % TFA) serve as suitable mobile and stationary phases, respectively. Flow rate should be maintained at 1 ml/min. Detection wavelengths will depend on the compounds under investigation; for observation of cinnamic acid decarboxylation to styrene, a detection wavelength of 265 nm functions well. Calibration curves should be prepared for each compound to enable sample quantification, as described in Messiha et al., (2023).

Additionally, LC-MS (or HPLC-MS) can be used to confirm product identities. Sample running and analysis methods will vary considerably depending on the instruments used, mobile and stationary phases, and the compounds under investigation. As such, discussion relating to LC-MS method development is outside the scope of this chapter.

9. Preparation of whole-cell UbiD-biocatalysts

Given the complexity associated with the purification of active UbiD enzymes, *E. coli* cells expressing both UbiX and the UbiD can be used as whole-cell biocatalysts. Such an approach may also improve cofactor economy by eliminating the requirement for expensive cofactors (e.g. NADPH and ATP) or their corresponding recycling systems, whilst still producing comparable product yields (Titchiner et al., 2022).

9.1 Equipment
- Orbital shaking incubator
- Centrifuge and rotor
- UV–Visible spectrophotometer
- Crimp-sealable mass spectrometry vials
- HPLC instrument equipped with UV detector with corresponding analysis software
- Suitable HPLC column (i.e. C18)

9.2 Buffers and reagents
- 1 L of autoclaved LB medium
- Antibiotic stocks at the appropriate working concentrations (filter sterilised through a 0.2 μM syringe filter)
- 0.25 M IPTG (filter sterilised through a 0.2 μM syringe filter)
- 1 M $MnCl_2$ (filter sterilised through a 0.2 μM syringe filter)
- UbiD substrate (i.e. cinnamic acid in the case of AnFdc1)
- Dimethyl sulfoxide (DMSO)
- HPLC/LC-MS grade solvents (i.e. water and acetonitrile with an appropriate ion pairing agent such as trifluoracetic acid)

9.3 Procedure
Transform UbiD/UbiX constructs into *E. coli* BL21(DE3) cells. For a suitable negative control, also transform an empty plasmid backbone into *E. coli* BL21(DE3) cells. Pre-culture a single co-transformant colony in between 5–50 ml of sterile LB broth (depending on the number of samples to prepare) containing the appropriate antibiotics, and induce cultures with IPTG as described earlier in the chapter. Following incubation, dilute cells in 50 mM KPi (pH 6.5) and 150 mM NaCl to give a final absorbance of 30 at OD_{600} nm. If the OD_{600} is below this, centrifuge the cultures at ~6300 g and resuspend the resulting pellets in a volume of buffer at 10-fold lower than that of the initial culture. Whole-cell biocatalyst suspensions can be aliquoted and stored at −20 °C until required, or used immediately. Transfer whole-cell biocatalyst samples (at a typical volume of 500 μl) into tightly sealable glass vials, and add 5 mM of substrate (prepared in DMSO, not exceeding 2 % *v/v*). Incubate the samples at 30 °C with shaking at 250 rpm for a total of 18 h. Quench reactions at different time points with equal volumes of acetonitrile, leave to stand for a minimum of 5 min and centrifuge at ~16000 × *g* to precipitate cell debris. Remove the supernatant, centrifuge for a second time (also at ~16000 × *g*), transfer into suitable HPLC vials and analyse as described elsewhere in the chapter.

10. UbiD-substrate screening through monitoring flavin spectral shift

Flavin-bound UbiD (as opposed to prFMN bound) can be used to screen compound binding by observing substrate induced shifts in the oxidised flavin UV–visible spectrum. This approach provides a quick, high-throughput method of identifying potential UbiD substrates for prFMN-mediated (de)carboxylation.

10.1 Equipment
- PD-10 desalting column
- UV-Visible spectrophotometer
- UV-range cuvettes (disposable or quartz)

10.2 Buffers and reagents
- Purified UbiD protein
- FMN
- Stocks of KCl, $MnCl_2$ and $MgCl_2$ prepared in suitable buffer.
- 50 mM NaPi pH 7.5 (or suitable alternative)
- Stocks of putative UbiD substrates to be screened

10.3 Reconstitution of UbiD with FMN
To improve flavin occupancy in the UbiD enzyme, protein stocks should be reconstituted with an excess of FMN. Adjust the concentration of the UbiD to 200 μM, before adding 500 μM of FMN. Add KCl, $MnCl_2$ and $MgCl_2$, each to final concentrations of 1 mM. Incubate the mixtures for 20 min at room temperature. Unbound FMN should then be removed via desalting into a suitable buffer, as described elsewhere in the chapter.

10.4 Spectral shift assay
Assays should be prepared within UV-range cuvettes and should contain ~30 μM of purified, flavin-bound UbiD protein. After recording an initial, enzyme-only spectrum, putative substrates should be added to a final concentration of 2 mM (not exceeding 5% v/v, 2% v/v if the substrate is dissolved in DMSO) and incubated for a period of 2 min. Following incubation, a second spectrum should be recorded.

Flavin-bound UbiD should exhibit a spectral peak at approximately ~450 nm, depending on the UbiD under investigation. Detectable spectral perturbations (exceeding 0.5 nm) following the addition of putative substrate indicates modification of the flavin environment, suggestive of an interaction between the compound and the UbiD-bound prFMN cofactor. Careful titration can allow determination of a K_d for the relevant compound by plotting spectral perturbation versus ligand concentration. Control reactions with free flavin should be performed where spectral perturbations are observed.

11. Experimental strategies to drive carboxylation

Since the process of carboxylation is thermodynamically unfavourable at ambient CO_2 levels, leveraging UbiD enzymes as carboxylation catalysts often requires the use of alternative strategies to drive the generic UbiD (de)carboxylase reaction in the carboxylative direction. This can be achieved through increasing the experimental [CO_2] (a "push" strategy), or by linking the reaction to an irreversible carboxylic acid consuming reaction (a "pull" strategy) (Fig. 5).

An experimental "push" strategy can be achieved by dissolving high concentrations (typically between 0.5–3 M) of bicarbonate in reaction assays, or by utilising pressurised CO_2 (30 bar) (Payer et al., 2017). This is perhaps the most widely adopted approach at driving the reaction equilibrium towards carboxylation, with examples present throughout the literature (reviewed in Glueck et al., 2010; Payer et al., 2019).

Alternatively, carboxylation reactions can be integrated into a biosynthetic cascade in which the resulting carboxylic acid undergoes a subsequent transformation. One successful demonstration of such a "pull" strategy involved the coupling of AnFdc with carboxylic acid reductase (CAR), the latter of which catalyses the irreversible conversion of aromatic carboxylic acids to the corresponding aldehydes. The resulting one-pot biocatalytic cascade was capable of efficient styrene functionalisation via CO_2 fixation under ambient conditions (Aleku et al., 2020).

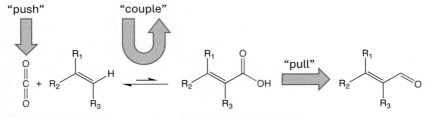

Fig. 5 Strategies to drive the generic UbiD (de)carboxylase reaction in the carboxylative direction.

While both "push" and "pull" strategies have proven effective, the former is not readily applicable in vivo and the latter is restricted to yielding non-carboxylic products. As such, biocatalytic systems that "couple" carboxylation to an irreversible exothermic reaction and facilitate the accumulation of carboxylic acid products in vivo are of biotechnological interest. The multi-component enzyme phenyl phosphate carboxylase reportedly "couples" carboxylation to dephosphorylation as part of a putative phenol degradation pathway, although the mechanics of this are yet to be fully described (Schühle & Fuchs, 2004).

12. Conclusion

While direct evidence of prFMN-dependency has only been demonstrated in a limited number of cases, the entire UbiD family likely exclusively depends on prFMN for activity. While the distinctive catalytic role of prFMN^{iminium} is now well described in case of the model AnFdc enzyme, the exact mechanisms of oxidative maturation, light-induced isomerisation and aromatic acid (de)carboxylation are yet to be fully elucidated. The substrate scope of some UbiD enzymes is well characterised and expanded upon through evolution, but the natural substrate of many UbiD homologues remains unknown. Enzyme engineering, through either the rational or directed approaches, offers an attractive means to expand upon the existing UbiD substrate repertoire (Payne et al., 2019; Payne et al. 2021; Saaret et al., 2021; Duţă et al., 2022; Roberts et al., 2023).

Despite these successes, wider application of UbiD enzymes towards the production of high-value compounds remains hampered by the reliable supply of the catalytically active prFMN^{iminium} species. As discussed, prFMN^{iminium} can be transformed into inactive prFMN species following exposure to both light and oxygen, whilst incorporation levels of prFMN within heterologously produced UbiD enzymes can often be unsatisfactory. While in vitro reconstitution of *apo*-UbiD with in vitro produced prFMN species remains a viable method of producing *holo*-UbiD enzymes, the ability of a specific UbiD enzyme to effectively regulate oxidation of the cofactor to the active, prFMN^{iminium} species remains hard to predict (Bailey et al., 2018; Gahloth et al., 2024; DiRocco et al., 2024). The recent characterisation of PhdC suggests prFMN maturases may perform a fundamental role in the production of active, *holo*-UbiD enzymes; other PhdC-like proteins await further characterisation (DiRocco et al., 2024).

As such, production and purification methods will likely require adaptation when investigating a novel UbiD enzyme or variant.

Finally, considering the diverse chemistry catalysed by prenylated flavin and the widespread presence of UbiX in many microbes, it is an intriguing possibility that prFMN may participate in other types of reactions. While this prospect is exciting, evidence supporting non-UbiD linked prFMN biochemistry is yet to materialise.

References

Aleku, G. A., Saaret, A., Bradshaw-Allen, R. T., Derrington, S. R., Titchiner, G. R., Gostimskaya, I., ... Leys, D. (2020). Enzymatic C–H activation of aromatic compounds through CO2 fixation. *Nature chemical biology, 16*(11), 1255–1260.

Arunrattanamook, N., & Marsh, E. N. G. (2018). Kinetic characterization of prenyl-flavin synthase from Saccharomyces cerevisiae. *Biochemistry, 57*(5), 696–700.

Bailey, S. S., Payne, K. A., Fisher, K., Marshall, S. A., Cliff, M. J., Spiess, R., ... Leys, D. (2018). The role of conserved residues in Fdc decarboxylase in prenylated flavin mononucleotide oxidative maturation, cofactor isomerization, and catalysis. *Journal of Biological Chemistry, 293*(7), 2272–2287.

Bailey, S. S., Payne, K. A., Saaret, A., Marshall, S. A., Gostimskaya, I., Kosov, I., ... Leys, D. (2019). Enzymatic control of cycloadduct conformation ensures reversible 1, 3-dipolar cycloaddition in a prFMN-dependent decarboxylase. *Nature Chemistry, 11*(11), 1049–1057.

Balaikaite, A., Chisanga, M., Fisher, K., Heyes, D. J., Spiess, R., & Leys, D. (2020). Ferulic acid decarboxylase controls oxidative maturation of the prenylated flavin mononucleotide cofactor. *ACS Chemical Biology, 15*(9), 2466–2475.

Beveridge, R., Migas, L. G., Payne, K. A., Scrutton, N. S., Leys, D., & Barran, P. E. (2016). Mass spectrometry locates local and allosteric conformational changes that occur on cofactor binding. *Nature Communications, 7*(1), 12163.

Bloor, S., Michurin, I., Titchiner, G. R., & Leys, D. (2023). Prenylated flavins: Structures and mechanisms. *The FEBS Journal, 290*(9), 2232–2245.

DiRocco, D. J., Roy, P., Mondal, A., Datar, P. M., & Marsh, E. N. G. (2024). An enzyme catalyzing the oxidative maturation of reduced prenylated-FMN to form the active coenzyme. *ACS Catalysis, 14*, 10223–10233.

Duță, H., Filip, A., Nagy, L. C., Nagy, E. Z. A., Tőtős, R., & Bencze, L. C. (2022). Toolbox for the structure-guided evolution of ferulic acid decarboxylase (FDC). *Scientific Reports, 12*(1), 3347.

Ferguson, K. L., Arunrattanamook, N., & Marsh, E. N. G. (2016). Mechanism of the novel prenylated flavin-containing enzyme ferulic acid decarboxylase probed by isotope effects and linear free-energy relationships. *Biochemistry, 55*(20), 2857–2863.

Gahloth, D., Fisher, K., Marshall, S., & Leys, D. (2024). The prFMNH2-binding chaperone LpdD assists UbiD decarboxylase activation. *Journal of Biological Chemistry, 300*(2).

Gahloth, D., Fisher, K., Payne, K. A., Cliff, M., Levy, C., & Leys, D. (2022). Structural and biochemical characterization of the prenylated flavin mononucleotide-dependent indole-3-carboxylic acid decarboxylase. *Journal of Biological Chemistry, 298*(4).

Glueck, S. M., Gümüs, S., Fabian, W. M., & Faber, K. (2010). Biocatalytic carboxylation. *Chemical Society Reviews, 39*(1), 313–328.

Keller, R. K., & Thompson, R. (1993). Rapid synthesis of isoprenoid diphosphates and their isolation in one step using either thin layer or flash chromatography. *Journal of Chromatography A, 645*(1), 161–167.

Leys, D. (2018). Flavin metamorphosis: Cofactor transformation through prenylation. *Current Opinion in Chemical Biology, 47*, 117–125.

Lira, L. M., Vasilev, D., Pilli, R. A., & Wessjohann, L. A. (2013). One-pot synthesis of organophosphate monoesters from alcohols. *Tetrahedron Letters, 54*(13), 1690–1692.

Marshall, S. A., Fisher, K., Cheallaigh, A. N., White, M. D., Payne, K. A., Parker, D. A., ... Leys, D. (2017). Oxidative maturation and structural characterization of prenylated FMN binding by UbiD, a decarboxylase involved in bacterial ubiquinone biosynthesis. *Journal of Biological Chemistry, 292*(11), 4623–4637.

Marshall, S. A., Payne, K. A., Fisher, K., Titchiner, G. R., Levy, C., Hay, S., & Leys, D. (2021). UbiD domain dynamics underpins aromatic decarboxylation. *Nature Communications, 12*(1), 5065.

Marshall, S. A., Payne, K. A., Fisher, K., White, M. D., Ní Cheallaigh, A., Balaikaite, A., ... Leys, D. (2019). The UbiX flavin prenyltransferase reaction mechanism resembles class I terpene cyclase chemistry. *Nature Communications, 10*(1), 2357.

Meganathan, R. (2001). Ubiquinone biosynthesis in microorganisms. *FEMS Microbiology Letters, 203*(2), 131–139.

Mergelsberg, M., Willistein, M., Meyer, H., Stärk, H. J., Bechtel, D. F., Pierik, A. J., & Boll, M. (2017). Phthaloyl-coenzyme A decarboxylase from Thauera chlorobenzoica: The prenylated flavin-, K+-and Fe2+-dependent key enzyme of anaerobic phthalate degradation. *Environmental Microbiology, 19*(9), 3734–3744.

Messiha, H. L., Scrutton, N. S., & Leys, D. (2023). High-titer bio-styrene production afforded by whole-cell cascade biotransformation. *ChemCatChem, 15*(5), e202201102.

Mondal, A., Roy, P., Carrannanto, J., Datar, P. M., DiRocco, D. J., Hunter, K., & Marsh, E. N. G. (2024). Surveying the scope of aromatic decarboxylations catalyzed by prenylated-flavin dependent enzymes. *Faraday Discussions, 024*(252), 208–222.

Payer, S. E., Faber, K., & Glueck, S. M. (2019). Non-oxidative enzymatic (de) carboxylation of (hetero) aromatics and acrylic acid derivatives. *Advanced Synthesis & Catalysis, 361*(11), 2402–2420.

Payer, S. E., Marshall, S. A., Bärland, N., Sheng, X., Reiter, T., Dordic, A., ... Glueck, S. M. (2017). Regioselective para-carboxylation of catechols with a prenylated flavin dependent decarboxylase. *Angewandte Chemie International Edition, 56*(44), 13893–13897.

Payne, K. A., Marshall, S. A., Fisher, K., Cliff, M. J., Cannas, D. M., Yan, C., ... Leys, D. (2019). Enzymatic carboxylation of 2-furoic acid yields 2, 5-furandicarboxylic acid (FDCA). *ACS Catalysis, 9*(4), 2854–2865.

Payne, K. A., Marshall, S. A., Fisher, K., Rigby, S. E., Cliff, M. J., Spiess, R., ... Leys, D. (2021). Structure and mechanism of Pseudomonas aeruginosa PA0254/HudA, a prFMN-dependent pyrrole-2-carboxylic acid decarboxylase linked to virulence. *Acs Catalysis, 11*(5), 2865–2878.

Payne, K. A., White, M. D., Fisher, K., Khara, B., Bailey, S. S., Parker, D., ... Leys, D. (2015). New cofactor supports α, β–unsaturated acid decarboxylation via 1, 3-dipolar cycloaddition. *Nature, 522*(7557), 497–501.

Roberts, G. W., & Leys, D. (2022). Structural insights into UbiD reversible decarboxylation. *Current Opinion in Structural Biology, 75*, 102432.

Roberts, G. W., Fisher, K., Jowitt, T., & Leys, D. (2023). Stability engineering of ferulic acid decarboxylase unlocks enhanced aromatic acid decarboxylation. *Current Research in Chemical Biology, 3*, 100043.

Saaret, A., Villiers, B., Stricher, F., Anissimova, M., Cadillon, M., Spiess, R., ... Leys, D. (2021). Directed evolution of prenylated FMN-dependent Fdc supports efficient in vivo isobutene production. *Nature Communications, 12*(1), 5300.

Schühle, K., & Fuchs, G. (2004). Phenylphosphate carboxylase: A new CC lyase involved in anaerobic phenol metabolism in Thauera aromatica. *Journal of Bacteriology, 186*(14), 4556–4567.

Shepherd, S. A., Karthikeyan, C., Latham, J., Struck, A. W., Thompson, M. L., Menon, B. R., ... Micklefield, J. (2015). Extending the biocatalytic scope of regiocomplementary flavin-dependent halogenase enzymes. *Chemical Science, 6*(6), 3454–3460.

Titchiner, G. R., Marshall, S. A., Miscikas, H., & Leys, D. (2022). Biosynthesis of pyrrole-2-carbaldehyde via enzymatic CO_2 fixation. *Catalysts, 12*(5), 538.

Wang, P. H., Khusnutdinova, A. N., Luo, F., Xiao, J., Nemr, K., Flick, R., ... Yakunin, A. F. (2018). Biosynthesis and activity of prenylated FMN cofactors. *Cell Chemical Biology, 25*(5), 560–570.

White, M. D., Payne, K. A., Fisher, K., Marshall, S. A., Parker, D., Rattray, N. J., ... Leys, D. (2015). UbiX is a flavin prenyltransferase required for bacterial ubiquinone biosynthesis. *Nature, 522*(7557), 502–506.

Zhu, X., Li, H., Ren, J., Feng, Y., & Xue, S. (2023). Engineering the biosynthesis of prFMN promotes the conversion between styrene/CO_2 and cinnamic acid catalyzed by the ferulic acid decarboxylase Fdc1. *Catalysts, 13*(6), 917.

CHAPTER SEVEN

Enzyme cascades for *in vitro* and *in vivo* FMN prenylation and UbiD (de)carboxylase activation under aerobic conditions

Anna N. Khusnutdinova[a,b], Khorcheska A. Batyrova[b], Po-Hsiang Wang[c], Robert Flick[d], Elizabeth A. Edwards[d], and Alexander F. Yakunin[a,d,*]

[a]Centre for Environmental Biotechnology, School of Environmental and Natural Sciences, Bangor University, Bangor, United Kingdom
[b]Institute of Basic Biological Problems, Federal Research Center "Pushchino Scientific Center for Biological Research, Russian Academy of Sciences" (FRC PSCBR RAS), Pushchino, Moscow Region, Russia
[c]Graduate Institute of Environmental Engineering, National Central University, Taoyuan, Taiwan
[d]Department of Chemical Engineering and Applied Chemistry, University of Toronto, Toronto, Canada
*Corresponding author. e-mail address: a.iakounine@bangor.ac.uk

Contents

1.	Introduction	152
2.	*In vitro* biosynthesis of prFMN and UbiD activation	156
	2.1 Enzyme cascades for *in vitro* FMN prenylation and apoUbiD activation	157
	2.2 Prenol phosphorylation to DMAP and FMN reduction using enzyme cascades	159
	2.3 UbiD activation using prFMN cascades	160
	2.4 Equipment	162
	2.5 Materials	162
	2.6 Procedure: recombinant protein expression and purification	163
	2.7 Procedure: *in vitro* prFMN biosynthesis using enzyme cascades	164
	2.8 Procedure: *in vitro* activation of purified UbiD using prFMN cascades	165
	2.9 Notes	165
3.	Procedure: *in vivo* FMN prenylation and UbiD activation using recombinant *E. coli* cells	166
	3.1 Equipment	167
	3.2 Materials	168
	3.3 Procedure	169
	3.4 Notes	170
4.	Conclusions	170
	Acknowledgments	171
	References	171

Methods in Enzymology, Volume 708
ISSN 0076-6879, https://doi.org/10.1016/bs.mie.2024.10.015
Copyright © 2024 Elsevier Inc. All rights are reserved, including those for text and data mining, AI training, and similar technologies.

Abstract

Microbial carboxylases and decarboxylases play important roles in the global carbon cycle and have many potential applications in biocatalysis and synthetic biology. The widespread family of reversible UbiD-like (de)carboxylases are of particular interest because these enzymes are active against a diverse range of substrates. Several characterized UbiD enzymes have been shown to catalyze reversible (de)carboxylation of aromatic and aliphatic substrates using the recently discovered prenylated FMN (prFMN) cofactor, which is produced by the associated family of UbiX FMN prenyltransferases. However, discovery and investigation of novel UbiD (de)carboxylases are delayed by our limited knowledge and the experimental complexities associated with FMN prenylation and UbiD activation resulting in the production of inactive recombinant UbiD enzymes. Therefore, there is a need for developing robust methods for efficient *in vitro* and *in vivo* FMN prenylation and UbiD activation for heterologous production of active UbiD enzymes. In this chapter, we present two protocols for *in vitro* and *in vivo* FMN prenylation and UbiD activation under aerobic conditions using enzyme cascades with regenerating systems and recombinant *E. coli* cells.

Abbreviations

DMAP	dimethylallylmonophosphate.
DMAPP	dimethylallylpyrophosphate.
DMSO	dimethyl sulfoxide.
FDH	formate dehydrogenase.
Fre	flavin reductase.
FMN	flavin mononucleotide.
IPTG	isopropyl-β-D-1-thiogalactoryranoside.
O/N	overnight.
polyp	polyphosphate.
prFMN	prenylated flavin mononucleotide.
RFK	riboflavin kinase.
RT	room temperature.
SDS-PAGE	sodium dodecyl sulfate polyacrylamide gel electrophoresis.

1. Introduction

Reversible carboxylases catalyzing carboxylation and decarboxylation reactions represent important enzymes in the global carbon cycle and natural metabolism (Bierbaumer et al., 2023; Erb, 2011). These enzymes use different catalytic mechanisms and play diverse physiological functions including carbon assimilation, anaplerosis, biosynthesis of organic metabolites, detoxification, and redox balancing (Bierbaumer et al., 2023; Erb, 2011; Glueck, Gumus, Fabian, & Faber, 2010; Kourist et al., 2014). Currently, over 110 (de) carboxylase families with different substrate preferences and cofactors are already known, and novel enzymes are expected to be discovered in the coming years (Bierbaumer et al., 2023; Erb, 2011; Li, Huo, Pulley, & Liu, 2012). The large widespread family of prenylated FMN dependent UbiD (de) carboxylases currently comprises up to 35,000 sequences in the InterPro Pfam database (IPR002830) from various archaea, bacteria, and fungi (Marshall, Payne, & Leys, 2017; Mondal et al., 2024). The UbiD family of reversible (de)carboxylases was named after the *Escherichia coli* UbiD enzyme, which was proposed to decarboxylate a prenylated hydroxybenzoic acid in the biosynthesis of ubiquinone producing 2-polyprenylphenol (Marshall, Payne, et al., 2017). Sequence similarity network (SSN) analysis of available UbiD sequences using the Enzyme Similarity Network (ESN) tool revealed the presence of multiple (over 50) clusters of these proteins, most of which remain to be characterized (Mondal et al., 2024). Phylogenetic analyses of UbiD enzymes demonstrated that they form at least 10 groups with different substrate specificities including the biochemically and structurally characterized ferulic acid decarboxylase Fdc1 from *Aspergillus niger*, which catalyzes non-oxidative decarboxylation of cinnamic acid derivatives with a broad substrate tolerance toward α,β-unsaturated carboxylic acids including nonnatural substituted acrylic acids (Aleku et al., 2018; Payne et al., 2015). Other UbiD enzymes have been shown to catalyze decarboxylation of a wide range of aromatic, heterocyclic, and unsaturated substrates such as protocatechuic acid (AroY), gallic acid (LpdC), vanillic acid (VdcC), 2,5-furandicarboxylic acid (HmfF), pyrrole-2-carboxylate, phthalate, and phenazine-1-carboxylate (PhdA), as well as carboxylation of styrene to cinnamate (Fdc1) (Aleku et al., 2021; Datar & Marsh, 2021; Marshall Payne et al., 2017; Zhu et al., 2023). The remarkably broad substrate scope, excellent stereoselectivity, and robustness of several characterized UbiD enzymes makes them highly attractive biocatalysts for various decarboxylation and carboxylation applications (Aleku et al., 2018; Bloor, Michurin, Titchiner, & Leys, 2023; Payer, Faber, & Glueck, 2019).

Previous biochemical and structural studies with the ferulic acid decarboxylase Fdc1 from *A. niger* demonstrated that its catalytic activity relies on the presence of a novel cofactor, the recently discovered prenylated FMN (prFMN), produced by the UbiX family of FMN prenyltransferases (EC 2.5.1.129) using reduced FMN and dimethylallylmonophosphate (DMAP) or dimethylallylpyrophosphate (DMAPP) as substrates (Marshall, Payne, Fisher, White, et al., 2019; Payne et al., 2015; White et al., 2015) (Fig. 1). Purified UbiX protein PA4019 from *Pseudomonas aeruginosa* attaches a dimethylallyl moiety (from DMAP) to the N5 and C6 atoms of reduced

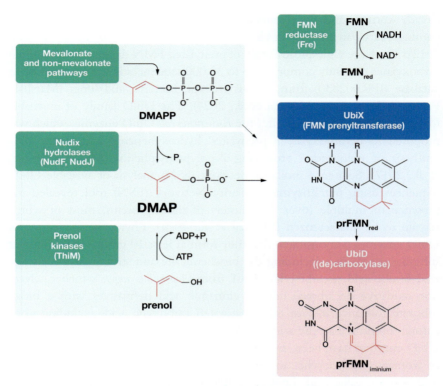

Fig. 1 Schematic diagram showing pathways and bacterial enzymes involved in the biosynthesis of DMAPP, DMAP and prFMN. In most organisms, DMAPP is naturally generated via two alternative biosynthetic pathways, the mevalonate (MVA) and non-mevalonate (MEP) pathways, that are lengthy, tightly regulated, and highly specific. DMAP can be produced from DMAPP by various Nudix hydrolases (e.g. NudF and NudJ in E. coli). In addition, DMAP can be produced by phosphorylating exogenous prenol using the hydroxyethylthiazole kinase ThiM (a non-natural reaction) in E. coli cells. FMN prenylation also requires reduced FMN (FMN$_{red}$) as a substrate, which is formed by the NAD(P)H-dependent flavin reductase Fre. Using DMAP (DMAPP in some microorganisms) and FMN$_{red}$ as substrates, the UbiX FMN prenyltransferase produces prFMN$_{red}$, which is transferred to UbiD followed by oxidative maturation to prFMN$_{iminium}$.

FMN, adding a fourth nonaromatic ring to FMN and producing reduced prenylated FMN (prFMNH$_2$) (White et al., 2015). The molecular mechanisms of these reactions were revealed by the crystal structures of PA4019 in complex with DMAP+FMN, covalent intermediate (N5-C1′ adduct), and product (prFMN), which include breaking the phosphate-C1′ bond of DMAP (initiated by FMN reduction), formation of the FMN N5-C1′ bond followed by adduct reorientation and formation of the dimethylallyl C3′ -C6 bond and $prFMN_{reduced}$ ($prFMNH_2$), which is quickly oxidized under aerobic conditions to a purple-colored semiquinone-like species ($prFMN_{radical}$) (White et al., 2015). Biochemical and structural studies of Fdc1 and UbiX proteins (PA4019 and AF1214 from *Archaeoglobus fulgidus*) also identified several other forms of prFMN including $prFMN_{ketimine}$, C1′-hydroxylated prFMN, protonated $prFMN$-C-$4a_{radical}$, C2′-hydroxylated prFMN, C1′-C2′-dihydroxylated prFMN, and the catalytically active $prFMN_{iminium}$ (Marshall, Fisher, et al., 2017; Payne et al., 2015; Wang et al., 2018; White et al., 2015). Furthermore, several biochemical studies revealed that some UbiX proteins (*e.g.* Pad1 from *Saccharomyces cerevisiae*, AnUbiX from *Aspergillus niger*, and *E. coli* K12 UbiX) prefer DMAPP as the prenyl group donor, whereas the *E. coli* O157:H7 can use both DMAP and DMAPP for FMN prenylation (Arunrattanamook & Marsh, 2018; Marshall, Payne, Fisher, White, et al., 2019) (Fig. 1).

It has been shown that inactive Fdc1 preparations can be activated under anaerobic conditions by the addition of purified $prFMN_{iminium}$ or reduced prFMN ($prFMNH_2$) followed by oxidative cofactor maturation under aerobic conditions, as well as by anaerobic incubation with purified heterologous UbiX-prFMN complexes (*e.g.* PA4019-$prFMNH_2$, PA4019-$prFMN_{radical}$, or AF1214-$prFMN_{radical}$) (Bailey et al., 2019; Balaikaite et al., 2020; Marshall, Fisher, et al., 2017; Wang et al., 2018; White et al., 2015). In addition, purified Fdc1 was activated under aerobic conditions after incubation with UbiX-$prFMN_{radical}$ complexes from different bacteria, with KLP0198 from *Klebsiella pneumonia* and TSTM2747 from *Salmonella typhimurium* producing the highest activation, whereas the UbiD protein ENC0058 from *Enterobacter cloacae* preferred RP0927 from *Rhodopseudomonas palustris* and TA1100 from *Thermoplasma acidophilum* for activation (Batyrova et al., 2020). Therefore, the efficiency of direct activation of apo-UbiD by UbiX-prFMN complexes also depends on the origin of both UbiD and UbiX proteins. Furthermore, two recent studies on the activation of UbiD-like enzymes gallate decarboxylase LpdC from *Lactobacillus plantarum* and phenazine-1-carboxylate decarboxylase PhdA from *Mycolicibacterium fortuitum* identified two cognate unrelated proteins LpdD and PhdC, respectively, as auxiliary enzymes (chaperones), assisting UbiD activation and facilitating the oxidative maturation of reduced prFMN and its transfer from UbiX to UbiD (DiRocco et al., 2024; Gahloth, Fisher, Marshall, & Leys, 2024). Finally, several *E. coli*-based *in vivo* systems for FMN prenylation and UbiD activation have been demonstrated, which involve co-expressing UbiD (Fdc1 or PhdA) with different UbiX and, for some UbiDs, with the prFMN maturase PhdC under aerobic conditions (Batyrova et al., 2020; DiRocco et al., 2024; Payne et al., 2015). To stimulate prFMN production by *E. coli* cells, the growth medium can be supplemented with 1 mM prenol, 0.2 mM riboflavin, and 1 mM $MnCl_2$ (Marshall, Payne, Fisher, Gahloth, et al., 2019; Wang et al., 2018).

The high interest in the UbiD family of (de)carboxylases arises from their biochemical diversity and broad substrate scope, suggesting many potential applications in biotechnology (Bloor et al., 2023; Leys & Scrutton, 2016; Roberts & Leys, 2022). This requires large-scale heterologous production of pure holoUbiDs loaded with prFMN for detailed biochemical and structural studies. However, recombinant expression and purification of novel UbiDs usually produce inactive apo-proteins, which need to be reconstituted with prFMN and activated. Furthermore, the discovery and experimental exploration of novel UbiD enzymes are hampered by the complexity and limitations of the FMN prenylation system, including the formation of several prFMN forms not associated with UbiD activity (Marshall, Payne, et al., 2017). In addition, preparations of purified prFMN are not commercially available, whereas commercial chemicals for *in vitro* FMN prenylation (*e.g.* DMAP) are expensive (Batyrova et al., 2020; Wang et al., 2018). Finally, some of the characterized UbiD enzymes (*e.g.* Fdc1) have been shown to be light-sensitive (due to light-induced prFMN isomerization) or oxygen-sensitive (AroY), whereas others can be both light- and oxygen-sensitive (HmfF and AnInD) (Bailey et al., 2018; Payer et al., 2017; Saaret, Balaikaite, & Leys, 2020). Here, we describe the robust enzyme cascades for *in vitro* FMN prenylation and UbiD activation using regenerating systems with inexpensive reagents, followed by the protocol for *in vivo* UbiD/UbiX co-expression and activation using recombinant *E. coli* strains under aerobic conditions.

2. *In vitro* biosynthesis of prFMN and UbiD activation

The first studies demonstrating *in vitro* Fdc1 activation by prFMN were performed using the UbiX-prFMN$_{reduced}$ complex (PA4019-prFMN$_{reduced}$) incubated with apoFdc1 under anaerobic conditions followed by oxidative maturation of the reconstituted Fdc1 producing the reactive prFMN$_{iminium}$ form (Payne et al., 2015; White et al., 2015). Later, it was found that apoFdc1 can be activated by the addition of purified prFMN$_{iminium}$ following dithionite reduction and air oxidation (Wang et al., 2018). In addition, Fdc1 activation was also observed under aerobic conditions (without DT reduction) in the presence of purified prFMN$_{iminium}$ or C1′-ene-prFMN$_{iminium}$, but the observed Fdc1 activity was 2–3 times lower than that after activation with DT reduction (Wang et al., 2018). However, the described protocols for the purification of protein-free prFMN$_{iminium}$ from the UbiX-prFMN$_{radical}$ complex produce small amounts of the cofactor, which is also unstable and should be used immediately (Wang et al., 2018). Therefore, *in vitro* UbiD activation using various UbiX-prFMN complexes appears to be a preferable approach for producing holo-UbiDs, because prFMN forms are more stable when they remain in complex with UbiX (Wang et al., 2018). Previous studies revealed that some UbiX proteins retain the produced prFMN cofactor when they are expressed and purified from *E. coli* cells grown with the addition of prenol and riboflavin, suggesting that these UbiX proteins can be used for cofactor stabilization, storage, and UbiD activation (Batyrova et al., 2020; Wang et al., 2018). Previously, we described *in vitro* methods for prFMN synthesis using dithionite, FMN, purified UbiX, and synthetic DMAP or DMAP generated by prenol phosphorylation (Khusnutdinova et al., 2019).

Here, we are presenting the expanded protocol for *in vitro* FMN prenylation and apoUbiD activation with purified UbiX and enzyme cascades supporting the production of UbiX-prFMN complexes using inexpensive substrates (FMN, prenol, polyphosphate, and formate). With these enzyme cascades, FMN prenylation and UbiD activation can be performed using both apoUbiX proteins and UbiX in complex with FMN. As shown in Fig. 2, the proposed protocol includes two enzyme cascades and regenerating systems producing DMAP and reduced FMN. The DMAP cascade is based on the polyphosphate-dependent ATP-regenerating system with a PPK2 family polyP kinase (CHU0107) supporting prenol phosphorylation activity of the *E. coli* ThiM. The FMN reduction cascade is composed of the *E. coli* flavin reductase Fre coupled with the FDH/formate-based NADH regenerating system. In the presence of FMN excess (0.5 mM) and NADH-regenerating system, the FMN reducing activity of Fre also quickly reduces the concentration of dissolved oxygen in the reaction mixture (*via* FMN$_{red}$ reoxidation by oxygen) facilitating the complete reduction of FMN indicated by solution discoloration.

2.1 Enzyme cascades for *in vitro* FMN prenylation and apoUbiD activation

The following section describes the protocols for *in vitro* FMN prenylation and apoUbiD activation, which include two enzyme cascades

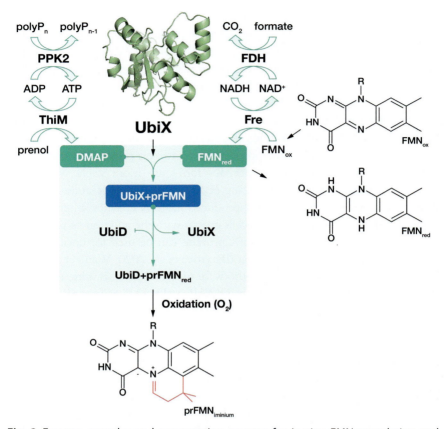

Fig. 2 Enzyme cascades and regenerating systems for *in vitro* FMN prenylation and UbiD activation under aerobic conditions. Prenol phosphorylation to DMAP (catalyzed by ThiM) is supported by the PPK2/polyphosphate-based ATP-regenerating system containing polyP, ADP, and PPK2 kinase (CHU0107 from *Cytophaga hutchinsonii*). FMN reduction is catalyzed by the NAD(P)H-dependent flavin reductase Fre, coupled with the FDH/formate-based NADH-regenerating system containing the formate dehydrogenase FDH from *Pseudomonas* sp. strain 101. In the presence of excess FMN, this regenerating system also quickly reduces the concentration of dissolved oxygen in the reaction mixture (*via* reoxidation of FMN$_{red}$ by oxygen) resulting in the complete reduction of FMN. Purified UbiX (AF1214 or PA4019) then uses DMAP and FMN$_{red}$ for FMN prenylation producing prFMN$_{red}$, which can be transferred directly to apoUbiD in the reaction mixture.

catalyzing prenol phosphorylation to DMAP (1) and FMN reduction (2) followed by prenylation of reduced FMN by UbiX (3) and transfer of prFMN from UbiX to apoUbiD and formation of holoUbiD (4) (Fig. 2). These reactions produce prFMN and holoUbiD using inexpensive substrates (FMN, prenol, polyphosphate, and formate) and several enzymes including purified UbiXs and UbiDs, as well as the *E. coli* hydroxyethylthiazole kinase ThiM and FMN reductase Fre, the PPK2 family polyphosphate kinase CHU0107 from *Cytophaga hutchinsonii*, and the *Pseudomonas sp.* formate dehydrogenase FDH (Fig. 2).

2.2 Prenol phosphorylation to DMAP and FMN reduction using enzyme cascades

As revealed by the first studies on FMN prenylation by UbiXs, most of these enzymes use DMAP and reduced FMN (FMN_{red}) as substrates (Payne et al., 2015; Saaret et al., 2020; White et al., 2015). The proposed enzyme cascade for producing DMAP is based on the *E. coli* hydroxyethylthiazole kinase ThiM catalyzing the non-natural reaction of prenol phosphorylation to DMAP (Wang et al., 2018). In the proposed DMAP production cascade (Fig. 2), the efficiency of prenol conversion to DMAP by ThiM has been increased more than four times using an ATP-regenerating system based on the PPK2-family polyP kinase CHU0107 from *Cytophaga hutchinsonii*, which uses inorganic polyP as a phosphate donor for ADP phosphorylation to ATP (K_m 0.35 mM and 0.29 mM for ADP and polyP, respectively) (Nocek et al., 2018). PolyP is ubiquitous in nature and has numerous functions, such as a universal source of energy (ATP) and phosphate, as well as protein stabilization (a chaperone) (Gray et al., 2014), making polyP-based ATP-regenerating systems especially attractive for applications in enzyme cascade systems. In addition to the wild type CHU0107, a truncated version of this enzyme (L285Stop, 1–284 aa) can be used for ATP regeneration, which was found to be at least two times more active than wild type CHU0107 (Nocek et al., 2018).

The second enzyme cascade in the proposed protocol generates reduced FMN (FMN_{red}) and includes the *E. coli* FMN reductase Fre (Fig. 2), which shows high activity and affinity to FMN and NADH (K_m 2.2 μM and 9.0 μM, respectively) (Fieschi, Niviere, Frier, Decout, & Fontecave, 1995). In the *E. coli* genome, the *fre* gene is located immediately downstream of *ubiD*, suggesting a functional relationship of Fre with FMN prenylation. To increase the reductant pool and reduce the cost of chemicals (NADH), we implemented a formate-supported NADH regenerating system using the formate dehydrogenase FDH from *Pseudomonas* sp. strain 101 (Batyrova et al., 2020). Under aerobic conditions and in the presence of excess free FMN (0.5 mM), this regenerating system also creates redox cycling in the reaction mixture, reducing the concentration of dissolved oxygen and facilitating the complete reduction of FMN (indicated by solution discoloration).

Since the UbiX prenyltransferase AF1214 from *A. fulgidus* showed high expression levels in *E. coli* cells, as well as high activity and affinity for prFMN (Wang et al., 2018), this enzyme was used for optimization of FMN prenylation and Fdc1 activation in the proposed enzyme cascades (Fig. 2). Furthermore, the UbiX proteins KLP0198 from *Klebsiella pneumoniae* and TSTM2747 from *Salmonella typhimurium* can be used for FMN prenylation and Fdc1 activation instead of AF1214, because these enzymes were 3–4 times more active in Fdc1 activation as revealed in our recent study (Batyrova et al., 2020). The efficiency of apoUbiD activation also depends on the nature of UbiX and UbiD proteins used. For example, for Fdc1 from *A. niger* the UbiX proteins KLP0198 and TSTM2747 were more active than AF1214 and PA4019, whereas the UbiD-like protocatechuic acid decarboxylase ENC0058 from *Enterobacter cloacae* showed the highest activity after activation by RP0927 from *Rhodopseudomonas palustris* or TA1100 from *Thermoplasma acidophilum* (Batyrova et al., 2020). The proposed prFMN cascade demonstrates high conversion rates of FMN to prFMN ($prFMN_{radical}$ and $prFMN_{red}$) under aerobic conditions using UbiX preparations with or without bound FMN. Unlike the dithionite-based FMN prenylation protocol, the proposed prFMN synthesis cascade does not require the use of expensive DMAP or anaerobic conditions and can be used for direct UbiD activation.

2.3 UbiD activation using prFMN cascades

In general, purified apoUbiD proteins (*e.g.* Fdc1) can be activated by anaerobic reconstitution with protein-free prFMN ($prFMN_{red}$ or $prFMN_{iminium}$) followed by oxidative cofactor maturation under aerobic conditions (Khusnutdinova et al., 2019; Payne et al., 2015; Wang et al., 2018; White et al., 2015). Furthermore, Fdc1 was activated by anaerobic incubation with purified UbiX-prFMN complexes ($PA4019$-$prFMN_{red}$ or $AF1214$-$prFMN_{red}$) with subsequent oxidative maturation under aerobic conditions (Marshall, Payne, Fisher, Gahloth, et al., 2019; Wang et al., 2018; White et al., 2015). In addition, purified UbiDs (Fdc1 and ENC0058 from *E. cloacae*) were found to be activated under aerobic conditions after incubation with oxygen-stable UbiX-$prFMN_{radical}$ complexes from different bacteria (Batyrova et al., 2020). In this section, we describe two protocols (30-min and overnight) for *in vitro* UbiD activation using prFMN cascades under aerobic conditions. In these experiments, we used the UbiX protein AF1214 from *A. fulgidus* and Fdc1 from *A. niger*, but these protocols can also be used with different UbiX and UbiD proteins (Batyrova et al., 2020). The 30-min UbiD activation protocol involves overnight (O/N) pre-incubation of the prFMN cascade for FMN prenylation followed by the addition of purified apoUbiD and activation during a 30 min incubation. As shown in Fig. 3, this protocol produced activated Fdc1 preparations with specific activities up to 1.1 U/mg (with cinnamic acid as substrate), which are comparable to Fdc1 activation using affinity purified AF1214-prFMN complex (up to 1.2 U/mg) (Fig. 3). In the second protocol (O/N), freshly prepared prFMN cascade is mixed with purified UbiD and incubated O/N for simultaneous FMN prenylation and UbiD activation. This protocol produced Fdc1 with higher specific activities (up to 2.2 U/mg), which were close to those after activation by purified (protein-free) prFMN extracted from the DT/DMAP reaction or from the prFMN cascade (2.0–2.3 U/mg) (Fig. 3).

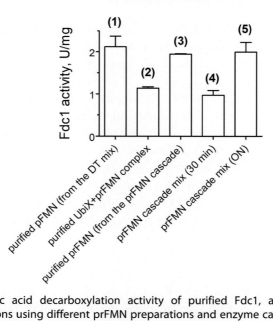

Fig. 3 Cinnamic acid decarboxylation activity of purified Fdc1, activated under aerobic conditions using different prFMN preparations and enzyme cascades. Purified apoFdc1 was activated using several methods: (1), by prFMN$_{radical}$ extracted from the DT mix; (2), by 30 min incubation with the purified UbiX-prFMN$_{radical}$ complex (AF1214); (3), by prFMN$_{radical}$ extracted from the prFMN cascade; (4), by 30 min incubation with the prFMN cascade mix (pre-incubated overnight for FMN prenylation); (5), by overnight incubation with the prFMN cascade mix.

2.4 Equipment

1. Shaker–incubator for culture tubes (37 °C)
2. Platform shaker for 2 L plastic flasks (37 °C)
3. 2.5 L baffled plastic flasks for *E. coli* cultures (autoclaved)
4. Preparative refrigerated centrifuges with rotors for *E. coli* cell harvesting and lysate clearing
   (for 1 L and 50 ml centrifuge bottles, Avanti)
5. Sonicator for *E. coli* cell disruption (QSonica Q700 with preparative probes)
6. UV–Vis spectrophotometer (SpectraMax M2)
7. Bench-top centrifuge (14,000 rpm)
8. Standard laboratory equipment for affinity purification of recombinant proteins using Ni–NTA resin
9. Protein electrophoresis system (Mini-Protean, Bio-Rad)
10. −80 °C freezer

2.5 Materials

1. Standard *E. coli* growth medium (Terrific Broth or equivalent)
2. Antibiotics for *E. coli* cultivation (filtered through 0.3 μm syringe filters)
3. Prenol (3-methyl-2-buten-1-ol, Sigma-Aldrich), 0.1 M stock solution in water
4. FMN (Sigma-Aldrich), 50 mM solution in water
5. Riboflavin, 50 mM stock solution in water (pH 3.0)
6. Adenosine triphosphate (ATP), 0.1 M stock solution in water
7. Polyphosphate sodium salt (sodium hexametaphosphate from Sigma-Aldrich, $(NaPO_3)_n$, cat. # 305553), 0.1 M stock solution in water
8. Sodium formate, 1 M stock solution in water
9. NADH (or NAD^+), 0.1 M stock solution in water
10. Isopropyl-β-D-1-thiogalactopyranoside (IPTG), 0.4 M stock solution in water
11. *E. coli* BL21(DE3) strains, each expressing one of the following recombinant enzymes with an N-terminal 6His-tag (from T7 promoter-based vectors, *e.g.* P15TVL or pET30a): UbiX prenyl transferase (AF1214 from *Archaeoglobus fulgidus*, UniProt ID O29054; or PA4019 from *P. aeruginosa*, UniProt ID Q9HX08), prenol kinase (the *E. coli* hydroxyethylthiazole kinase ThiM, UniProt ID P76423), flavin reductase (the *E. coli* NAD(P)H-flavin reductase Fre, UniProt ID P0AEN1), formate dehydrogenase (FDH from *Pseudomonas* sp. strain 101, UniProt ID P33160), riboflavin kinase (RFK, the riboflavin

kinase FMN1 from *Neurospora crassa*, UniProt ID Q6M923), PPK2 polyphosphate kinase (CHU0107 from *Cytophaga hutchinsonii*, UniProt ID A0A6N4SMB5)

12. Liquid nitrogen
13. Centrifugal concentrators (filters) for protein concentration (MW cut-off 3 kDa)
14. Glass or plastic chromatographic columns for gravity-flow protein purification
15. Ni-NTA agarose resin for affinity purification of recombinant proteins with 6His-tags
16. Protein purification buffers for Ni-chelate affinity chromatography: binding buffer (BB: 50 mM HEPES-K, pH 7.5, 0.4 M NaCl, 5 mM imidazole, 5 % glycerol), washing buffer (WB: BB with 30 mM imidazole), and elution buffer (EB: BB with 500 mM imidazole)
17. Bradford reagent (Bio-Rad) for determining protein concentration in preparations

2.6 Procedure: recombinant protein expression and purification

1. Inoculate 10 ml of Terrific broth (TB) medium with appropriate antibiotics using a single *E. coli* colony from a freshly transformed plate and grow the starter culture O/N (12-16 h) at 37 °C with shaking at 200 rpm.
2. Use 10 ml of starter culture to inoculate 1 L of TB medium with appropriate antibiotics in 2.5-L baffled flasks and grow the cultures with shaking at 200 rpm aerobically at 37 °C to optical density (OD_{600}) 0.6-0.8.
3. Add IPTG (0.4 mM final concentration) and incubate at 37 °C (for 3-5 h) or at 16 °C O/N with shaking at 200 rpm (or at 37 °C for 3-5 h, if this suits you better).
4. Harvest cells by centrifugation at 5000 × g for 15 min
5. Discard supernatants and resuspend pellets in 30 ml of BB. Freeze cells in liquid nitrogen for storage (at −80 °C) or use them immediately for protein purification.
6. Transfer cell suspensions to sonication cups and place the cups in an ice-water bath.
7. Sonicate cells for 20 min (using 5 s on/off pulses) at 20 kHz.
8. Centrifuge sonicated cells at 40,000 × g for 20 min at 4 °C and transfer supernatants to 50 ml Falcon tubes.

9. Purify recombinant proteins using standard Ni-affinity chromatography (using washing buffer with 30 mM imidazole and elution buffer with 500 mM imidazole, recommended column volumes 10−15 ml).
10. Determine protein concentrations in eluted column fractions using Bradford reagent.
11. Analyze the level of protein purity in eluted fractions using denaturing (sodium dodecyl sulfate) polyacrylamide gel electrophoresis (SDS-PAGE).
12. Pool and concentrate selected protein fractions (to 3−5 mg/ml), flash freeze aliquots in liquid nitrogen and store at −80 °C.

2.7 Procedure: *in vitro* prFMN biosynthesis using enzyme cascades

1. Thaw purified UbiX (AF1214 or PA4019) and cascade enzymes: Fre, FDH, ThiM, CHU0107, and RFK (if riboflavin will be used instead of FMN).
2. Prepare 2-4x concentrated master mix for the FMN prenylation reaction mixture containing (final concentrations) 100 mM Tris-Cl (pH 7.5), 35 mM prenol, 0.5 mM FMN (or riboflavin), 10 mM $MgCl_2$, 10 mM ATP, 3 mM polyP, 50 mM Na-formate, 2 mM NADH, 0.04 mg/ml BSA, mix well (vortex).
3. Add to the FMN prenylation reaction mixture the following cascade enzymes: CHU0107 (final concentration 0.5 mg/ml), ThiM (1.2 mg/ml), UbiX (1.3 mg/ml), FDH (0.7 mg/ml), Fre (0.5 mg/ml), and RFK (0.5 mg/ml for reactions with riboflavin instead of FMN).
4. Incubate the reaction mixture under aerobic conditions (in parafilm-sealed vials) for 12 h at RT without shaking.
5. Oxidize the colorless reaction mixtures by vortexing for several minutes. The processed samples with produced prFMN should change to a brownish color (due to the combination of yellow FMN with magenta-colored UbiX-prFMN$_{radical}$). After this step, the produced cascade mix can be used either for direct UbiD activation (described below in Section 2.8) or for the extraction of protein-free prFMN cofactor (described below, steps 6-9) and UbiD activation.
6. Add ice-cold acetonitrile to final concentration 80 % (v/v) to reaction mixtures and mix well by vortexing.
7. Centrifuge the reaction mixture at maximal speed (14,000 rpm) on a bench-top centrifuge for 10 min and discard the supernatant containing unbound FMN.

8. Resuspend the pellet in 80 % (v/v) ice-cold acetone by vortexing, centrifuge at maximal speed (14,000 rpm) for 10 min and discard the supernatant.
9. Pellets can be stored at −80 °C or dissolved in water for prFMN extraction as described previously (Khusnutdinova et al., 2019).

2.8 Procedure: *in vitro* activation of purified UbiD using prFMN cascades

1. Before adding apoUbiD, the reacted prFMN cascades containing UbiX in complex with prFMN$_{red}$ (step 4, Section 2.7) should be oxidized by vigorous vortexing for 2—5 min to produce brownish UbiX-prFMN$_{radical}$ complexes.
2. Add purified apoUbiD (Fdc1 from Section 2.6) to the oxidized prFMN cascade solution (aiming toward 2-10 molar excess of UbiX).
3. Incubate under aerobic conditions at 37 °C for 30 min or O/N and collect aliquots for the spectrophotometric assay of Fdc1 activity using cinnamic acid as substrate (at 270 nm, as described previously by (Payne et al., 2015)).
4. Alternatively, add purified UbiD (Fdc1) to the freshly prepared FMN prenylation cascade (2-10 molar excess of UbiX) and incubate O/N at 37 °C followed by the Fdc1 activity assay with cinnamic acid as substrate (as in the previous step).
5. Make aliquots and flash freeze the activated UbiD solution in liquid nitrogen and store at −80 °C.

2.9 Notes

1. Instead of the wild type polyP kinase CHU0107, a truncated version of this enzyme (L285Stop, 1-284 aa) can also be used for ATP regeneration, as it was found to be at least two times more active than the wild type CHU0107 (Nocek et al., 2018).
2. The FMN reduction cascade can be extended by using riboflavin as substrate (instead of FMN) and adding RFK for riboflavin phosphorylation to FMN, but this system produces UbiDs with lower specific activities (Batyrova et al., 2020).
3. The proposed prenylation cascade can also be used to screen purified UbiX proteins for prenylation activity against various flavins.
4. Occasionally, purified recombinant AF1214 can precipitate after long storage on ice. To dissolve protein precipitates, the AF1214 preparation can be incubated briefly at room temperature.

5. After adding all enzymes to the reaction mixture, the solution color should quickly change to pale yellow due to FMN reduction.
6. Precipitated UbiX pellets with bound prFMN can be frozen and stored at −80 °C.
7. Discoloration of the reaction mixture is a convenient indicator of Fre activity and FMN reduction.
8. Since purified Fdc1 is light sensitive, direct light exposure should be minimized.
9. Fdc1 preparations obtained after O/N incubation at 37 °C can show significant variations in specific activity (probably due to long incubation at this temperature).
10. In our previous study (Wang et al., 2018), Fdc1 activation was stimulated by phosphate addition. The proposed cascade mix includes the polyP-based ATP-regenerating system containing significant amounts of free phosphate (and additional free phosphate is produced by UbiX during FMN prenylation).
11. In addition to AF1214 and PA4019, high FMN prenylation rates and efficient UbiD activation were also observed with other UbiX proteins: KLP0198 from *K. pneumonia* (UniProt ID Q462H4), TSTM2747 from *S. typhimurium* (Q8ZMG1), RP0927 from *R. palustris* (Q6NB98), and TA1100 from *T. acidophilum* (Q9HJ72) (Batyrova et al., 2020).
12. For the UbiD/UbiX systems associated with a prFMN-binding chaperone, a mediator protein facilitating the transfer and maturation of prFMN$_{red}$ from UbiX to UbiD (DiRocco et al., 2024; Gahloth et al., 2024), the corresponding proteins (*e.g.* LpdD from *Lactobacillus plantarum* or PhdC from *Mycolicibacterium fortuitum*) should be expressed, affinity purified (like described for UbiX), and included into the prFMN enzyme cascade (in equimolar concentration to UbiD).

3. Procedure: *in vivo* FMN prenylation and UbiD activation using recombinant *E. coli* cells

Although *E. coli* cells express the endogenous UbiX protein during growth (Yung et al., 2016), the levels of its expression and activity are insufficient for the activation of recombinant UbiDs, resulting in the production of inactive UbiD proteins (*e.g.* Fdc1 < 0.1 U/mg) (Batyrova et al., 2020). Previous studies revealed that purified UbiDs can be activated *in vitro* using heterologous UbiX proteins in complex with prFMN (Batyrova et al., 2020; Wang et al., 2018; White et al., 2015). In line with this, active preparations of Fdc1 (from *A. niger* or *S. cerevisiae*) were obtained from *E. coli* cells co-expressing Fdc1 with heterologous UbiX proteins from different bacteria including *E. coli*, *A. fulgidus* (AF1214), *K. pneumoniae* (KLP0198), and *S. typhimurium* (TSTM2747) (Batyrova et al., 2020; Marshall, Payne, Fisher, Gahloth, et al., 2019; Payne et al., 2015). Similar to the production of recombinant UbiX proteins in complex with prFMN, the expression of active UbiDs and FMN prenylation by *E. coli* cells can be stimulated by the addition of prenol (1.0 mM), riboflavin (0.1 mM), and Mn^{2+} (1 mM) to growing cultures (Batyrova et al., 2020; Marshall, Payne, Fisher, Gahloth, et al., 2019; Wang et al., 2018; Zhu et al., 2023). Although the initial study on *in vivo* FMN prenylation by the UbiX protein AF1214 expressed in *E. coli* cells recommended switching to anaerobic conditions (+DMSO) after IPTG induction to achieve significant levels of FMN prenylation (Wang et al., 2018), subsequent research demonstrated efficient FMN prenylation and Fdc1 activation under aerobic conditions when AF1214 was replaced by KLP0198 or TSTM2747 (Batyrova et al., 2020). Generally, the use of aerobic growth conditions for recombinant *E. coli* cultures would be preferable for *in vivo* UbiD activation due to higher yields of biomass and recombinant proteins compared to anaerobic conditions. Finally, in a more recent study, Zhu et al. reported on engineering the *E. coli* strain SC-6 with plasmid-based expression of four additional enzymes contributing to *in vivo* FMN prenylation (ThiM, Fre, RFK, and FDH), which produced highly active Fdc1 (from *S. cerevisiae*) when co-expressed with the UbiX protein PA4019 under aerobic conditions (Zhu et al., 2023).

This chapter describes a protocol for combined *in vivo* FMN prenylation and UbiD activation under aerobic conditions using *E. coli* cells co-expressing Fdc1 from *A. niger* with various UbiX, including TSTM2747 from *S. typhimurium*, AF1214 from *A. fulgidus*, and KLP0198 from *K. pneumoniae* (Fig. 4). With prenol addition and TSTM2747 as the UbiX prenyltransferase, the *E. coli* cultures produced activated Fdc1 preparations, which showed cinnamic acid decarboxylase activities up to 1.5 U/mg after purification (Batyrova et al., 2020).

3.1 Equipment
1. UV-Vis spectrophotometer (eg. SpectraMax M2)
2. Shaking incubator for culture tubes (37 °C)
3. Thermostat for bacterial plates (37 °C)

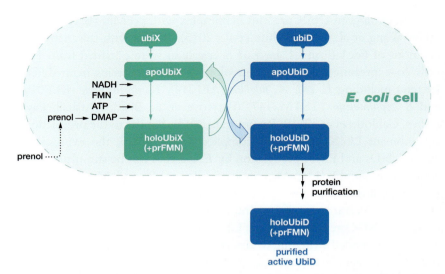

Fig. 4 Schematic diagram showing *in vivo* FMN prenylation and UbiD activation under aerobic conditions using *E. coli* cells co-expressing UbiX and UbiD. The addition of prenol (1 mM) to the growth medium stimulates FMN prenylation by UbiX and eventually UbiD activation inside *E. coli* cells. After cell harvesting and sonication, active holoUbiD proteins can be purified under aerobic conditions using affinity chromatography.

4. Platform shaker for 2.5-L plastic flasks with temperature control (37 °C/16 °C)
5. Baffled plastic flasks (2.5 L) for *E. coli* cultures (autoclaved)
6. Preparative refrigerated centrifuges with rotors for *E. coli* cell harvesting and lysate clearing for 1-L and 50-ml centrifuge bottles (Avanti)
7. Sonicator for *E. coli* cell disruption (QSonica Q700 with preparative probes)
8. Bench-top centrifuges
9. Standard equipment for affinity purification of recombinant proteins using Ni-NTA resin
10. Protein electrophoresis system (Mini-Protean, Bio-Rad)
11. −80 °C freezer

3.2 Materials

1. *E. coli* growth medium (Terrific Broth or equivalent)
2. Antibiotics for *E. coli* cultures (filtered through 2 μm syringe filters)
3. Prenol (3-methyl-2-buten-1-ol, Sigma Aldrich), 0.1 M stock solution in water
4. Isopropyl-β-D-1-thiogalactopranoside (IPTG), 0.4 M stock solution in water

FMN prenylation and UbiD activation 169

5. Cinnamic acid sodium salt, 0.1 M stock solution in water
6. *E. coli* BL21(DE3) strains expressing the *A. niger* Fdc1 with a C-terminal 6His-tag (from a pET30a plasmid, a gift from Dr. David Leys, Manchester University) co-expressed with the UbiX prenyltransferase TSTM2747 from *S. typhimurium* (UniProt ID Q8ZMG1, cloned into a pCDFDuet plasmid in the MCS2) or other UbiX proteins.
7. Liquid nitrogen
8. Centrifugal filters for protein concentration (MW cut-off 3 kDa)
9. Glass or plastic chromatographic columns for gravity flow protein purification
10. Ni–NTA resin for affinity protein purification
11. Standard protein purification buffers for Ni-chelate affinity chromatography: binding buffer (BB: 50 mM HEPES-K, pH 7.5, 0.4 M NaCl, 5 mM imidazole, 5 % glycerol), washing buffer (WB: BB with 30 mM imidazole), and elution buffer (EB: BB with 500 mM imidazole)
12. Bradford reagent (Bio-Rad) for determining protein concentration in preparations

3.3 Procedure

1. Using chemical transformation or electroporation, transform *E. coli* BL21(DE3) with selected plasmids for co-expressing Fdc1 from *A. niger* and TSTM2747 from *S. typhimurium* (or other UbiX proteins) (Fig. 4), incubate plates O/N at 37 °C and select isolated colonies.
2. Inoculate 10 ml of Terrific broth (TB) medium with appropriate antibiotics with a single colony obtained from the freshly transformed plate and grow the starter culture O/N (12-16 h) at 37 °C with shaking at 200 rpm.
3. Use 10 ml of the starter culture to inoculate TB medium (1 L) with appropriate antibiotics in 2.5-L baffled flasks and grow the cultures with shaking at 200 rpm aerobically at 37 °C to optical density $OD_{600} = 0.6$-0.8.
4. Add prenol (1 mM final concentration) and IPTG (0.4 mM final concentration), incubate at 37 °C (3-5 h) or at 16 °C (O/N) with shaking at 200 rpm.
5. Harvest *E. coli* cells by centrifugation at $5000 \times g$ for 15 min
6. Remove supernatants and resuspend cell pellets in 30 ml of BB. Freeze cells in liquid nitrogen for storage at −80 °C or immediately use them for sonication and protein purification.
7. For sonication, transfer cell suspensions to metal sonication cups and place the cups in an ice-water bath.
8. Sonicate cells for 20 min at 20 kHz (using pulses 5 s on/off).

9. Centrifuge sonicated samples at $40,000 \times g$ for 20 min at 4 °C and transfer supernatants to 50 ml Falcon tubes.
10. Purify Fdc1 using standard Ni-affinity chromatography (using washing buffer with 30 mM imidazole and elution buffer with 500 mM imidazole) (Fig. 4).
11. Determine protein concentrations in eluted column fractions using the Bradford reagent.
12. Analyze the level of protein purity in eluted fractions using SDS-PAGE.
13. Concentrate selected protein fractions (to 3-5 mg/ml), flash freeze in liquid nitrogen and store at −80 °C.

3.4 Notes

1. In addition to TSTM2747, other UbiX prenyltransferases can be used for the co-expression and activation of recombinant UbiDs (Fig. 4), including KPL0198 from *K. pneumonia*, PA4019 from *P. aeruginosa*, AF1214 from *A. fuldigus*, TA1100 from *T. acidophilum* and others, as analyzed previously (Batyrova et al., 2020).
2. For the UbiX/UbiD systems associated with prFMN-binding chaperones (*e.g.* LpdD from *L. plantarum* or PhdC from *M. fortuitum*), these proteins can be co-expressed with the corresponding UbiX and UbiD using a pCDFDuet plasmid followed by the *in vivo* UbiD activation protocol described above (Fig. 4). Alternatively, purified apoUbiDs can first be screened against a panel of different UbiX-prFMN$_{radical}$ complexes, and then the UbiX proteins with significant UbiD activation activity can be selected for cloning and *in vivo* UbiD activation.
3. Since purified UbiD proteins can be sensitive to light and/or oxygen (Bailey et al., 2018; Gahloth et al., 2022; Payer et al., 2017), it is desirable to minimize direct light exposure of protein fractions during protein purification and perform it in an anaerobic glove box (for O_2-sensitive proteins). In addition, storage in liquid nitrogen is recommended for purified O_2-sensitive UbiD proteins (for long term storage).

4. Conclusions

Exploration of biochemical diversity of the large family of UbiD-like carboxylases and decarboxylases requires recombinant production and purification of significant amounts of active UbiD proteins loaded with prFMN. In this chapter, we described the protocols for *in vitro* and *in vivo*

FMN prenylation and UbiD activation under aerobic conditions using enzyme cascades with regenerating systems or recombinant *E. coli* cells co-expressing selected UbiX and UbiD proteins. As demonstrated using the UbiD protein Fdc1 from *A. niger*, both *in vitro* enzyme cascades and *in vivo* protein co-expression in *E. coli* cells can produce holoFdc1 preparations with specific activities comparable to those after activation using protein-free prFMN or purified UbiX-prFMN complexes. Overall, these protocols can be carried out in a regular biochemical laboratory using inexpensive chemicals and standard equipment. It is anticipated that various *in vitro* and *in vivo* methods for UbiD activation might be required for characterizing novel UbiD enzymes and developing their applications in biocatalysis (Marshall, Payne, Fisher, Gahloth, et al., 2019). Hence, the proposed protocols will contribute to the development of novel UbiD activation methods, thereby facilitating the discovery of novel (de)carboxylases and their practical applications.

Acknowledgments

This work was supported by the NSERC Strategic Network grant Industrial Biocatalysis Network and by the FuturEnzyme project funded by the European Union's Horizon 2020 Research and Innovation Program under grant agreement101000327. We thank www.dizign.net for their assistance with figure design and preparation.

References

Aleku, G. A., Prause, C., Bradshaw-Allen, R. T., Plasch, K., Glueck, S. M., Bailey, S. S., ... Leys, D. (2018). Terminal alkenes from acrylic acid derivatives via non-oxidative enzymatic decarboxylation by ferulic acid decarboxylases. *ChemCatChem, 10*(17), 3736–3745.

Aleku, G. A., Roberts, G. W., Titchiner, G. R., & Leys, D. (2021). Synthetic enzyme-catalyzed CO_2 fixation reactions. *ChemSusChem, 14*, 1781–1804.

Arunrattanamook, N., & Marsh, E. N. G. (2018). Kinetic characterization of prenyl-flavin synthase from *Saccharomyces cerevisiae*. *Biochemistry, 57*(5), 696–700.

Bailey, S. S., Payne, K. A. P., Fisher, K., Marshall, S. A., Cliff, M. J., Spiess, R., ... Leys, D. (2018). The role of conserved residues in Fdc decarboxylase in prenylated flavin mononucleotide oxidative maturation, cofactor isomerization, and catalysis. *The Journal of Biological Chemistry, 293*(7), 2272–2287.

Bailey, S. S., Payne, K. A. P., Saaret, A., Marshall, S. A., Gostimskaya, I., Kosov, I., ... Leys, D. (2019). Enzymatic control of cycloadduct conformation ensures reversible 1,3-dipolar cycloaddition in a prFMN-dependent decarboxylase. *Nature Chemistry, 11*(11), 1049–1057.

Balaikaite, A., Chisanga, M., Fisher, K., Heyes, D. J., Spiess, R., & Leys, D. (2020). Ferulic acid decarboxylase controls oxidative maturation of the prenylated flavin mononucleotide cofactor. *ACS Chemical Biology, 15*(9), 2466–2475.

Batyrova, K. A., Khusnutdinova, A. N., Wang, P. H., Di Leo, R., Flick, R., Edwards, E. A., ... Yakunin, A. F. (2020). Biocatalytic *in vitro* and *in vivo* FMN prenylation and (de)carboxylase activation. *ACS Chemical Biology, 15*(7), 1874–1882.

Bierbaumer, S., Nattermann, M., Schulz, L., Zschoche, R., Erb, T. J., Winkler, C. K., ... Glueck, S. M. (2023). Enzymatic conversion of CO(2): From natural to artificial utilization. *Chemical Reviews, 123*(9), 5702–5754.

Bloor, S., Michurin, I., Titchiner, G. R., & Leys, D. (2023). Prenylated flavins: Structures and mechanisms. *The FEBS Journal, 290*(9), 2232–2245.

Datar, P. M., & Marsh, E. N. G. (2021). Decarboxylation of aromatic carboxylic acids by the prenylated-FMN-dependent enzyme phenazine-1-carboxylic acid decarboxylase. *ACS Catalysis, 11*, 11723–11732.

DiRocco, D. J., Roy, P., Mondal, A., Datar, P. M., & Marsh, E. N. G. (2024). An enzyme catalyzing the oxidative maturation of reduced prenylated-FMN to form the active enzyme. *ACS Catalysis, 14*, 10223–10233.

Erb, T. J. (2011). Carboxylases in natural and synthetic microbial pathways. *Applied and Environmental Microbiology, 77*(24), 8466–8477.

Fieschi, F., Niviere, V., Frier, C., Decout, J. L., & Fontecave, M. (1995). The mechanism and substrate specificity of the NADPH:flavin oxidoreductase from *Escherichia coli*. *The Journal of Biological Chemistry, 270*(51), 30392–30400.

Gahloth, D., Fisher, K., Marshall, S., & Leys, D. (2024). The prFMNH(2)-binding chaperone LpdD assists UbiD decarboxylase activation. *The Journal of Biological Chemistry, 300*(2), 105653.

Gahloth, D., Fisher, K., Payne, K. A. P., Cliff, M., Levy, C., & Leys, D. (2022). Structural and biochemical characterization of the prenylated flavin mononucleotide-dependent indole-3-carboxylic acid decarboxylase. *The Journal of Biological Chemistry, 298*(4), 101771.

Glueck, S. M., Gumus, S., Fabian, W. M., & Faber, K. (2010). Biocatalytic carboxylation. *Chemical Society Reviews, 39*(1), 313–328.

Gray, M. J., Wholey, W. Y., Wagner, N. O., Cremers, C. M., Mueller-Schickert, A., Hock, N. T., ... Jakob, U. (2014). Polyphosphate is a primordial chaperone. *Molecular Cell, 53*(5), 689–699.

Khusnutdinova, A. N., Xiao, J., Wang, P. H., Batyrova, K. A., Flick, R., Edwards, E. A., & Yakunin, A. F. (2019). Prenylated FMN: Biosynthesis, purification, and Fdc1 activation. *Methods in Enzymology, 620*, 469–488.

Kourist, R., Guterl, J.-K., Miyamoto, K., & Sieber, V. (2014). Enzymatic decarboxylation—An emerging reaction for chemicals production from renewable resources. *ChemCatChem, 6*, 689–701.

Leys, D., & Scrutton, N. S. (2016). Sweating the assets of flavin cofactors: New insight of chemical versatility from knowledge of structure and mechanism. *Current Opinion in Structural Biology, 41*, 19–26.

Li, T., Huo, L., Pulley, C., & Liu, A. (2012). Decarboxylation mechanisms in biological system. *Bioorganic Chemistry, 43*, 2–14.

Marshall, S. A., Fisher, K., Ni Cheallaigh, A., White, M. D., Payne, K. A., Parker, D. A., ... Leys, D. (2017). Oxidative maturation and structural characterization of prenylated FMN binding by UbiD, a decarboxylase involved in bacterial ubiquinone biosynthesis. *The Journal of Biological Chemistry, 292*(11), 4623–4637.

Marshall, S. A., Payne, K. A. P., Fisher, K., Gahloth, D., Bailey, S. S., Balaikaite, A., ... Leys, D. (2019). Heterologous production, reconstitution and EPR spectroscopic analysis of prFMN dependent enzymes. *Methods Enzymol, 620*, 489–508.

Marshall, S. A., Payne, K. A. P., Fisher, K., White, M. D., Ni Cheallaigh, A., Balaikaite, A., ... Leys, D. (2019). The UbiX flavin prenyltransferase reaction mechanism resembles class I terpene cyclase chemistry. *Nature Communications, 10*(1), 2357.

Marshall, S. A., Payne, K. A. P., & Leys, D. (2017). The UbiX-UbiD system: The biosynthesis and use of prenylated flavin (prFMN). *Archives of Biochemistry and Biophysics, 632*, 209–221.

Mondal, A., Roy, P., Carrannanto, J., Datar, P. M., DiRocco, D. J., Hunter, K., & Marsh, E. N. G. (2024). Surveying the scope of aromatic decarboxylations catalyzed by prenylated-flavin dependent enzymes. *Faraday Discussions*.

Nocek, B. P., Khusnutdinova, A. N., Ruszkowski, M., Flick, R., Burda, M., Batyrova, K., ... Yakunin, A. F. (2018). Structural insights into substrate selectivity and activity of bacterial polyphosphate kinases. *ACS Catalysis, 8*, 10746–10760.

Payer, S. E., Faber, K., & Glueck, S. M. (2019). Non-oxidative enzymatic (de)carboxylation of (hetero)aromatics and acrylic acid derivatives. *Advanced Synthesis & Catalysis, 361*(11), 2402–2420.

Payer, S. E., Marshall, S. A., Barland, N., Sheng, X., Reiter, T., Dordic, A., ... Glueck, S. M. (2017). Regioselective para-carboxylation of catechols with a prenylated flavin dependent decarboxylase. *Angewandte Chemie (International Ed. in English), 56*(44), 13893–13897.

Payne, K. A., White, M. D., Fisher, K., Khara, B., Bailey, S. S., Parker, D., ... Leys, D. (2015). New cofactor supports alpha,beta-unsaturated acid decarboxylation via 1,3-dipolar cycloaddition. *Nature, 522*(7557), 497–501.

Roberts, G. W., & Leys, D. (2022). Structural insights into UbiD reversible decarboxylation. *Current Opinion in Structural Biology, 75*, 102432.

Saaret, A., Balaikaite, A., & Leys, D. (2020). Biochemistry of prenylated-FMN enzymes. *Enzymes, 47*, 517–549.

Wang, P. H., Khusnutdinova, A. N., Luo, F., Xiao, J., Nemr, K., Flick, R., ... Yakunin, A. F. (2018). Biosynthesis and activity of prenylated FMN cofactors. *Cell Chemical Biology, 25*(5), 560–570 e566.

White, M. D., Payne, K. A., Fisher, K., Marshall, S. A., Parker, D., Rattray, N. J., ... Leys, D. (2015). UbiX is a flavin prenyltransferase required for bacterial ubiquinone biosynthesis. *Nature, 522*(7557), 502–506.

Yung, P. Y., Grasso, L. L., Mohidin, A. F., Acerbi, E., Hinks, J., Seviour, T., ... Lauro, F. M. (2016). Global transcriptomic responses of Escherichia coli K-12 to volatile organic compounds. *Scientific Reports, 6*, 19899.

Zhu, X., Li, H., Ren, J., Feng, Y., & Xue, S. (2023). Engineering the biosynthesis of prFMN promotes the conversion between styrene/CO_2 and cinnamic acid catalyzed by the ferulic acid decarboxylase Fdc1. *Catalysts, 13*, 917.

CHAPTER EIGHT

Cellular and biochemical approaches to define GGCX carboxylation of vitamin K-dependent proteins

Kathleen L. Berkner[a],* [iD] **, Kevin W. Hallgren[a], Mark A. Rishavy[a], and Kurt W. Runge[b]** [iD]

[a]Department of Cardiovascular and Metabolic Sciences, Lerner Research Institute, Cleveland Clinic Lerner College of Medicine at CWRU, Cleveland, OH, United States
[b]Department of Inflammation and Immunity, Lerner Research Institute, Cleveland Clinic Lerner College of Medicine at CWRU, Cleveland, OH, United States
*Corresponding author. e-mail address: berknek@ccf.org

Contents

1. Introduction	176
2. Cellular analysis of VKD protein carboxylation	179
2.1 Analyzing VKD protein carboxylation in 293 cells expressing endogenous GGCX	179
2.2 Generating cell lines edited to eliminate endogenous GGCX	183
2.3 Expression of r-GGCX variants	184
3. Biochemical analysis of VKD protein carboxylation	191
3.1 Assays to monitor GGCX catalysis	191
3.2 An assay to monitor the entire GGCX reaction	195
3.3 An assay to monitor GGCX processivity	198
4. Considerations	199
References	201

Abstract

The gamma-glutamyl carboxylase (GGCX) generates clusters of carboxylated glutamic acid residues in vitamin K-dependent (VKD) proteins, which is required for their diverse functions including hemostasis and regulation of calcification. The GGCX modifies a VKD protein using several substrates and cofactors, and has regulatory mechanisms like processivity that ensures full carboxylation of VKD proteins. The GGCX mechanism is incompletely understood. GGCX mutations cause disease: vitamin K clotting factor deficiency (VKCFD1) associated with severe bleeding and pseudoxanthoma elasticum (PXE)-like, where bleeding is mild but calcification is excessive. Why mutations cause disease has only been revealed for a few GGCX mutants. The chapter describes biochemical and cellular assays developed to define GGCX mechanism and determine why GGCX mutations cause disease. Multiple components are important to VKD protein

Methods in Enzymology, Volume 708
ISSN 0076-6879, https://doi.org/10.1016/bs.mie.2024.10.023
Copyright © 2024 Elsevier Inc. All rights are reserved, including those for text and data mining, AI training, and similar technologies.

carboxylation. GGCX requires reduced vitamin K generated by the vitamin K epoxide reductase (VKORC1), which is activated by redox protein(s). GGCX activity is also supported by the reductases VKORC1-like and ferroptosis suppressor protein-1 (FSP1). Carboxylation occurs in the endoplasmic reticulum during VKD protein secretion and is impacted by quality control mechanisms. This chapter emphasizes the importance in maintaining an appropriate stoichiometry of the components when studying recombinant (r-) r-GGCX function in cells. Also emphasized is the essentiality in using both cellular and biochemical approaches to study r-GGCX function, as cellular analysis can detect altered VKD protein carboxylation by r-GGCX mutants but cannot explain the r-GGCX defects. This point is illustrated by studies on a GGCX mutant that causes the PXE-like disease, where biochemical analysis was necessary for revealing mutant dysfunction, i.e. impaired processivity.

1. Introduction

The gamma-glutamyl carboxylase (GGCX) is an integral membrane enzyme that converts specific glutamic acid (Glu) residues to carboxylated Glus (Glas) in vitamin K-dependent (VKD) proteins (Berkner & Runge, 2022). Clusters of Glas are generated, which form a calcium-binding module that is essential for VKD protein function. This family of proteins includes clotting factors required for hemostasis, as well as nonhemostatic proteins with diverse roles such as regulation of calcification, growth control, inflammation, apoptosis and cell signaling (Berkner & Runge, 2004; Shearer & Newman, 2008). All VKD proteins are modified by a single GGCX (Wu, Cheung, Frazier, & Stafford, 1991), which is expressed in virtually all tissues in humans (Caspers et al., 2015). Disruption of VKD protein carboxylation causes disease. Naturally occurring GGCX mutations result in VKCFD1 (for vitamin K clotting factor deficiency) associated with severe bleeding (Dhouha Darghouth et al., 2009; D. Darghouth et al., 2006; Ghosh et al., 2022; Hao et al., 2021; Mutucumarana et al., 2000; Rishavy, Hallgren, Zhang, Runge, & Berkner, 2019; Watzka et al., 2014; Zhang & Ginsburg, 2004), and GGCX null mice die around birth due to hemorrhaging (Zhu et al., 2007). Some GGCX mutations cause the pseudoxanthoma elasticum (PXE)-like disease associated with excessive calcification in skin and mild bleeding (De Vilder, Debacker, & Vanakker, 2017; Ghosh et al., 2021; Li, Grange, et al., 2009; Li, Jiang, Pfendner, Varadi, & Uitto, 2009; Li, Schurgers, et al., 2009; Rishavy et al., 2022; Tie et al., 2016; Vanakker et al., 2007). The mechanisms by which GGCX mutations cause defective hemostasis and calcification are poorly understood. This chapter focuses on cellular and biochemical assays to understand the functional consequences of GGCX variants.

The GGCX reaction is complex, with catalysis and regulation involving multiple substrates and cofactors (Fig. 1A). The reduced vitamin K cofactor is generated from dietary vitamin K, which comprises phylloquinone and menaquinones that share a naphthoquinone nucleus but differ in isoprenyl groups with different lengths and degree of saturation (Shearer & Newman, 2014). GGCX selectively modifies VKD proteins because they contain a GGCX-interacting sequence adjacent to the Glu cluster in the Gla domain undergoing carboxylation (Fig. 1B), which in most cases is a propeptide cleaved after carboxylation. This sequence binds to GGCX at a site distinct from the catalytic site and is referred to as the exosite binding domain (EBD). The EBD is important to allosteric regulation, as EBD binding to GGCX stimulates Glu catalysis (Knobloch & Suttie, 1987). The EBD also mediates GGCX processivity, which refers to GGCX remaining bound to a VKD protein until the multiple Glu residues in the Gla domain become fully carboxylated (Stenina, Pudota, McNally, Hommema, & Berkner, 2001). The mechanism by which GGCX modifies VKD proteins is only partially understood, and how GGCX mutations disrupt carboxylation to cause disease has only been explained for a few mutations.

Multiple proteins facilitate VKD protein carboxylation. The GGCX reaction generates a vitamin K epoxide product that must be recycled to vitamin K hydroquinone for continuous carboxylation (Berkner & Runge, 2022) (Fig. 1C). The vitamin K epoxide is recycled to the hydroquinone form in two reactions, and the vitamin K epoxide reductase (VKORC1) performs both reactions (Rishavy et al., 2013). VKORC1 is expressed in all tissues in humans (Caspers et al., 2015), and mice lacking this reductase show lethal hemorrhaging around birth (Spohn et al., 2009). The VKORC1 paralog VKORC1-like also reduces vitamin K epoxide and supports VKD protein carboxylation in vivo (Hammed et al., 2013; Lacombe, Rishavy, Berkner, & Ferron, 2018); however, its role is not well understood. Both VKORC1 and VKORC1-like are inactivated during vitamin K epoxide reduction and are reactivated through the action of redox proteins that remain to be determined. VKORC1 is the target of anticoagulants like warfarin that suppress vitamin K regeneration, and bleeding complications from warfarin overdose are ameliorated by vitamin K administration, implicating a warfarin resistant reductase in supporting carboxylation. This reductase was recently identified as FSP-1, a protein known to regulate ferroptosis (Jin et al., 2023; Mishima et al., 2022).

The secretory process impacts VKD protein carboxylation. VKD proteins are carboxylated in the endoplasmic reticulum, and most undergo additional

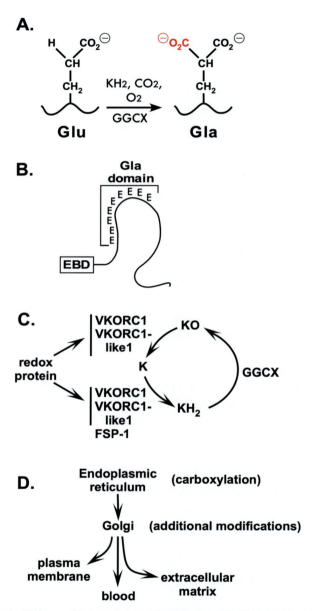

Fig. 1 Vitamin K-dependent protein carboxylation. (A) The gamma-glutamyl carboxylase (GGCX) uses oxygenation of vitamin K hydroquinone (KH$_2$) to generate carboxylated glutamic acid (Gla). (B) Vitamin K-dependent (VKD) proteins are targeted to GGCX via an exosite binding domain (EBD) adjacent to the Gla domain, which contains multiple glutamic acid residues (E) that undergo carboxylation. (C) Carboxylation produces vitamin K epoxide (KO), which is reduced to KH$_2$ in two

modifications in the Golgi (Fig. 1D). The VKD proteins are then secreted to the plasma membrane (e.g. proline rich Gla proteins) or into the circulation (e.g. clotting factors) or the extracellular matrix (e.g. Matrix Gla Protein). As described in the next section, quality control mechanisms that degrade poorly carboxylated VKD protein can impact the secretion of VKD proteins.

2. Cellular analysis of VKD protein carboxylation

Analyzing recombinant (r-) r-GGCX variants in cells provides a natural environment where GGCX and VKD proteins interact with other components required for carboxylation and the secretory machinery (Fig. 1, Introduction). Early studies used baculovirus-infected insect cells because they lack endogenous GGCX (Berkner & McNally, 1997; Roth et al., 1993); however, CRISPR-Cas9 editing now makes mammalian cells a better system. A test of many cell lines has shown that HEK293 cells (referred to as 293 cells) are the most efficient for VKD protein carboxylation (Berkner, 1993). This section describes methods for studying VKD protein carboxylation, eliminating endogenous GGCX by CRISPR-Cas9 editing, introducing r-GGCX variants into edited cells and characterizing r-GGCX 293 cells.

It is important to note that cellular analysis is an indirect approach for analyzing VKD protein carboxylation. Secreted proteins are monitored to assess carboxylation, which is necessary because the intracellular VKD protein population is a large mixture of forms. Specifically, in addition to the carboxylated product, the intracellular population includes uncarboxylated substrate, immature proteins that are unfolded or that lack disulfide bonds, and proteins in the Golgi undergoing other post translational modifications. Therefore, while the cellular studies are informative, conclusions about r-GGCX function need to consider the indirect nature of this approach.

2.1 Analyzing VKD protein carboxylation in 293 cells expressing endogenous GGCX

2.1.1 Reagents

Cell culture media: Dulbecco's Modified Eagle's Medium Nutrient F-12 Ham (DMEM:F12) is from Caisson, supplements (penicillin (100 units/ml),

---

steps by VKORC1 and VKORC1-like through a vitamin K quinone (K) intermediate. FSP-1 performs the single K to KH$_2$ reaction. (D) VKD proteins are carboxylated during secretion to either the plasma membrane or outside the cell. Examples of modifications in the Golgi are proteolytic removal of the EBD and glycosylation.

streptomycin (10 μg/ml), glutamine (2 mM), sodium pyruvate (1 mM)) are from Gibco, and 10 % fetal bovine serum (Charcoal/Dextran Treated) is from R and D Systems.

Serum free media: Identical to the cell culture media, except that serum is replaced by a cocktail of fetuin (500 μg/ml, from Sigma-Aldrich) plus insulin (10 μg/ml), transferrin (6 μg/ml), selenium (0.01 μg/ml) and HEPES (20 mM, pH 7.1) that are from Gibco.

Phosphate buffered saline (PBS).

Vitamin K: variable concentrations of phytonadione (Hospira).

Chemical lysis buffer: 25 mM sodium phosphate (pH 7.9), 25 mM potassium chloride, 20 % glycerol, 0.75 % CHAPS, and 2 mM phenylmethanesulfonyl fluoride.

Factor IX standard: human factor IX purified from plasma (Enzyme Research Laboratories).

Antibodies: anti-factor IX (monoclonal (Green Mountain Antibodies) and polyclonal (in house) antibodies); anti-GAPDH (Sigma); anti-Gla (BioMedica Diagnostics).

In choosing a VKD reporter to study carboxylation, an important consideration is the impact of quality control on the secreted VKD protein population. The carboxylation status can impact whether a VKD protein is secreted or degraded, and the response differs among individual VKD proteins. For example, factor IX is secreted irrespective of carboxylation status, while poorly carboxylated protein C and protein Z are degraded (Hallgren, Hommema, McNally, & Berkner, 2002; Souri, Iwata, Zhang, & Ichinose, 2009; Tokunaga, Wakabayashi, & Koide, 1995; W. Wu, Bancroft, & Suttie, 1997). Quality control mechanisms that filter poorly carboxylated VKD protein have the potential to confound analysis of a secreted protein; for example, a r-GGCX mutant may impair carboxylation but VKD protein degradation would make the defect less apparent in secreted protein. Secretion versus degradation was revealed in pulse-chase analyses that have only been performed on a few VKD proteins (Hallgren et al., 2002; Souri et al., 2009; Tokunaga et al., 1995). This subsection describes the analysis for a VKD protein secreted independent of carboxylation status, factor IX.

Analysis is shown for 293 cells expressing endogenous GGCX and r-factor IX, where fully carboxylated factor IX is generated (Hallgren et al., 2002; Rishavy et al., 2022). The generation of cells with r-VKD proteins has previously been described in a Methods in Enzymology chapter (Berkner, 1993). Analyses use cells incubated in vitamin K; however, to allow parallel analysis with cells lacking vitamin K, the cells are routinely passaged in serum treated with charcoal/dextran that removes vitamin K. Vitamin K cannot be eliminated by simply fluid changing the cells into media lacking vitamin K because intracellular stores of vitamin K are retained for several weeks (unpublished data). To study VKD protein carboxylation, cells are exchanged into serum free media, which eliminates bovine VKD proteins in serum that would interfere with analyses. Cells ($\sim 10^7$) in 10 cm plates are rinsed with PBS (2 ml), and then incubated in serum free media (5 ml) for varying times in either the presence or absence of vitamin K, and media containing secreted factor IX is harvested. The concentration of vitamin K varies with the specific experiment. The analysis shown below uses 50 ng/ml vitamin K, but ranges of 0.1–10 µg/ml have been used in dose response experiments. Cellular toxicity is observed at vitamin K levels above 10 µg/ml. Following incubation of cells (e.g. overnight in the example shown here), media is collected and filtered using a unit containing a 0.45 µm membrane. Cell lysates are prepared by rinsing cells with PBS (10 ml), followed by incubation with chemical lysis buffer and centrifugation (860xg, 5 min). All manipulations are performed at 4 °C, and samples are stored at −80 °C.

Factor IX expression is monitored in a Western using an antibody that recognizes both carboxylated and uncarboxylated factor IX (Fig. 2A and B). The antibody used in Fig. 2 was a polyclonal antibody generated in house; however, commercially available polyclonal or monoclonal antibodies could be used, after testing for reactivity with carboxylated and uncarboxylated forms. Quantitation of factor levels using purified plasma factor IX indicates similar levels of factor IX secretion in the presence or absence of vitamin K, consistent with pulse–chase analysis showing that factor IX is secreted independent of carboxylation (Hallgren et al., 2002). Factor IX secreted from cells containing or lacking vitamin K are qualitatively different because carboxylation impacts other post-translational modifications (e.g. glycosylation, (Berkner, 1993; Yan et al., 1990)). Glycosylation differences also account for the difference in gel migration of secreted factor IX versus plasma factor IX (Fig. 2A), which we found does not affect activity.

Carboxylation is monitored using an antibody that recognizes Gla residues (Fig. 2C; the antibody is described in more detail below). The antibody is specific for factor IX secreted from cells containing vitamin K and does not react with media from progenitor 293 cells, which allows the specific analysis of carboxylation in factor IX 293 cells. The anti-Gla and anti-factor IX signals for secreted factor IX are quantitated using plasma factor IX as a standard (Fig. 2A and C), and the Gla versus factor IX ratios for samples and standard assess the extent of carboxylation.

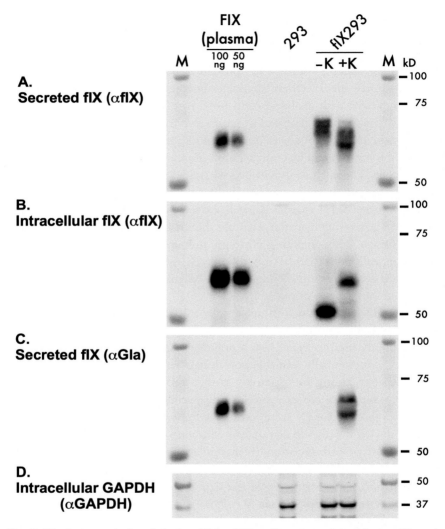

Fig. 2 **Western analysis of factor IX in 293 cells expressing r-factor IX and endogenous GGCX.** Factor IX (fIX) expression and carboxylation in cells containing or lacking vitamin K (K) are analyzed using antibodies against fIX (A, B) or Gla (C), respectively. GAPDH is analyzed as a loading control (D). Purified fIX isolated from plasma (Enzyme Research Laboratories) is used to quantitate the fIX levels in cells. In panel B, the difference in gel migration of intracellular fIX in cells cultured with (+K) or without (-K) is due to 12 negative charges added to fIX during carboxylation. The intracellular factor IX differs in molecular weight from secreted fIX (in panel A) due to post-translational modifications (e.g. glycosylation) that occur in the Golgi after carboxylation in the endoplasmic reticulum.

2.2 Generating cell lines edited to eliminate endogenous GGCX

2.2.1 Reagents

Cell culture media: identical to the composition indicated in the previous subsection.

Transfection reagent (Mirus Biosciences).

DNeasy Blood and Tissue Kit (Qiagen).

EnGen Mutation Detection Kit (New England Biolabs).

Chemical lysis buffer: identical to the buffer indicated in the previous subsection.

GGCX is expressed in virtually all tissues, and r-GGCX analysis in cells therefore requires editing to eliminate endogenous GGCX. Editing by CRISPR-Cas9 can be performed with the pCas-Guide vector (Origene), which expresses both the Cas9 enzyme and CRISPR guide RNA (gRNA) that targets cutting to a specific site (Doudna & Charpentier, 2014). Targeting gRNAs are commercially available (e.g. IDTdna.com), and the companies have identified sites in coding sequences and determined off-target sites elsewhere in the genome. Cut sites are chosen by three criteria: (1) The site is at or 5' to sequences encoding regions essential for protein function; (2) Sites in the coding sequences of other genes are not targeted; (3) The cut site is located before the last intron (preferably last several introns) to cause frameshift mutations that can induce nonsense-mediated decay of the mRNA. DNAs encoding the top four gRNAs are individually cloned into the pCas-Guide vector using Gibson Assembly (Gibson et al., 2009; Rabe & Cepko, 2020), and plasmids bearing the gRNAs are validated by restriction enzyme mapping and sequence analysis. Only one GGCX isoform has been identified, and multiple g-RNAs have been successful in generating GGCX-edited cells.

The gDNAs are tested for editing efficiency by transiently transfecting plasmids (2.5 µg) into 10^7 293 cells. Control untransfected cells are processed in parallel. After 3 days, cells are harvested using EDTA (0.5 mM, 1 ml), centrifuged (860xg, 5 min) and rinsed twice with PBS (10 ml) with repeated centrifugation, all at 4 °C. Cell extracts are then prepared in chemical lysis buffer and DNA prepared using the Qiagen DNeasy Blood and Tissue Kit following the manufacturer's instructions. DNA concentration is determined by Nanodrop, followed by PCR amplification using GGCX primers that generate an approximate 1 kb product. The PCR product is denatured, rehybridized and digested with T7 endonuclease, which recognizes and cuts

DNA mismatches (Fig. 3A), using the EnGen Mutation Detection Kit and manufacturer's instructions. A negative control lacking T7 endonuclease is processed in parallel. The PCR products are gel electrophoresed, and then quantitated (Gel Doc EZ Imager, Biorad) to determine editing efficiency. Fig. 3B shows an example of DNA samples from cells with or without editing.

The gDNA showing the highest editing efficiency is then used to generate a cell line eliminated in GGCX expression. 293 cells transiently transfected as described above are seeded at a low cell density to generate well separated clones, and 2–3 dozen clones are isolated using trypsin and cloning cylinders. Individual clones are seeded into duplicate 24 wells, and the T7 assay is performed once cells in a well reach near confluency. Performing the assay at this stage of growth minimizes the number of clones that need to be processed further. Cells showing editing are scaled up in duplicate to $\sim 10^7$ cells that are either frozen or used to prepare lysates to validate editing.

Three assays are used to validate editing. Cells are first tested in an activity assay that monitors peptide carboxylation (described in the biochemical analysis section below) and by Western analysis (Fig. 3C and D). A subset of clones are then subjected to sequence analysis. Genomic DNA is amplified by PCR, the PCR product is cloned into pCR4-TOPO (Invitrogen), and multiple transformants are sequenced. These analyses have revealed that 293 cells are triploid for GGCX (Rishavy et al., 2019), and a single transfection is sufficient for eliminating all three copies of the GGCX gene.

2.3 Expression of r-GGCX variants

Carboxylation requires multiple proteins (Fig. 1, Introduction), and therefore an important consideration in expressing r-GGCX variants is the stoichiometry of components. While 293 cells are capable of secreting fully carboxylated VKD proteins (Fig. 2 and Hallgren et al., 2002; Rishavy et al., 2022), altering the relative levels of carboxylation components disrupts the normal process. For example, r-factor IX expressed at high levels is secreted as mostly uncarboxylated and inactive protein (Kaufman, Wasley, Furie, Furie, & Shoemaker, 1986). Overexpression of r-VKORC1 that generates the reduced vitamin K cofactor required by GGCX increases the amount of carboxylated VKD protein, but only by a modest amount (2–3 fold) (Hallgren, Qian, Yakubenko, Runge, & Berkner, 2006). Overexpression of r-GGCX does not improve carboxylation and instead suppresses the

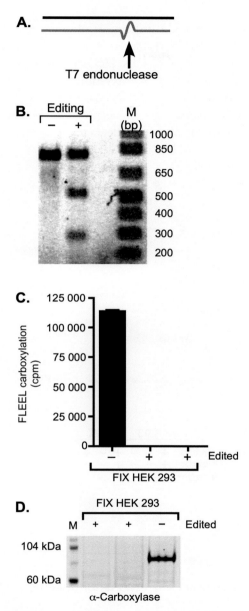

Fig. 3 **CRISPR-Cas9 editing of 293 cells.** Editing is detected in PCR-amplified DNA using T7 endonuclease (A), which cleaves the edited site (B). An example of targeting and primer sequences used in editing are described in Rishavy et al., 2019. Edited sites are validated by an activity assay that monitors carboxylation of a peptide (FLEEL) (C) and by Western analysis (D). *Panels C and D are reprinted from Figs. 3D and 3E from Rishavy et al., J. Thromb. Haemost. 2019, 17:1053–1063, with permission from Elsevier.*

secretion of carboxylated VKD protein, as observed with both stable and transient expression of GGCX (Hallgren et al., 2002; Rishavy et al., 2022). As shown in Fig. 4, when r-factor IX 293 cells are stably transfected with r-GGCX, an increase in r-GGCX expression decreases the levels of both intracellular (Fig. 4A) and secreted (Fig. 4B) factor IX. Pulse-chase studies show that r-GGCX overexpression retains the factor IX in a factor IX-GGCX complex that is ultimately degraded (Hallgren et al., 2002).

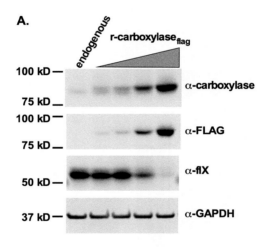

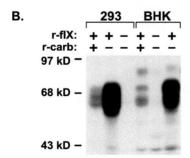

Fig. 4 r-GGCX overexpression suppresses factor IX expression. (A) Western analysis of r-factor IX (fIX) 293 cells stably transfected with r-carboxylase$_{FLAG}$ (r-GGCX$_{FLAG}$) shows that intracellular fIX levels decrease as r-GGCX$_{FLAG}$ expression increases. (B) Secreted r-fIX levels are also suppressed by high levels of r-GGCX (indicated as r-carb) in both 293 and baby hamster kidney (BHK) cells. *Panel A is reprinted from Supplemental Fig. 3 in Rishavy et al., J. Thromb. Haemost. 2019, 17:1053–1063, with permission from Elsevier. Panel B is reprinted from Fig. 1C in Hallgren et al., Biochemistry, 2002, 41:15045–15055, with permission from the American Chemical Society.*

Ensuring the appropriate stoichiometry of carboxylation components to obtain meaningful results necessitates stable expression of r-GGCX at a level similar to that of endogenous GGCX in 293 cells, which fully carboxylates VKD proteins (Hallgren et al., 2002; Rishavy et al., 2022). Transient transfection complicates analysis because it generates a cell population expressing a wide range of r-GGCX levels (e.g. 100-fold, Fig. 5A), which substantially varies the stoichiometry of r-GGCX versus other carboxylation components in individual cells. The consequence is a VKD protein population that is mixed in the extent of carboxylation, as described here using wild type r-GGCX as an example. Cells expressing r-GGCX at levels similar to endogenous GGCX will generate fully carboxylated VKD protein. However, cells expressing higher GGCX levels will produce poorly carboxylated VKD products because GGCX requires reduced vitamin K generated by VKORC1, and r-GGCX excess over VKORC1 will decrease catalysis and consequent carboxylated VKD protein. High levels of r-GGCX may also result in factor IX degradation as observed with stably transfected cells (Fig. 4) (Hallgren et al., 2002; Rishavy et al., 2019). Cells expressing low levels of r-GGCX will also secrete a mixed population of VKD protein. Some protein will be fully carboxylated because VKORC1 is in excess over r-GGCX; however, some protein will be uncarboxylated because the low levels of r-GGCX compared to VKD protein will be insufficient for carboxylation, similar to what has been observed with factor IX overexpression (Kaufman et al., 1986). The results from transient transfection will therefore be an average of a mixed VKD protein population with different extents of carboxylation, which will significantly complicate interpreting the results.

Fig. 5B–D shows an example of cells transiently expressing r-GGCX, which are compared to progenitor cells expressing r-factor IX and endogenous GGCX. The level of r-GGCX expression, which is the average of cells expressing a range of concentrations (Fig. 5A), is similar to that of endogenous GGCX (Fig. 5B). However, the levels of factor IX secretion and carboxylation are significantly lower in the transiently transfected r-GGCX cells (Fig. 5C and D), and intracellular analysis indicates factor IX and GGCX degradation (data not shown). The results are similar to data on clonal isolates expressing high levels of r-GGCX (Fig. 4), and indicate that transient expression introduces ambiguities that impact the results. Therefore, to obtain meaningful results, we use stably transfected r-GGCX cells where variants are expressed at levels similar to endogenous GGCX or r-wild type GGCX in 293 cells that generate fully carboxylated VKD protein (Hallgren et al., 2002; Rishavy et al., 2022).

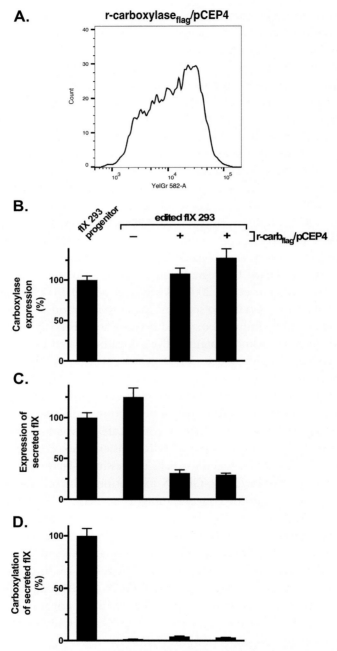

Fig. 5 Transient r-GGCX expression suppresses factor IX secretion and carboxylation. (A) Factor IX (flX) 293 cells edited to eliminate endogenous carboxylase (GGCX) and transiently transfected with a plasmid encoding r-wild type GGCX in the pCEP4 vector

Vitamin K-dependent protein carboxylation

189

Reagents.

Cell culture media: described in the section on analyzing VKD protein carboxylation.

Expression vectors: vectors with cassettes that allow expression of a gene of interest and a selectable marker.

Selection agents: Geneticin (500 µg/ml, Gibco) for neomycin selection and puromycin (2.5 µg/ml, Gibco) for puromycin selection.

Transfection reagent: Lipofectin (Invitrogen).

Chemical lysis buffer: described in the section on analyzing VKD protein carboxylation.

Sequential transfection of recombinant GGCX and VKD proteins is used because this approach has been more reliable than cotransfection of both proteins in obtaining cells with the appropriate stoichiometry of carboxylation components. The cell lines can be generated using either order of transfection of plasmids encoding GGCX or VKD proteins. A typical screen uses a vector containing cassettes that encode r-GGCX and resistance to neomycin (e.g. pCMV6-AC (Origene)) and 5 µg is transfected into 293 cells lacking endogenous GGCX. Throughout the screen, cells are cultured in the absence of vitamin K using charcoal/dextran treated serum in the media. Cells are exchanged into selection media containing Geneticin (500 µg/ml) 3 days post-transfection and seeded at a low density that results in well separated colonies. Individual clones are isolated using trypsin and cloning cylinders, and the cells are then scaled up in numbers. To limit overall effort, cells are analyzed early in the screen ($\sim 10^6$ cells in 6-well plates) by preparing cell extracts using chemical lysis buffer (200 µl), followed by Western analysis using either an antibody against the GGCX C-terminus or an epitope appended to GGCX. Appropriate antibodies are commercially available. The clonal isolates show significant variability in expression (e.g. as shown in Fig. 6), and clones expressing r-GGCX at levels like that of endogenous GGCX in 293 cells are selected for further growth. Cells are scaled up to $\sim 10^7$ cells in duplicate 10 cm plates, which are either frozen or analyzed for expression. Analyses include a Western and an activity assay that monitors peptide carboxylation (described in the section on biochemical analysis). An

---

(r-carb$_{flag}$/pCEP4) show a wide range of expression as revealed by FACS analysis (Rishavy et al., 2019). The cells were monitored for GGCX expression (B), fIX secretion (C) and carboxylation (D) in parallel with 293 cells expressing endogenous GGCX and r-fIX (fIX 293 progenitor) (Rishavy et al., 2019). *Panels are reprinted from Supplemental Fig. 2 in Rishavy et al., J. Thromb. Haemost. 2019, 17:1053–1063, with permission from Elsevier.*

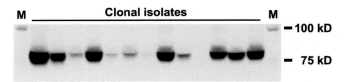

Fig. 6 A screen of stably transfected r-GGCX 293 cells lacking endogenous GGCX. Clonal isolates expressing r-GGCX are screened in a Western using an antibody against the C-terminus of GGCX. Clones with r-GGCX levels similar to that of endogenous GGCX in 293 cells are chosen for further analysis.

appropriate clone is then used in generating cells that also express a VKD protein. Cells are transfected with a vector containing cassettes that encode r-VKD protein and resistance to puromycin (e.g. pCMV6-A-Puro (Origene)). The goal is to isolate cell lines that express r-GGCX and r-VKD protein at levels which, in the case of wild type GGCX, produce fully carboxylated VKD protein. Cells are then characterized, using 2–3 clones to ensure that similar results are obtained with independent clones.

Characterization includes monitoring Glu carboxylation to Gla, and both quantitative and semi-quantitative methods have been used. Quantitative approaches require affinity purification of the protein, followed by amino acid analysis or mass spectrometry to monitor Gla production (Berkner, 1993; Hallgren, Zhang, Kinter, Willard, & Berkner, 2013). Because these methods are difficult in the case of high throughput experiments, an alternative approach has been developed using antibodies that detect carboxylation. The Western blot in Fig. 2, for example, uses a pan-specific anti-Gla antibody that was generated against a consensus sequence in hemostatic VKD proteins (Brown, Stenberg, Persson, & Stenflo, 2000), which is commercially available. Antibodies have also been developed against specific VKD proteins, i.e. the epitope is a peptide comprising the Gla domain (Schurgers et al., 2005; Viegas et al., 2014). The principle behind this approach is that carboxylation induces a conformational change in the Gla domain that elicits antibody reactivity. These antibodies are referred to as conformation specific, and a major assumption is that they recognize fully carboxylated VKD protein. However, even a few Gla residues may result in antibody recognition, and the number of Gla residues required for inducing the conformation change is unknown. The antibody approach for monitoring carboxylation is semi-quantitative, which needs to be considered when interpreting results.

Using a VKD reporter that can be assayed for activity is of value in analyzing r-GGCX function. Factor IX, for example, can be monitored in a clotting assay that monitors the ability of factor IX secreted from cells to

Vitamin K-dependent protein carboxylation

restore clotting in factor IX–deficient serum. Activity is quantitated by comparing the values to pure factor IX and plasma factor IX standards, and this assay has been described in detail in a previous Methods in Enzymology chapter (Berkner, 1993). When this assay was applied to the factor IX 293 cells that secrete fully carboxylated protein (Fig. 2), factor IX was found to have 100% activity (Berkner & McNally, 1997; Rishavy et al., 2022).

Analysis of a naturally occurring GGCX mutant that causes the PXE–like disease, V255M, illustrates the value of an activity assay in understanding r-GGCX function. Fig. 7 shows Western analysis of 293 cells lacking GGCX and coexpressing r-factor IX and either r-wild type or r-V255M GGCX. Both cell lines secrete similar levels of factor IX, however cells expressing r-V255M secrete significantly more Gla (Fig. 7A and B). The results are not due to differences in intracellular accumulation of factor IX, as similar levels are observed in cell lysates (Fig. 7C). While V255M appears to function better in carboxylating factor IX, the activity of factor IX is poor (Fig. 7D). As described in the following section, biochemical analyses revealed that V255M causes the PXE–like disease due to impaired processivity that results in partially carboxylated VKD proteins.

3. Biochemical analysis of VKD protein carboxylation

Cellular analyses can reveal changes in VKD protein carboxylation by a r-GGCX mutant but cannot explain why carboxylation is altered. Biochemical analyses are therefore essential for understanding the consequence of GGCX mutations on function. Presented below are assays we have developed that monitor the carboxylation of natural VKD substrates and mimic how carboxylation occurs in vivo. This section also gives examples of insight into GGCX function that resulted from biochemical approaches, e.g. revealing GGCX allostery and processivity, and explaining the defect in the V255M mutant that causes the PXE–like disease.

3.1 Assays to monitor GGCX catalysis

Reagents.

Core reaction cocktail: BES (pH 6.9), DTT (2.5 mM), CHAPS (0.16% w/v), NaCl (500 mM), phosphatidyl choline (0.16% w/v), vitamin K hydroquinone (KH$_2$, 200 μM; prepared as previously described (Rishavy & Berkner, 2008)) and [^{14}C]-CO$_2$ (1.3 mM; the stock solution from Perkin Elmer is 17 mM).

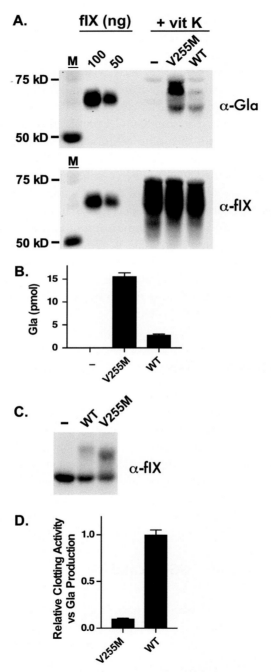

Fig. 7 Western analysis of a GGCX mutant (V255M) that causes the PXE-like disease. (A) Factor IX (fIX) secreted from cells edited to eliminate endogenous GGCX (−) and stably expressing mutant (r-V255M) or r-wild type (WT) GGCX was analyzed for

Glu-containing peptide: Phe-Leu-Glu-Glu-Leu (FLEEL, Anaspec).

EBD-containing peptide: an ~18 amino acid sequence derived from a VKD protein.

PFTE boiling chips (Saint Gobain Performance Plastics).

GGCX-VKD protein complexes: isolated from cells as described below.

The simplest GGCX assay monitors carboxylation of a small Glu-containing peptide derived from a VKD protein (e.g. FLEEL, Fig. 8A). GGCX samples are incubated at 20 °C in a final volume of 200 µl with core reaction cocktail containing FLEEL (2.5 mM) and EBD peptide (10 µM). The reaction time varies depending upon the experiment and is linear for at least 8 h. Reactions are quenched by the addition of trichloroacetic acid (1 ml, 10 %), and then kept on ice for at least one hour, and up to 24 h. The samples are centrifuged (18,000 xg 15 min, 4 °C), and 1 ml is transferred to a scintillation vial containing a boiling chip. Vials are placed on a hot plate and the samples are then boiled in a fume hood until the volume is reduced to ~100 µl. After cooling, 10 ml of scintillation fluid is added to each vial, followed by scintillation counting. A negative control with sample incubated in the absence of vitamin K is included and these values are subtracted as background counts.

The assay is rapid and convenient for monitoring GGCX editing in cells (e.g. Fig. 3C) or r-GGCX expression in cell screens. The assay has also provided mechanistic insights. For example, it revealed EBD allosteric regulation of Glu catalysis (arrow in Fig. 8A) (Knobloch & Suttie, 1987) and identified a GGCX residue (His160) essential for the coupling of vitamin K epoxidation to Glu-carboxylation that ensures efficient carboxylation (Rishavy & Berkner, 2008; Rishavy, Hallgren, & Berkner, 2011).

A second assay for catalysis monitors the carboxylation of full-length VKD protein in a complex with GGCX. The complex is generated in cells, which are cultured in the absence of vitamin K so that the VKD protein is uncarboxylated. Both mammalian cells and baculovirus-infected insect cells have been used as a source of the GGCX-VKD protein complexes. The assay monitors multiple reactions that result in full carboxylation of the Gla

carboxylation (anti-Gla) and expression (anti-fIX). (B) Quantitation of secreted fIX revealed higher levels of Gla production in the r-V255M cells. (C) Intracellular fIX expression was also analyzed. (D) The activity of factor IX secreted from cells was determined by a clotting assay. *Panels A, B and D are reprinted from Figures 4D, 4E and 4F from Rishavy et al. Blood, 2022 140:1710–1722, with permission from Elsevier.*

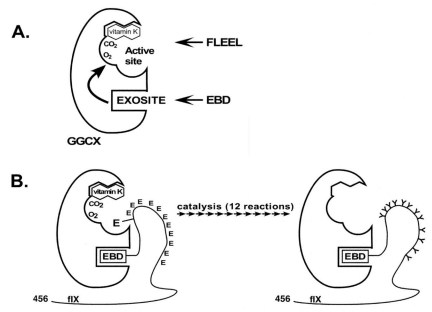

Fig. 8 Assays to monitor GGCX catalysis. (A) GGCX catalysis is monitored using peptides derived from the Gla domain (e.g. FLEEL) and exosite binding domain (EBD) of VKD proteins. (B) Complexes between GGCX and VKD proteins (e.g. factor IX (fIX)) are also used to monitor catalysis. The reaction is distinct from the assay in panel A because the EBD remains bound to GGCX throughout the multiple Glu (E) to Gla (Y) reactions.

domain (e.g. 12 for factor IX, Fig. 8B), and is distinctly different from the peptide assay: Specifically, in the peptide assay multiple carboxylations occur as the small peptides enter and exit the active site (Fig. 8A). In contrast, in the protein assay the VKD protein remains bound throughout multiple reactions that fully carboxylate the Gla domain (Fig. 8B).

The source of complexes used in the reaction is either cell extracts (i.e. lysates or microsomes prepared as previously described (Berkner, 1993)) or complex immunopurified using antibody immobilized to resin. Antibodies that react with either GGCX (or an epitope appended to GGCX) or VKD protein are used, depending upon the experiment. The anti-VKD protein antibody should recognize sequences well removed from the EBD-Gla domain region, and suitable antibodies are commercially available. GGCX-VKD protein samples are incubated in core reaction cocktail (400 μl final volume), and timed aliquots are quenched by the addition of SDS-PAGE loading buffer. The aliquots are heated at 70 °C for 10 min, and are centrifuged (860xg, 5 min, 4 °C) in the case of samples immunopurified on antibody resin. Samples are then subjected to gel electrophoresis along with [^{14}C]-standards and quantitation by PhosphorImager analysis. Conversion of Glus to Glas incorporates negative charge into VKD proteins, which in the case of factor IX causes a shift in gel migration that allows the course of the reaction to be monitored (Fig. 9A). Quantitation shows similar rates of carboxylation for complexes that are either free in cell extracts or immobilized on resin (Fig. 9B).

The rate of complex carboxylation is nonlinear (Fig. 9B), with the last few Glu residues requiring a much longer time to become carboxylated. Consequently, the time to fully carboxylate the Gla domain is very slow (e.g. 10–15 min for factor IX), which as discussed below is relevant to achieving full carboxylation in vivo. The assay has been informative in showing that full carboxylation of VKD protein is not sufficient for GGCX release, which is accelerated by the presence of exogenous VKD protein (Hallgren et al., 2002).

3.2 An assay to monitor the entire GGCX reaction

Reagents.

Core reaction cocktail: same as described in the subsection on GGCX catalysis.

VKD substrate: peptide containing sequences comprising the EBD and Gla domain.

The GGCX reaction is the sum of multiple steps, i.e. binding of VKD protein to GGCX, catalysis of multiple Glu residues, and release of the carboxylated VKD protein (Fig. 10A). The reaction is assayed using either cell extracts that contain GGCX but not VKD proteins or affinity-purified GGCX obtained using anti-GGCX or anti-epitope tag antibody. The VKD substrate is a peptide with the EBD sequence attached to all or part of the Gla domain, which can be obtained from companies able to synthesize long peptides. GGCX and VKD substrate are incubated at 20 °C in core reaction cocktail (final volume of 400 μl), and aliquots withdrawn at timed intervals are quenched with SDS-PAGE loading dye. Samples are gel electrophoresed along with [^{14}C]-standards, followed by PhosphorImager analysis that converts the signals to pmol Gla. These values are then compared to pmol of GGCX to determine the length of time for each VKD protein to undergo the entire reaction of binding, catalysis and release (Fig. 10A).

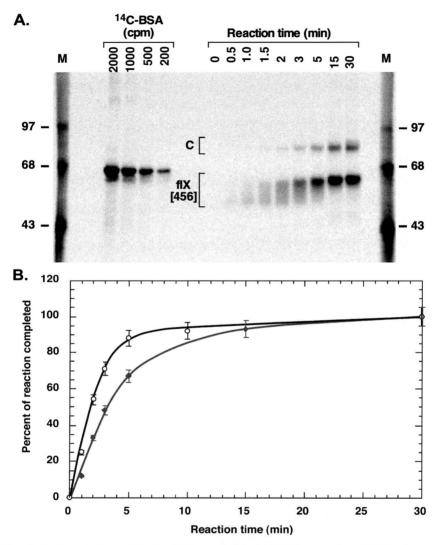

Fig. 9 Carboxylation of factor IX in a factor IX-GGCX complex. (A) The image shows [^{14}C]-CO_2 incorporation into complexes of factor IX (fIX) and GGCX (C), which is also a VKD protein that becomes carboxylated (Berkner & Pudota, 1998; Hallgren et al., 2013). (B) Quantitation is shown for fIX-GGCX complexes either free in cell extracts (open circle) or immobilized on anti-fIX resin (closed circle). *Panels are reprinted from Figs. 2 and 3 in Stenina et al. Biochemistry, 2001 40:10301–10309 with permission from the American Chemical Society.*

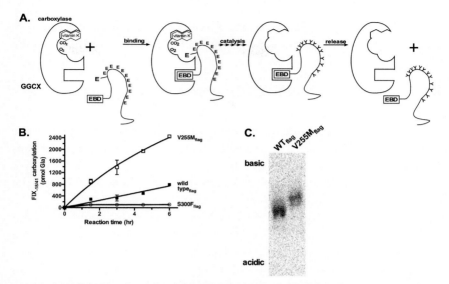

Fig. 10 **Biochemical analysis of the entire GGCX reaction.** (A) The GGCX reaction comprises binding of VKD proteins, catalytic conversion of multiple Glu (E) residues to Glas (Y), and release of carboxylated VKD product. The substrate shown (fIX$_{18/41}$) contains the factor IX (fIX) exosite binding domain (EBD) linked to the entire Gla domain. (B) FIX$_{18/41}$ is reacted with wild type and mutant carboxylases. (C) Reaction products are analyzed by isoelectric focusing. Gla generation adds negative charge to VKD proteins, making them more acidic. The V255M product is less acidic than that generated by wild type GGCX, indicating undercarboxylation. *Panels are reprinted from Figs. 2A, 2B, 2G from Rishavy et al. Blood, 2022 140:1710–1722, with permission from Elsevier.*

Release is the rate limiting step in carboxylation (Hallgren et al., 2002; Rishavy et al., 2022), and comparing the time required for the entire reaction versus catalysis is of interest for whether a VKD protein is released before full carboxylation. We found that the time for the entire reaction (i.e. binding, catalysis and release) for the VKD proteins factor IX and Matrix Gla Protein is only 5–6 fold slower than the time to catalyze all Glu residues in the Gla domains (Rishavy et al., 2022). Importantly, the relative rates of catalysis versus release could vary in vivo. For example, vitamin K controls the rate of catalysis and the levels of vitamin K are low in vivo and vary significantly with diet, which could impact full carboxylation. GGCX mutants can also impact the rates of catalysis versus the entire reaction. The V255M mutant that causes the PXE-like disease showed a faster rate for the entire reaction (Fig. 10B) and slower catalysis (not shown) than wild type GGCX (Rishavy et al., 2022). The consequence was undercarboxylation of VKD protein by V255M, as revealed by isoelectric focusing (Fig. 10C)

(Rishavy et al., 2022). Therefore, the number of VKD substrates carboxylated by V255M was higher than wild type carboxylase, but the quality of the products was poorer.

3.3 An assay to monitor GGCX processivity

Reagents.

Reaction cocktail: same as the core reaction cocktail described in the subsection on GGCX catalysis, except for the use of lower concentrations of vitamin K hydroquinone (10 μM).

GGCX-VKD protein complexes: isolated from cells as described in the subsection on catalysis.

Challenge VKD protein: a VKD substrate that is distinguishable from VKD protein in the GGCX-VKD protein complex; the substrate is either a synthetic peptide or a protein generated in cells.

GGCX processivity is monitored using a challenge assay that mimics in vivo conditions, i.e. where carboxylation of VKD protein-GGCX complexes occurs in the presence of a pool of uncarboxylated VKD substrate. VKD protein-GGCX complexes are reacted in the presence of a challenge VKD substrate that is distinguishable from VKD protein in the complex and present in large excess over the complex, and carboxylation of both VKD forms is monitored (Fig. 11A). The VKD protein-GGCX complex is generated in cells and affinity-purified using an anti-factor IX antibody immobilized to resin that reacts with factor IX in the complex but not with the challenge VKD substrate. The VKD protein-GGCX complex (~50 nM) and challenge substrate (1 μM) are incubated in reaction cocktail (final volume 350 μl) at 20 °C. Aliquots removed at timed intervals are quenched with SDS-PAGE loading dye and then subjected to SDS-PAGE electrophoresis along with [^{14}C]-standards. PhosphorImager analysis then quantitates the amount of [^{14}C]-CO_2 incorporation into challenge substrate and VKD protein.

VKD protein carboxylated by wild type GGCX is unaffected by the presence of challenge protein and becomes fully carboxylated, which established GGCX processivity (Stenina et al., 2001). Wild type GGCX carboxylates the challenge VKD protein after VKD protein in the complex (Figs. 11B and 12A). In contrast, the V255M mutant that causes the PXE-like disease carboxylates both VKD substrates at the same time (Fig. 12B), and carboxylates factor IX in the complex more poorly than wild type GGCX (Fig. 12A and B). Therefore, V255M processivity is impaired, which can account for partial VKD protein carboxylation (Fig. 10C) and

Vitamin K-dependent protein carboxylation 199

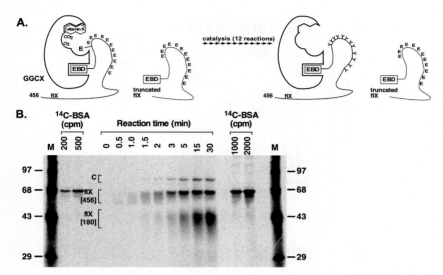

Fig. 11 Monitoring GGCX processivity. (A) A GGCX-VKD protein complex is reacted in the presence of a challenge substrate that is distinguishable from VKD protein in the complex and in excess of the complex. Glu (E) carboxylation to Gla (Y) in each VKD form is then monitored. The example shown is full length factor IX (fIX, 456 residues) attached to the exosite binding domain (EBD). The challenge substrate is EBD-factor IX truncated after amino acid 180. (B) Carboxylation of VKD protein in the complex (fIX [456]), challenge substrate (fIX[180]) and GGCX (C) are monitored by gel electrophoresis and PhosphorImager analysis. *Panel B is reprinted from Fig. 5 in Stenina et al. Biochemistry, 2001 40:10301–10309 with permission from the American Chemical Society.*

poor VKD clotting activity in cells (Fig. 7D) and explains why V255M causes the PXE-like disease in the patient (Rishavy et al., 2022). It is worth noting that an independent study on the V255M mutant that only tested VKD protein carboxylation in cells could not detect a defect in V255M (Hao et al., 2021). The methods described in this chapter that revealed V255M dysfunction illustrate the power in using both biochemical and cellular approaches to determine the function of r-GGCX variants.

4. Considerations

As covered in this chapter, both biochemical and cellular approaches are required to define GGCX function. Biochemical analyses are direct and reveal information not obtainable using the cellular approach that indirectly studies carboxylation by analyzing secreted VKD proteins. Biochemical

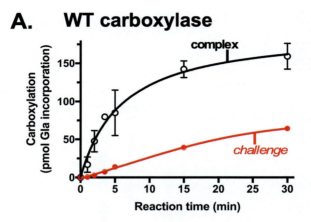
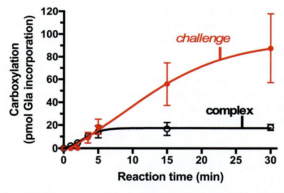

Fig. 12 The V255M GGCX mutant shows disrupted processivity. (A) Wild type (WT) and (B) V255M carboxylases were tested in the processivity assay described in Fig. 11A, which revealed impaired V255M processivity (Rishavy et al., 2022). *Panels are reprinted from Figs. 3F and 3G from Rishavy et al. Blood, 2022 140:1710–1722, with permission from Elsevier.*

analyses have revealed important insights that include defining why VKD proteins are selectively targeted for carboxylation, showing that GGCX uses an unusual mechanism in activating vitamin K to drive Glu carboxylation, and identifying functional GGCX domains that inform on why naturally occurring mutations cause disease. Biochemical analyses also revealed GGCX allostery important to efficient carboxylation and GGCX processivity that ensures full carboxylation of the multiple Glu residues in VKD proteins. This information is essential for interpreting cellular studies.

Advantages of the cellular approach are that the activities of VKD proteins can be monitored and GGCX function can be studied in the presence of other carboxylation components as well as the secretory machinery. Examples of important information revealed by cellular studies are that quality control mechanisms regulate degradation versus secretion of poorly carboxylated VKD proteins, that altering the stoichiometry of VKD proteins or GGCX disrupts carboxylation, and that VKD proteins show differential carboxylation. The use of both cellular and biochemical approaches has often been highly informative, as exemplified in this chapter with the GGCX mutant V255M where both approaches were needed to understand why the mutant causes the PXE-like disease. Another example is the impact of vitamin K reduction on GGCX activity: biochemical analyses have revealed multiple reductases that generate the reduced vitamin K GGCX cofactor and cellular studies are important in assessing their relative contribution to carboxylation. A combined approach that uses both biochemical and cellular approaches is therefore key for understanding VKD protein carboxylation.

References

Berkner, K. L. (1993). Expression of recombinant vitamin K-dependent proteins in mammalian cells: Factors IX and VII. *Methods in Enzymology, 222*, 450–477.

Berkner, K. L., & McNally, B. A. (1997). Purification of vitamin K-dependent carboxylase from cultured cells. *Methods in Enzymology, 282*, 313–333.

Berkner, K. L., & Pudota, B. N. (1998). Vitamin K-dependent carboxylation of the carboxylase. *Proceedings of the National Academy of Sciences of the United States of America, 95*(2), 466–471. Retrieved from: http://www.ncbi.nlm.nih.gov/cgi-bin/Entrez/referer?http://www.pnas.org/cgi/content/full/95/2/466.

Berkner, K. L., & Runge, K. W. (2004). The physiology of vitamin K nutriture and vitamin K-dependent protein function in atherosclerosis. *Journal of Thrombosis and Haemostasis: JTH, 2*(12), 2118–2132. Retrieved from: http://www.ncbi.nlm.nih.gov/entrez/query.fcgi?cmd=Retrieve&db=PubMed&dopt=Citation&list_uids=15613016.

Berkner, K. L., & Runge, K. W. (2022). Vitamin K-dependent protein activation: Normal gamma-glutamyl carboxylation and disruption in disease. *International Journal of Molecular Sciences, 23*(10), https://doi.org/10.3390/ijms23105759.

Brown, M. A., Stenberg, L. M., Persson, U., & Stenflo, J. (2000). Identification and purification of vitamin K-dependent proteins and peptides with monoclonal antibodies specific for gamma -carboxyglutamyl (Gla) residues. *The Journal of Biological Chemistry, 275*(26), 19795–19802. Retrieved from: http://www.ncbi.nlm.nih.gov/cgi-bin/Entrez/referer?http://www.jbc.org/cgi/content/full/275/26/19795.

Caspers, M., Czogalla, K. J., Liphardt, K., Muller, J., Westhofen, P., Watzka, M., & Oldenburg, J. (2015). Two enzymes catalyze vitamin K 2,3-epoxide reductase activity in mouse: VKORC1 is highly expressed in exocrine tissues while VKORC1L1 is highly expressed in brain. *Thrombosis Research, 135*(5), 977–983. https://doi.org/10.1016/j.thromres.2015.01.025.

Darghouth, D., Hallgren, K. W., Issertial, O., Bazaa, A., Berkner, K. L., Rosa, J.-P., & Favier, R. (2009). Compound heterozygosity of a W493C substitution and R704/ premature stop codon within the {gamma}-glutamyl carboxylase in combined vitamin K-dependent coagulation factor deficiency in a french family. *Blood (ASH Annual Meeting Abstracts), 114*(22), 1302. Retrieved from: http://abstracts.hematologylibrary. org/cgi/content/abstract/ashmtg;114/22/1302.

Darghouth, D., Hallgren, K. W., Shtofman, R. L., Mrad, A., Gharbi, Y., Maherzi, A., ... Rosa, J.-P. (2006). Rosa Compound heterozygosity of novel missense mutations in the gamma-carboxylase gene causes hereditary combined vitamin K-dependent coagulation factor deficiency. *Blood, 108*(6), 1925–1931. Retrieved from: http://www.ncbi.nlm.nih.gov/entrez/ query.fcgi?cmd=Retrieve&db=PubMed&dopt=Citation&list_uids=16720838.

De Vilder, E. Y., Debacker, J., & Vanakker, O. M. (2017). GGCX-associated phenotypes: An overview in search of genotype-phenotype correlations. *International Journal of Molecular Sciences, 18*(2), 34. https://doi.org/10.3390/ijms18020240.

Doudna, J. A., & Charpentier, E. (2014). Genome editing. The new frontier of genome engineering with CRISPR-Cas9. *Science (New York, N. Y.), 346*(6213), 1258096. https://doi.org/10.1126/science.1258096.

Ghosh, S., Kraus, K., Biswas, A., Muller, J., Buhl, A. L., Forin, F., ... Oldenburg, J. (2021). GGCX mutations show different responses to vitamin K thereby determining the severity of the hemorrhagic phenotype in VKCFD1 patients. *Journal of Thrombosis and Haemostasis: JTH, 19*(6), 1412–1424. https://doi.org/10.1111/jth.15238.

Ghosh, S., Kraus, K., Biswas, A., Muller, J., Forin, F., Singer, H., ... Czogalla-Nitsche, K. J. (2022). GGCX variants leading to biallelic deficiency to gamma-carboxylate GRP cause skin laxity in VKCFD1 patients. *Human Mutation, 43*(1), 42–55. https://doi.org/10. 1002/humu.24300.

Gibson, D. G., Young, L., Chuang, R. Y., Venter, J. C., Hutchison, C. A., 3rd, & Smith, H. O. (2009). Enzymatic assembly of DNA molecules up to several hundred kilobases. *Nature Methods, 6*(5), 343–345. https://doi.org/10.1038/nmeth.1318.

Hallgren, K. W., Hommema, E. L., McNally, B. A., & Berkner, K. L. (2002). Carboxylase overexpression impairs factor IX secretion: Implications for the release of vitamin K-dependent proteins. *Biochemistry, 41*(50), 15045–15055. Retrieved from: http://www.ncbi.nlm.nih.gov/ entrez/query.fcgi?cmd=Retrieve&db=PubMed&dopt=Citation&list_uids=12475254.

Hallgren, K. W., Qian, W., Yakubenko, A. V., Runge, K. W., & Berkner, K. L. (2006). r-VKORC1 expression in factor IX BHK cells increases the extent of factor IX carboxylation but is limited by saturation of another carboxylation component or by a shift in the rate-limiting step. *Biochemistry, 45*(17), 5587–5598. Retrieved from: http://www.ncbi.nlm.nih. gov/entrez/query.fcgi?cmd=Retrieve&db=PubMed&dopt=Citation&list_uids=16634640.

Hallgren, K. W., Zhang, D., Kinter, M., Willard, B., & Berkner, K. L. (2013). Methylation of gamma-carboxylated Glu (Gla) allows detection by liquid chromatography-mass spectrometry and the identification of Gla residues in the gamma-glutamyl carboxylase. *Journal of Proteome Research, 12*(6), 2365–2374. https://doi.org/10.1021/pr3003722.

Hammed, A., Matagrin, B., Spohn, G., Prouillac, C., Benoit, E., & Lattard, V. (2013). VKORC1L1, an enzyme rescuing the vitamin K 2,3-epoxide reductase activity in some extrahepatic tissues during anticoagulation therapy. *The Journal of Biological Chemistry, 288*(40), 28733–28742. https://doi.org/10.1074/jbc.M113.457119.

Hao, Z., Jin, D., Chen, X., Schurgers, L. J., Stafford, D. W., & Tie, J. K. (2021). Gamma-glutamyl carboxylase mutations differentially affect the biological function of vitamin K-dependent proteins. *Blood, 137*(4), 533–543. https://doi.org/10.1182/blood.2020006329.

Jin, D. Y., Chen, X., Liu, Y., Williams, C. M., Pedersen, L. C., Stafford, D. W., & Tie, J. K. (2023). A genome-wide CRISPR-Cas9 knockout screen identifies FSP1 as the warfarin-resistant vitamin K reductase. *Nature Communications, 14*(1), 828. https://doi. org/10.1038/s41467-023-36446-8.

Kaufman, R. J., Wasley, L. C., Furie, B. C., Furie, B., & Shoemaker, C. B. (1986). Expression, purification, and characterization of recombinant gamma-carboxylated factor IX synthesized in Chinese hamster ovary cells. *The Journal of Biological Chemistry, 261*(21), 9622–9628.

Knobloch, J. E., & Suttie, J. W. (1987). Vitamin K-dependent carboxylase. Control of enzyme activity by the "propeptide": Region of factor X. *The Journal of Biological Chemistry, 262*(32), 15334–15337.

Lacombe, J., Rishavy, M. A., Berkner, K. L., & Ferron, M. (2018). VKOR paralog VKORC1L1 supports vitamin K-dependent protein carboxylation in vivo. *JCI Insight, 3*(1), https://doi.org/10.1172/jci.insight.96501.

Li, Q., Grange, D. K., Armstrong, N. L., Whelan, A. J., Hurley, M. Y., Rishavy, M. A., ... Uitto, J. (2009). Mutations in the GGCX and ABCC6 genes in a family with pseudoxanthoma elasticum-like phenotypes. *Journal of Investigative Dermatology, 129*(3), 553–563. Retrieved from: http://www.ncbi.nlm.nih.gov/entrez/query.fcgi?cmd=Retrieve&db=PubMed&dopt=Citation&list_uids=18800149.

Li, Q., Jiang, Q., Pfendner, E., Varadi, A., & Uitto, J. (2009). Pseudoxanthoma elasticum: Clinical phenotypes, molecular genetics and putative pathomechanisms. *Experimental Dermatology, 18*(1), 1–11. https://doi.org/10.1111/j.1600-0625.2008.00795.x.

Li, Q., Schurgers, L. J., Smith, A. C., Tsokos, M., Uitto, J., & Cowen, E. W. (2009). Co-existent pseudoxanthoma elasticum and vitamin K-dependent coagulation factor deficiency: Compound heterozygosity for mutations in the GGCX gene. *The American Journal of Pathology, 174*(2), 534–540. https://doi.org/10.2353/ajpath.2009.080865.

Mishima, E., Ito, J., Wu, Z., Nakamura, T., Wahida, A., Doll, S., ... Conrad, M. (2022). A non-canonical vitamin K cycle is a potent ferroptosis suppressor. *Nature, 608*(7924), 778–783. https://doi.org/10.1038/s41586-022-05022-3.

Mutucumarana, V. P., Stafford, D. W., Stanley, T. B., Jin, D. Y., Solera, J., Brenner, B., ... Wu, S. M. (2000). Expression and characterization of the naturally occurring mutation L394R in human gamma-glutamyl carboxylase. *Journal of Biological Chemistry, 275*(42), 32572–32577. Retrieved from: http://www.ncbi.nlm.nih.gov/cgi-bin/Entrez/referer?http://www.ncbi.nlm.nih.gov/htbin-post/Omim/getmim%3ffield=medline_uid&search=10934213.

Rabe, B. A., & Cepko, C. (2020). A simple enhancement for gibson isothermal assembly. 2020.2006.2014.150979 *bioRxiv*. https://doi.org/10.1101/2020.06.14.150979.

Rishavy, M. A., & Berkner, K. L. (2008). Insight into the coupling mechanism of the vitamin K-dependent carboxylase: mutation of histidine 160 disrupts glutamic acid carbanion formation and efficient coupling of vitamin K epoxidation to glutamic acid carboxylation. *Biochemistry, 47*(37), 9836–9846. Retrieved from: http://www.ncbi.nlm.nih.gov/entrez/query.fcgi?cmd=Retrieve&db=PubMed&dopt=Citation&list_uids=18717596.

Rishavy, M. A., Hallgren, K. W., & Berkner, K. L. (2011). The vitamin K-dependent carboxylase generates gamma-carboxylated glutamates by using CO_2 to facilitate glutamate deprotonation in a concerted mechanism that drives catalysis. *The Journal of Biological Chemistry, 286*(52), 44821–44832. https://doi.org/10.1074/jbc.M111.249177.

Rishavy, M. A., Hallgren, K. W., Wilson, L. A., Hiznay, J. M., Runge, K. W., & Berkner, K. L. (2022). GGCX mutants that impair hemostasis reveal the importance of processivity and full carboxylation to VKD protein function. *Blood, 140*(15), 1710–1722. https://doi.org/10.1182/blood.2021014275.

Rishavy, M. A., Hallgren, K. W., Wilson, L. A., Usubalieva, A., Runge, K. W., & Berkner, K. L. (2013). The vitamin K oxidoreductase is a multimer that efficiently reduces vitamin K epoxide to hydroquinone to allow vitamin K-dependent protein carboxylation. *The Journal of Biological Chemistry, 288*(44), 31556–31566. https://doi.org/10.1074/jbc.M113.497297.

Rishavy, M. A., Hallgren, K. W., Zhang, H., Runge, K. W., & Berkner, K. L. (2019). Exon 2 skipping eliminates gamma-glutamyl carboxylase activity, indicating a partial splicing defect in a patient with vitamin K clotting factor deficiency. *Journal of Thrombosis and Haemostasis: JTH, 17*(7), 1053–1063. https://doi.org/10.1111/jth.14456.

Roth, D. A., Rehemtulla, A., Kaufman, R. J., Walsh, C. T., Furie, B., & Furie, B. C. (1993). Expression of bovine vitamin K-dependent carboxylase activity in baculovirus-infected insect cells. *Proceedings of the National Academy of Sciences of the United States of America, 90*(18), 8372–8376.

Schurgers, L. J., Teunissen, K. J., Knapen, M. H., Kwaijtaal, M., van Diest, R., Appels, A., ... Vermeer, C. (2005). Novel conformation-specific antibodies against matrix gamma-carboxyglutamic acid (Gla) protein: Undercarboxylated matrix Gla protein as marker for vascular calcification. *Arteriosclerosis, Thrombosis, and Vascular Biology, 25*(8), 1629–1633. https://doi.org/10.1161/01.ATV.0000173313.46222.43.

Shearer, M. J. (2008). Newman Metabolism and cell biology of vitamin K. *Thrombosis and Haemostasis, 100*(4), 530–547. Retrieved from: https://www.ncbi.nlm.nih.gov/pubmed/18841274.

Shearer, M. J., & Newman, P. (2014). Recent trends in the metabolism and cell biology of vitamin K with special reference to vitamin K cycling and MK-4 biosynthesis. *Journal of Lipid Research, 55*(3), 345–362. https://doi.org/10.1194/jlr.R045559.

Souri, M., Iwata, H., Zhang, W. G., & Ichinose, A. (2009). Unique secretion mode of human protein Z: Its Gla domain is responsible for inefficient, vitamin K–dependent and warfarin-sensitive secretion. *Blood, 113*(16), 3857–3864. https://doi.org/10.1182/blood-2008-07-171884.

Spohn, G., Kleinridders, A., Wunderlich, F. T., Watzka, M., Zaucke, F., Blumbach, K., ... Oldenburg, J. (2009). VKORC1 deficiency in mice causes early postnatal lethality due to severe bleeding. *Thrombosis and Haemostasis, 101*(6), 1044–1050. Retrieved from: http://www.ncbi.nlm.nih.gov/pubmed/19492146.

Stenina, O., Pudota, B. N., McNally, B. A., Hommema, E. L., & Berkner, K. L. (2001). Tethered processivity of the vitamin K-dependent carboxylase: Factor IX is efficiently modified in a mechanism which distinguishes Gla's from Glu's and which accounts for comprehensive carboxylation in vivo. *Biochemistry, 40*(34), 10301–10309.

Tie, J. K., Carneiro, J. D., Jin, D. Y., Martinhago, C. D., Vermeer, C., & Stafford, D. W. (2016). Characterization of vitamin K-dependent carboxylase mutations that cause bleeding and nonbleeding disorders. *Blood, 127*(15), 1847–1855. https://doi.org/10.1182/blood-2015-10-677633.

Tokunaga, F., Wakabayashi, S., & Koide, T. (1995). Warfarin causes the degradation of protein C precursor in the endoplasmic reticulum. *Biochemistry, 34*(4), 1163–1170. Retrieved from: http://www.ncbi.nlm.nih.gov/pubmed/7827066.

Vanakker, O. M., Martin, L., Gheduzzi, D., Leroy, B. P., Loeys, B. L., Guerci, V. I., ... De Paepe, A. (2007). Pseudoxanthoma elasticum-like phenotype with cutis laxa and multiple coagulation factor deficiency represents a separate genetic entity. *The Journal of Investigative Dermatology, 127*(3), 581–587. Retrieved from: http://www.ncbi.nlm.nih.gov/entrez/query.fcgi?cmd=Retrieve&db=PubMed&dopt=Citation&list_uids=17110937.

Viegas, C. S., Herfs, M., Rafael, M. S., Enriquez, J. L., Teixeira, A., Luis, I. M., ... Simes, D. C. (2014). Gla-rich protein is a potential new vitamin K target in cancer: Evidences for a direct GRP-mineral interaction. *BioMed Research International, 2014*, 340216. https://doi.org/10.1155/2014/340216.

Watzka, M., Geisen, C., Scheer, M., Wieland, R., Wiegering, V., Dorner, T., ... Oldenburg, J. (2014). Bleeding and non-bleeding phenotypes in patients with GGCX gene mutations. *Thrombosis Research, 134*(4), 856–865. https://doi.org/10.1016/j.thromres.2014.07.004.

Wu, S. M., Cheung, W. F., Frazier, D., & Stafford, D. W. (1991). Cloning and expression of the cDNA for human gamma-glutamyl carboxylase. *Science (New York, N. Y.)*, *254*(5038), 1634–1636.

Wu, W., Bancroft, J. D., & Suttie, J. W. (1997). Structural features of the kringle domain determine the intracellular degradation of under-g-carboxylated prothrombin: Studies of chimeric rat/human prothrombin. *Proceedings of the National Academy of Sciences of the United States of America*, *94*(25), 13654–13660. Retrieved from: http://www.ncbi.nlm.nih.gov/cgi-bin/Entrez/referer?http://www.pnas.org/cgi/content/full/94/25/13654.

Yan, S. C., Razzano, P., Chao, Y. B., Walls, J. D., Berg, D. T., McClure, D. B., & Grinnell, B. W. (1990). Characterization and novel purification of recombinant human protein C from three mammalian cell lines. *Biotechnology (N. Y.)*, *8*(7), 655–661.

Zhang, B., & Ginsburg, D. (2004). Familial multiple coagulation factor deficiencies: New biologic insight from rare genetic bleeding disorders. *Journal of Thrombosis and Haemostasis: JTH*, *2*(9), 1564–1572. https://doi.org/10.1111/j.1538-7836.2004.00857.x.

Zhu, A., Sun, H., Raymond, R. M., Jr., Furie, B. C., Furie, B., Bronstein, M., ... Ginsburg, D. (2007). Fatal hemorrhage in mice lacking {gamma}-glutamyl carboxylase. *Blood*, *109*(12), 5270–5275. Retrieved from: http://www.ncbi.nlm.nih.gov/entrez/query.fcgi?cmd=Retrieve&db=PubMed&dopt=Citation&list_uids=17327402.

CHAPTER NINE

Assessment of gamma-glutamyl carboxylase activity in its native milieu

Xuejie Chen, Darrel W. Stafford, and Jian-Ke Tie*

Department of Biology, University of North Carolina at Chapel Hill, Chapel Hill, NC, United States
*Corresponding author. e-mail address: jktie@email.unc.edu

Contents

1. Introduction	208
2. Cell-based assessment of VKD carboxylation activity using ELISA	211
2.1 Materials and equipment	213
2.2 Procedure	213
2.3 Notes	218
3. Functional study of naturally occurring GGCX mutations in the cellular milieu	219
3.1 Materials and equipment	221
3.2 Procedure	221
3.3 Notes	226
4. Assessment of vitamin K epoxidase activity in the cellular milieu	227
4.1 Materials and equipment	228
4.2 Procedure	228
4.3 Notes	231
5. Summary and conclusions	231
Acknowledgments	232
References	232

Abstract

Gamma-glutamyl carboxylase (GGCX), a polytopic membrane protein found in the endoplasmic reticulum, catalyzes the posttranslational modification of a variety of vitamin K-dependent (VKD) proteins to their functional forms. GGCX uses the free energy from the oxygenation of reduced vitamin K to remove the proton from the glutamate residue to drive VKD carboxylation. During the process of carboxylation, reduced vitamin K is oxidized to vitamin K epoxide. Therefore, GGCX is a dual-function enzyme that possesses both glutamate carboxylation and vitamin K epoxidation activities. Genetic variations in GGCX are mainly associated with bleeding disorders referred to as combined VKD coagulation factors deficiency. Comorbid non-bleeding phenotypes are also observed in patients carrying GGCX mutations. Our current knowledge concerning GGCX's function has been obtained mainly from *in vitro* experimentation under artificial conditions, which limits its use in interpreting the

---

Methods in Enzymology, Volume 708
ISSN 0076-6879, https://doi.org/10.1016/bs.mie.2024.10.011
Copyright © 2024 Elsevier Inc. All rights are reserved, including those for text and data mining, AI training, and similar technologies.

clinical phenotypes associated with GGCX genotypes. In this chapter, we describe the background, establishment, and application of mammalian cell-based assays for both the carboxylation and epoxidation activities of GGCX. We provide detailed procedures for making the reporter cell lines, creating CRISPR-Cas9-mediated gene-knockout reporter cell lines, and using these cell lines for functional studies of GGCX and its naturally occurring mutations. Combined with different reporter proteins, this cell-based strategy has been successfully used for the functional study of vitamin K-related enzymes, high-throughput screening of VKD carboxylation inhibitors, and genome-wide CRISPR-Cas9 knockout library screening of the unknown enzymes associated with vitamin K reduction.

1. Introduction

Gamma–glutamyl carboxylase (GGCX, EC 4.1.1.90), also known as vitamin K-dependent (VKD) carboxylase, is an integral membrane glycoprotein that resides in the endoplasmic reticulum (Carlisle & Suttie, 1980). GGCX catalyzes the posttranslational modification of specific glutamate (Glu) residues to gamma–carboxyglutamate (Gla) in VKD proteins using reduced vitamin K (vitamin K hydroquinone, KH_2) as a cofactor. Carboxylation is required for the biological functions of numerous VKD proteins involved in a broad range of physiological processes including blood coagulation, vascular calcification, bone metabolism, signal transduction, and cancer cell proliferation (Chatrou, Reutelingsperger, & Schurgers, 2011; Kaesler, Schurgers, & Floege, 2021; G. Wu et al., 2018). The enzymatic activity of GGCX was first discovered in the 1970s when studying the biosynthesis of coagulation factor prothrombin (Girardot, Delaney, & Johnson, 1974; Shah & Suttie, 1974). It was found that radioactive $^{14}CO_2$ was incorporated into prothrombin within two hours in rats and that the amount of $^{14}CO_2$ incorporation was increased three to four-fold by the administration of vitamin K. It was concluded that prothrombin biosynthesis involves the carboxylation of Glu residues in prothrombin precursor by a membrane-bound enzyme. The identification of the gene encoding GGCX (Wu, Cheung, Frazier, & Stafford, 1991) and purification of GGCX protein to near homogeneity (Wu, Morris, & Stafford, 1991) make it possible to study the function of GGCX at the molecular level.

During the process of VKD carboxylation, the γ-proton of the glutamic acid is abstracted by a strong base *via* a "base strength amplification" pathway (Dowd, Hershline, Ham, & Naganathan, 1995) followed by the

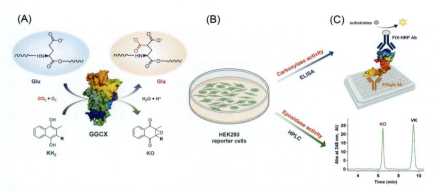

Fig. 1 **Schematic diagram of the cell-based GGCX activity assay.** (A) GGCX catalyzes the posttranslational modification of specific glutamate (Glu) residues to gamma-carboxyglutamate (Gla) residues in vitamin K-dependent proteins and simultaneously converts reduced vitamin K (KH_2) to vitamin K epoxide (KO). (B) The HEK293 reporter cells are incubated with vitamin K or other conditions before functional analysis. (C) The carboxylase and epoxidase activity of GGCX are measured from cell culture medium and cell pellet by ELISA and HPLC, respectively.

addition of carbon dioxide. In this "base strength amplification" mechanism, GGCX uses the free energy of KH_2 oxygenation to transform a weak base to a strong base to remove the proton from the Glu residue at the gamma-carbon and thus drive carboxylation. Concomitant with carboxylation is the oxidation of KH_2 to vitamin K epoxide (KO) (Suttie, Geweke, Martin, & Willingham, 1980; Willingham & Matschiner, 1974). Under normal conditions, for each molecule of Glu carboxylated, one molecule of KH_2 is oxidized to KO (Larson, Friedman, & Suttie, 1981; Wood & Suttie, 1988). The formation of KO during the carboxylation reaction has been called an epoxidation reaction. Therefore, GGCX is a dual-function enzyme that possesses both glutamate carboxylation and vitamin K epoxidation activities (Fig. 1).

The enzymatic activity of GGCX was originally evaluated *in vitro* by measuring the incorporation of radioactive $^{14}CO_2$ into the endogenous microsomal protein substrate, prothrombin precursor (Esmon, Sadowski, & Suttie, 1975). As crude microsomal extracts from normal animals contain a small amount of endogenous protein substrate, treating animals with warfarin or feeding them vitamin K-deficient diets was shown to be beneficial in obtaining substrate-enriched microsomal preparations for GGCX function studies (Vermeer, Soute, De Metz, & Hemker, 1982). The use of the synthetic short peptide rather than the endogenous protein precursor as the substrate was a milestone for GGCX *in vitro* activity assay.

Suttie and Hageman reported that the pentapeptide FLEEV (corresponding to residues 5–9 of bovine prothrombin) has identical conditions for carboxylation as those for prothrombin (Suttie & Hageman, 1976). These authors also compared the carboxylation of a variety of short synthetic peptide substrates with the endogenous protein precursors (Rich, Lehrman, Kawai, Goodman, & Suttie, 1981; Suttie, Lehrman, Geweke, Hageman, & Rich, 1979). In general, none of the synthetic substrates were carboxylated as effectively as the endogenous protein substrates, and the apparent Km value for the peptide substrate was found to be in the millimolar range.

As GGCX interacts with coagulation factor precursors mainly through an 18-amino acid propeptide which significantly increases small peptide substrate carboxylation (Knobloch & Suttie, 1987), Ulrich et al. used a 28-residue peptide based on residues −18 to + 10 in prothrombin (proPT28) as the peptide substrate of GGCX (Ulrich, Furie, Jacobs, Vermeer, & Furie, 1988). This 28-residue peptide was efficiently carboxylated with a Km value that was three orders of magnitude lower than that of the penta-peptide FLEEL. Additionally, a 59-residue peptide containing the pro-peptide sequence and the whole Gla domain of factor IX (proFIX59) was also proved to be an efficient substrate for *in vitro* carboxylation study with an apparent Km value (0.55 μM) approximately five times lower than that of proPT28 (Wu, Soute, Vermeer, & Stafford, 1990). Furthermore, it has been shown that decarboxylated bone Gla protein and matrix Gla protein are also good substrates for GGCX function studies (Engelke, Hale, Suttie, & Price, 1991; Vermeer, Soute, Hendrix, & de Boer-van den Berg, 1984). Nevertheless, these assays are based on measuring the incorporation of radioactive $^{14}CO_2$ into different peptide substrates using chemically reduced KH_2 as the cofactor. It is worth noting that Kaesler et al. developed a nonradioactive assay for the GGCX function study that used a fluorescein isothiocyanate-labeled hexapeptide FLEELK that can be readily separated and detected in its unmodified and carboxylated form by reversed-phase high-performance liquid chromatography (HPLC) with fluorescence detection (Kaesler et al., 2012).

The epoxidase activity of GGCX was originally assessed by measuring the formation of radioactive KO in the carboxylation reaction using silica-based thin-layer chromatography to separate the reactant and product (Sadowski, Schnoes, & Suttie, 1977; Willingham & Matschiner, 1974). It was later replaced by using non-labeled vitamin K in the carboxylation reaction, and K vitamins in the reaction mixture were separated by HPLC using a Permaphase ODS column and methanol-methyl cyanide-water (2: 2: l, v/v) as the eluting solvent (Elliott, Odam, & Townsend, 1976; Elliott, Townsend, & Odam, 1980). The K vitamins eluted from the column were quantitated by UV absorbance at 254 nm. This approach has since been used as a standard epoxidase activity assay aside from minor modifications on sample extraction, reversed-phase HPLC column, and the mobile phase composition.

Despite significant progress in our understanding of how GGCX modifies its substrates, our knowledge concerning GGCX's function has been obtained mainly from *in vitro* experimentation under artificial conditions. Thus, we lack an understanding of how GGCX carboxylates the natural VKD protein substrates in the native milieu and how the naturally occurring GGCX mutations differentially affect the functions of various VKD proteins leading to distinct clinical phenotypes. In the past decades, we have developed several cell-based systems to study the functions of vitamin K cycle enzymes in their native milieu (Tie & Stafford, 2017). In this chapter, we describe the detailed procedure of the cell-based assays for determining GGCX's carboxylation and epoxidation activities.

2. Cell-based assessment of VKD carboxylation activity using ELISA

As VKD carboxylation was first discovered in the biosynthesis of coagulation factors, most of the early cell-based studies on GGCX were focused on the carboxylation and/or production of functional clotting factors. For example, to test the propeptide's contribution to GGCX binding and VKD protein carboxylation in the cellular environment, Galeffi and Brownlee stably expressed wild-type FIX and its propeptide mutant in dog kidney cells (Galeffi & Brownlee, 1987). Carboxylated FIX secreted in the cell culture medium was enriched by barium citrate precipitation and quantitated by ELISA. Results from this study support the hypothesis that the propeptide region of the clotting factor is required for its carboxylation, and the arginine at position −4 is required for the correct processing of the propeptide. By sequential immunoaffinity purification and characterization of purified FIX secreted from Chinese hamster ovary (CHO) cells stably expressing FIX, Bristol et al. reported that fully carboxylated profactor IX is biologically inactive unless its propeptide is removed (Bristol, Freedman, Furie, & Furie, 1994). Using the same approach, it has been shown that when two different VKD proteins (FIX and protein C) were co-expressed in the same cell line (human embryonic kidney 293, HEK293), both proteins could be completely carboxylated (Lingenfelter & Berkner, 1996). Additionally, by characterizing purified prothrombin secreted from CHO cells expressing different variants of prothrombin, Furie et al. reported that the propeptide of prothrombin is sufficient to direct VKD carboxylation on the adjacent Glu–rich region (Furie et al., 1997). Furthermore, Camire et al. purified and characterized factor X secreted from HEK293 cells that stably express wild–type factor X and chimeric factor X with its signal peptide and propeptide replaced by that of prothrombin (Camire, Larson, Stafford, & High, 2000). Results from this study suggest that the production of functional carboxylated factor X was significantly improved when a weaker propeptide was attached to factor X.

These cell–based studies provide invaluable information on VKD carboxylation in the cellular environment. However, significant work was involved in each of these approaches, which includes establishing a stable cell line for expressing the VKD protein, collecting a large amount of cell culture medium for protein purification, and characterizing the carboxylation status of the purified proteins. Therefore, these approaches are not compatible with routine GGCX functional analysis for a relatively large number of samples. To establish an efficient and easy-to-use cell–based approach for studying VKD carboxylation, we created a reporter protein by harnessing the commercially available monoclonal antibody that specifically recognizes fully carboxylated FIX and the characteristics of endoplasmic reticulum-associated degradation of uncarboxylated protein C. We stably expressed protein C with its Gla domain replaced by that of FIX (FIXgla-PC) as the reporter protein in HEK293 cells and directly measured the carboxylation efficiency of the reporter protein in cell culture medium using a conventional sandwich-based ELISA (Tie, Jin, Straight, & Stafford, 2011). The linear response of the carboxylated reporter protein (FIXgla-PC) in the ELISA assay spans about three orders of magnitude (0.24 to 125 ng/mL) (Tie & Stafford, 2017). When the 96-well plate is used in cell culture, this assay can easily handle large numbers of samples, and the final ELISA assay can be completed within one day. For example, we have successfully adapted this approach for high-throughput screening of a drug library to identify drugs that cause bleeding disorders by off-targeting the vitamin K cycle (Chen et al., 2020). With different VKD proteins as the reporter protein, this cell-based assay has been extended to study hepatic and extrahepatic VKD carboxylation (Hao et al., 2021). Additionally, we have successfully adapted this cell-based system for genome-wide CRISPR-Cas9 knockout screening for the long-sought-after enzyme – warfarin-resistant vitamin K reductase

(Jin et al., 2023). It is worth noting that other groups have also extended this cell-based assay to study VKD carboxylation using different reporter proteins with various signal readout approaches (Fregin et al., 2013; Ghosh et al., 2021; Haque, McDonald, Kulman, & Rettie, 2014). The following section details the process of the cell-based assay for VKD carboxylation using FIX as a reporter protein.

2.1 Materials and equipment

- phCMV1 mammalian expression vector (AMSBIO, Catalog #P003100).
- Human embryonic kidney 293 (HEK293) cell line (ATCC, Catalog #CRL-1573).
- 1 × DPBS (Dulbecco's phosphate-buffered saline) (Gibco, Catalog #14190-144).
- DMEM/F-12 cell culture medium (Dulbecco's Modified Eagle Medium/Nutrient Mixture F-12) (Gibco, Catalog #11330-032).
- Fetal bovine serum (FBS) (Sigma–Aldrich, Catalog #F2442).
- Penicillin streptomycin (Corning, Catalog #30-002 CL).
- Xfect transfection reagent (Takara Bio, Catalog #631317).
- Geneticin™ (G418) (Gibco, Catalog # 10131035).
- Vitamin K (10 mg/mL, injectable emulsion) (Hospira, Catalog #NDC 0409-9158).
- Recombinant coagulation factor IX (Pfizer, BeneFIX).
- Mouse anti-carboxylated FIX gla domain monoclonal antibody (Green Mountain Antibodies, Catalog #GMA-001).
- Affinity-purified sheep anti-FIX polyclonal antibody (Affinity Biologicals, Catalog #SAFIX-AP).
- HRP conjugated goat anti-FIX polyclonal antibody (Affinity Biologicals, Catalog #GAFIX-APHRP).
- ABTS 1-component peroxidase substrate kit (Sera Care, Catalog # 5120-0043).
- UV–visible spectrophotometer (Molecular Devices, SpectraMax ABS Plus).
- 96-well microplate, high-binding (Greiner Bio-One, Catalog #655081).
- Forma Scientific CO_2 water jacketed incubator (ThermoFisher Scientific).
- GraphPad Prism Software v10.

2.2 Procedure

2.2.1 Construction of factor IX expression vector

- Clone human FIX cDNA into phCMV1 mammalian expression vector following the standard molecular cloning approach (see note 1).

2.2.2 Dose-response curve (kill curve) titration of G418 for HEK293 cells

- HEK 293 cells are maintained in complete growth medium (DMEM/F-12 medium with 10 % FBS and 1 × penicillin streptomycin) at 37 °C with 5 % CO_2.
- Seed cells into a 12-well plate so that cells will reach approximately 25 % confluency on the day of treatment.
- The next day, replace the culture medium with a medium containing varying concentrations of G418 (0, 50, 100, 200, 400, 600, 800 µg/mL).
- Replace the G418 selective media every 3-4 days and observe cell growth under an inverted microscope every other day.
- Determine the lowest concentration of G418 that begins to give massive cell death in 5 days (kills > 60 % cells) and kills all cells within 14 days.
- Use the selected lowest G418 concentration (in our case, 400 µg/mL) for making the stably transfected cell line.

2.2.3 Creation of the stable HEK293-FIX reporter cell line

- One day before transfection, seed HEK 293 cells into a 12-well plate in 1 mL complete growth medium so that the cells will reach 50-70 % confluency at the time of transfection.
- Dilute 1 µg FIX containing plasmid DNA with the Xfect reaction buffer to a final volume of 50 µL. Mix well by vortexing for 5 s at high speed (see note 2).
- Add 0.3 µL Xfect transfection reagent to the diluted plasmid DNA solution and mix well by vortexing for 10 s at high speed.
- Incubate for 10 min at room temperature to allow the formation of nanoparticle complexes.
- Add 0.5 mL complete growth medium to the above nanoparticle complex solution. Mix well by repeatedly pipetting the solution up and down for several times.
- Gently replace the 12-well plate medium with the above nanoparticle complex-containing medium.
- Incubate the plate at 37 °C with 5 % CO_2 for 4 h.
- Add another 0.5 mL fresh complete growth medium and incubate the plate at 37 °C with 5 % CO_2 overnight.
- The following day, detach the transfected cell by trypsin digestion, seed the trypsinized cells into 10 cm plates at 100-fold, 1000-fold, and 10,000-fold dilutions, and culture the cells with medium containing G418 at the pre-determined concentration from Section 2.2.2.

- Change the selective medium every 3-4 days until the G418-resistant colonies can be identified.
- Pick up single colonies under inverted microscope and expand them in 96-well plates for further characterization.

2.2.4 Selection of factor IX over-expression cell colonies by ELISA
- TBST buffer: 20 mM Tris-HCl pH 7.5, 150 mM NaCl, 0.1 % Tween 20.
- FIX standard: recombinant factor IX (Pfizer, BeneFIX) was dissolved in the buffer as the manufacturer recommended and serially diluted with the complete growth medium for ELISA standard (see note 3).
- FIX samples: cell culture medium from single colonies was used directly for FIX quantitation.
- Dilute sheep anti-FIX polyclonal antibody to 2 µg/mL with 50 mM carbonate/bicarbonate coating buffer (pH 9.6) (1.59 g Na_2CO_3 and 2.93 g $NaHCO_3$ in 1000 mL deionized water).
- Add 100 µL diluted antibody per well to a 96-well ELISA plate. Cover the plate with adhesive plastic and incubate overnight at 4 °C.
- Wash the plate 3 times with TBST buffer to remove the unbound antibodies.
- Block the plate with 200 µL blocking buffer (1 % BSA in TBST) per well and incubate the plate at room temperature for one hour or overnight at 4 °C.
- Remove the blocking buffer and directly add 100 µL sample or FIX standard per well. Incubate the plate for 2 h at room temperature.
- Wash the plate 3 times with TBST buffer to remove the unbound samples or FIX standard.
- Add 100 µL HRP-conjugated anti-FIX antibody (1 µg/mL in TBST buffer) per well. Incubate the plate at room temperature for one hour.
- Wash the plate 3 times with TBST buffer to remove the unbound HRP-conjugated antibodies.
- Dispense 100 µL ABTS peroxidase substrates solution per well and read the absorbance at a wavelength between 405-410 nm (see note 4).
- The colony with the highest FIX expression is selected as the stable reporter cell line (HEK293-FIX) for further study (see note 5).

2.2.5 Vitamin K titration using transiently expressed reporter protein
- Although the stable reporter cell line gives consistent and reliable results for vitamin K titration, it is a time-consuming process to get the stable

reporter cell line. When less samples (<30) need to be analyzed, transiently expressed reporter cells can be used instead of making a stable reporter cell line.

- One day before transfection, seed HEK293 cells in complete growth medium into a 96-well plate so that the cells will reach 50-70 % confluency at the time of transfection.
- Dilute 2 μg FIX containing plasmid DNA with the Xfect reaction buffer to a final volume of 100 μL. Mix well by vortexing for 5 s at high speed.
- Add 0.6 μL Xfect transfection reagent to the diluted plasmid DNA solution and mix well by vortexing for 10 s at high speed. Incubate for 10 min at room temperature to allow the formation of nanoparticle complexes.
- Spin down for one second to collect the contents at the bottom of the tube and add 1 mL of complete growth medium.
- Gently replace the cell culture medium in each well with 60 μL of the DNA-Xfect nanoparticle containing transfection medium that was prepared in the preceding step. Incubate the plate at 37 °C with 5 % CO_2 for 4 h (see note 6).
- Prepare a 2- or 3-fold serial dilution of vitamin K using the complete growth medium. Add 60 μL vitamin K-containing medium to each well and gently rock the plate back and forth to mix.
- Incubate the plate at 37 °C with 5 % CO_2 for 24-48 h.
- The cell culture medium is directly used for quantitation of the carboxylated and total FIX secreted in the cell culture medium.
- For total FIX quantitation, follow the procedure as described in Section 2.2.4.
- For carboxylated FIX quantitation, the ELISA plate is coated with anti-carboxylated FIX gla domain antibody (2 μg/mL in 50 mM carbonate/bicarbonate coating buffer, pH 9.6).
- After the plate is washed and blocked, add 20 μL $CaCl_2$ solution (10 mM $CaCl_2$ in 100 mM Tris-HCl, pH 7.5,) and 80 μL sample or FIX standard to each well (see note 7). Shake the plate a few times in the plate reader to mix. Incubate the plate for 2 h at room temperature.
- Wash the plate 3 times with TBST buffer containing 2 mM $CaCl_2$ to remove the unbound samples or FIX standard.
- Add 100 μL HRP-conjugated anti-FIX antibody (1 μg/mL in TBST buffer containing 2 mM $CaCl_2$) per well. Incubate the plate for one hour at room temperature.

Assessment of gamma-glutamyl carboxylase activity in its native milieu 217

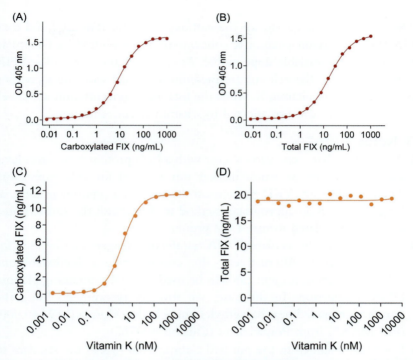

Fig. 2 Assessment of vitamin K-dependent carboxylation activity with ELISA using FIX as the reporter protein. The standard curves of carboxylated FIX (A) and total FIX (B) were determined by ELISA using commercial FIX protein and plotted using a Four-parameter logistic (4PL) regression model by GraphPad Prism. The amount of carboxylated FIX (C) and total FIX (D) secreted in the cell culture medium with increasing concentrations of vitamin K were measured by ELISA and plotted using a variable slope model with GraphPad Prism.

- Wash the plate 3 times with TBST buffer containing 2 mM $CaCl_2$ to remove the unbound antibodies.
- Add 100 μL ABTS peroxidase substrates solution per well and read the absorbance at a wavelength between 405 and 410 nm.

2.2.6 Data analysis using GraphPad Prism
- The standard curve of carboxylated FIX (Fig. 2A) and total FIX (Fig. 2B) is plotted using a Four-parameter logistic (4PL) regression model (see note 8).
- The carboxylated and total FIX protein concentrations in the samples treated with different vitamin K concentrations can be calculated/interpolated based on their standard curves, respectively.

- The dose-response curves for the carboxylated FIX (Fig. 2C) and total FIX (Fig. 2D) *versus* vitamin K concentrations are plotted by GraphPad Prism using the variable slope model. As expected, the carboxylated FIX concentration in the cell culture medium increased with the increasing concentrations of vitamin K, while the total FIX concentration in the cell culture medium was not affected by vitamin K concentrations.

2.3 Notes

(1) Any mammalian expression vector with a high protein expression level and an antibiotic selection marker can be used for FIX expression. Theoretically, any VKD protein can be used as a reporter protein, as long as there is an appropriate method to distinguish the carboxylated and uncarboxylated forms of the protein.

(2) Any commercially available DNA transfection reagent can be used for the transfection. Alternatively, the cost-effective polyethylenimine (PEI) transfection reagent can also be used for FIX-containing plasmid DNA transfection. In this case, we strongly recommend using the transfection grade linear polyethylenimine hydrochloride (MW 40,000) (PEI MAX®) from Polysciences (Catalog #24765).

(3) It is recommended to use purified carboxylated FIX as the standard in the ELISA assay. However, cell culture medium collected from FIX expression mammalian cells treated with vitamin K could also be used as the standard in the ELISA assay, as long as the carboxylated FIX concentration is known in the medium.

(4) We prefer to read the ELISA plate continuously without stopping the color development. This is especially beneficial for obtaining a wider linear range of the standard curve by fitting the data obtained at different reading times. The maximum absorbance value of the highest reading should not exceed the absorbance value of 2.0.

(5) To obtain a relatively pure reporter cell line, we recommend performing a second round of single colony pickup by diluting the first round selected colony in 10 cm culture plates as described in Section 2.2.3. Alternatively, it is not necessary to establish a stable reporter cell line if not many samples need to be analyzed. In this case, the reporter protein can be transiently expressed in HEK293 cells for functional study, as described in Section 2.2.5.

(6) To prevent cell loss, medium removal or replacement must be gently performed. Alternatively, if a stable reporter cell line is used, seed the reporter cells on the previous day and proceed to the following step directly.

(7) As the anti-carboxylated FIX gla domain antibody is a conformation-specific calcium-dependent antibody, it is important to include 2 mM $CaCl_2$ in all buffers in the following steps.
(8) If GraphPad Prism is not available, data can be plotted manually using the logit-log curve-fitting model, as previously described (Tie & Stafford, 2017).

3. Functional study of naturally occurring GGCX mutations in the cellular milieu

The cloning of the GGCX gene (Wu, Cheung, et al., 1991) makes it possible to study the correlation between GGCX genotypes and clinical phenotypes. Genetic screenings of patients with vitamin K-related disorders have identified over 40 naturally occurring mutations in GGCX (Brenner et al., 1998; De Vilder, Debacker, & Vanakker, 2017). Due to defects in the carboxylation of VKD coagulation factors, these mutations are mainly associated with bleeding disorders referred to as combined vitamin K-dependent coagulation factors deficiency type 1 (VKCFD1). Comorbid non-bleeding phenotypes of cardiac, dermatologic, ophthalmologic, and orthopedic origin were also observed in VKCDF1 patients, which are linked to defects in the carboxylation of extrahepatic VKD proteins, such as osteocalcin, matrix Gla protein, and Gla-rich protein (Galunska, Yotov, Nikolova, & Angelov, 2024). As GGCX is a dual-function enzyme that requires multiple substrates for modifying various structurally and functionally distinct VKD proteins, the characterization of GGCX mutations has been challenging.

The clinically reported VKCFD1 cases could be traced back to the 1950s (Newcomb, Matter, Conroy, Demarsh, & Finch, 1956), long before the GGCX gene was discovered. The first GGCX mutation identified from VKCFD1 patients was the homozygous conversion of a leucine codon (CTG) to an arginine codon (CGG) at residue 394 (Brenner et al., 1998). Functional study of the L394R mutant was initially performed by expressing the wild-type and the mutant protein in *Drosophila* cells and using the cell lysate for GGCX activity assay by measuring the incorporation of radioactive $^{14}CO_2$ to the pentapeptide substrate FLEEL (Brenner et al., 1998). This approach was later improved by expressing the enzymes in insect cells and determining GGCX activity using affinity-purified protein (Mutucumarana et al., 2000), which has since been a standard approach for studying the function of GGCX mutations. Despite the advantages of using purified enzyme rather than cell lysate (Darghouth et al., 2006) or crude microsomes (Li et al., 2009), the disadvantage of this approach is that it requires expressing and purifying a relatively large amount of protein for each GGCX mutation studied. Importantly, results obtained from these studies were mainly from *in vitro* activity assays using the small peptide substrate FLEEL in the presence of detergent, phospholipids, and high salt concentrations. Thus, these results have limited use for interpreting the multiple distinct clinical phenotypes caused by GGCX mutations. For example, the apparent affinities of coagulation factors' propeptides for GGCX vary over 100-fold *in vitro* (Stanley, Jin, Lin, & Stafford, 1999). Nevertheless, these coagulation factors appear to be fully carboxylated under physiological conditions. Additionally, it is not clear why some GGCX mutations cause bleeding disorders while others cause non-bleeding syndromes (Watzka et al., 2014). Furthermore, the defect in the carboxylation of coagulation factors in patients carrying GGCX mutations can be ameliorated by the administration of high doses of vitamin K (Napolitano, Mariani, & Lapecorella, 2010). However, it is not clear why the defect in the carboxylation of matrix Gla protein (MGP), a strong inhibitor of vascular calcification, by the same GGCX mutation, cannot be rescued by vitamin K - either *in vivo* or *in vitro* (Tie et al., 2016).

In an attempt to assess GGCX function using animal models, the endogenous GGCX gene was knocked out in mice by gene targeting (Zhu et al., 2007). Unfortunately, all homozygous GGCX-deficient mice succumbed to massive intra-abdominal hemorrhage shortly after birth. Although mice with liver-specific GGCX deficiency have been successfully created (Azuma et al., 2014), manipulating GGCX variants and their substrates in the mouse model appears to be impracticable.

Unlike in animals, VKD carboxylation has no effect on mammalian cell survival. Consequently, we sought to develop a GGCX-deficient reporter cell line for the functional study of the naturally occurring GGCX mutations in the cellular environment (Tie et al., 2016). We used CRISPR-Cas9-mediated genome editing to knock out the endogenous GGCX from the aforementioned FIXgla-PC/HEK293 reporter cells. The assay is based on the ability of the exogenously expressed GGCX or its mutants to carboxylate the VKD reporter protein in HEK293 cells. This approach has been extended to study the effect of GGCX mutations on the carboxylation of hepatic and extrahepatic VKD proteins using the corresponding VKD proteins as the reporter proteins (Hao et al., 2021). It has also been shown that other VKD coagulation factors could be used as the reporter proteins in the GGCX-deficient HEK293 cells to study GGCX mutations found in VKCFD1 patients (Ghosh et al., 2021). The following section details creating the GGCX-deficient reporter cell line with FIX as the reporter protein and using the established reporter cell line for the functional study of GGCX mutations.

3.1 Materials and equipment

- Mammalian expression vector pcDNA3.1/hygro(+) (ThermoFisher Scientific, Catalog #V87020).
- Chimeric guide RNA (gRNA) and human codon-optimized SpCas9 expression vector pX330-U6-Chimeric-BB-CBh-hSpCas9 (A gift from Dr. Feng Zhang. Addgene plasmid #42230).
- Rabbit anti-GGCX polyclonal antibody (Proteintech, Catalog #16209-1-AP).
- Mouse anti-GAPDH monoclonal antibody (Proteintech, Catalog #60004-1-Ig).
- HRP-conjugated goat anti-rabbit antibody (Jackson ImmunoResearch Laboratories, Catalog #111-035-003).
- HRP-conjugated goat anti-mouse antibody (Jackson ImmunoResearch Laboratories, Catalog #115-035-003).
- RIPA lysis buffer (Pierce, Catalog #89900).
- QuickExtract™ DNA Extraction Solution (Lucigen, Catalog #QE09050)
- Advantage™ 2 PCR enzyme system (Takara Bio, Catalog #639207).
- 1 × cOmplete protease inhibitor cocktail (Roche, Catalog #11873580001).
- SurePAGE Bis-Tris precast gels (GenScript, Catalog #M00653).
- Skim milk powder (BD, Catalog #232100).
- Amersham ECL western blotting detection reagent (Cytiva, Catalog #RPN2106).
- Celltrics™ 30 μm sterilized filter (Sysmex, Catalog #04-004-2320).
- Protein electrophoresis and blotting system (Bio-Rad).
- UV–visible spectrophotometer (Molecular Devices, SpectraMax ABS Plus).

3.2 Procedure

3.2.1 CRISPR-Cas9 knockout of GGCX from HEK293-FIX reporter cells

- To knock out the endogenous GGCX of HEK293-FIX reporter cells (see note 1), we used CRISPR-Cas9-mediated genome editing. The gRNA (5'-CACAGCTTATCCGGTAGCAC-3') targeting exon 4 of

GGCX is selected for gene targeting, as we have shown that it is the most efficient gRNA in the GeCKO v2 CRISPR library (Addgene catalog # 1000000048) for GGCX targeting (Tie et al., 2016).

- Synthesize the partially complementary gRNA oligonucleotides with 4nt overhangs compatible for cloning into the pX330-U6-Chimeric-BB-CBh-hSpCas9 vector (see note 2).
  Forward oligo: 5'-**CACCG**CACAGCTTATCCGGTAGCAC-3'
  Reverse oligo: 5'-**AAAC**GTGCTACCGGATAAGCTGTG**C**-3'
- Phosphorylate and anneal the forward and reverse oligos and clone the gRNA fragment into *Bbs*I digested pX330 vector (see note 3).
- Transiently transfect the gRNA containing pX330 vector into the HEK293-FIX cells using a 12-well plate as described in Section 2.2.3.
- Forty-eight hours post-transfection, detach the transfected cells with trypsin and use a 30 μm sterilized CellTrics™ filter to prepare a diluted single-cell suspension in the complete growth medium.
- Seed the single-cell suspension into ten 96-well plates with less than five cells per well. Incubate the plates at 37 °C with 5 % CO_2.
- Check cell growth every 3-4 days and replace the culture medium as needed.

3.2.2 Selection of GGCX knockout HEK293-FIX reporter cells

- When cell clusters are visible under an optical microscope in most of the wells, replace the culture medium with 150 μL complete growth medium containing 10 μM vitamin K.
- Twenty-four hours later, take 50 μL medium with a multi-channel pipette to determine the carboxylated FIX by ELISA, as described in Section 2.2.5 (see note 4).
- Select wells that have significantly reduced production of carboxylated FIX based on the absorbance at 405-410 nm. Check the selected wells under an inverted microscope to exclude those that do not have reporter cells (see note 5).
- Mark the wells with the lowest absorbance at 405-410 nm with visible cell clusters. Take 50 μL medium from these wells and determine the total FIX level, as described in Section 2.2.4. Select 12 wells with the highest ratio of total to carboxylated FIX.
- Detach cells from the selected wells by trypsin digestion and make single-cell suspension as described in Section 3.2.1. Seed the cells from each well at two different cell densities into two 10 cm cell culture dishes.

Incubate the dishes at 37 °C with 5 % CO_2 for single colony pickup, as described in Section 2.2.3 (see note 6).

- When cell clusters of the picked-up single colonies are visible in most of the wells, replace the culture medium with 150 µL complete growth medium containing 10 µM vitamin K.
- Twenty-four hours later, take 50 µL medium to determine the total FIX and carboxylated FIX individually by ELISA, as described above.
- Select 10 colonies with the highest ratio of total to carboxylated FIX for further characterization (see note 7).

3.2.3 Characterization of GGCX knockout HEK293-FIX reporter cells

- To further confirm GGCX gene knockout in the reporter cells, first, clone wild-type GGCX cDNA into the mammalian expression vector pcDNA3.1 using the standard molecular cloning approach.
- Transiently express GGCX in the GGCX knockout HEK293-FIX reporter cells in a 96-well plate using Xfect transfection reagent, as described in Section 2.2.5.
- Four hours later, add an equal volume of complete growth medium containing 20 µM vitamin K to the transfected cells.
- Incubate the plate at 37 °C with 5 % CO_2 for forty-eight hours. Determine carboxylated FIX in the cell culture medium from GGCX-deficient HEK293-FIX reporter cells with and without GGCX transfection. A robust signal is expected for the carboxylated FIX in the GGCX transfection cells compared with the non-transfected cells, as shown in Fig. 3A. However, the uncarboxylated reporter protein FIX is not affected by GGCX transfection (Fig. 3B).
- Next, examine GGCX protein expression in the GGCX knockout HEK293-FIX cells by western blot analysis.
- Seed the GGCX-knockout HEK293-FIX reporter cells in a 24-well plate.
- Wash the cells twice with 1 × PBS to remove the medium when cells reach ~90 % confluency.
- Harvest and lyse cells on ice for 30 min with RIPA lysis buffer containing 1 × cOmplete™ protease inhibitor cocktail.
- Separate proteins in the cell lysate by SDS-PAGE using a 4-12 % SurePAGE Bis-Tris precast gel and transfer protein bands onto a PVDF membrane following a standard transfer protocol.
- Block the PVDF membrane by 5 % skim milk for 2 h at room temperature. Probe the GGCX and the loading control GAPDH protein

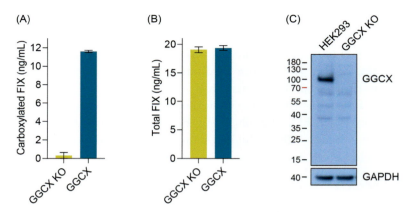

Fig. 3 Characterization of GGCX knockout HEK293-FIX reporter cell line. (A) Carboxylated reporter protein FIX secreted in the cell culture medium from GGCX knockout HEK293-FIX reporter cells (GGCX KO) and these cells exogenously express GGCX by transient transfection. (B) Total reporter protein FIX in the cell culture medium from cells as in Fig. 3A. (C) Endogenous GGCX expression in HEK293 cells and these cells with GGCX knockout.

bands using rabbit anti-GGCX and mouse anti-GAPDH antibodies as the primary antibodies, respectively.
- After washing off the unbound primary antibodies, incubate the PVDF membrane with HRP-conjugate goat anti-rabbit and goat anti-mouse antibodies, respectively.
- After washing off the unbound HRP-conjugated secondary antibodies, visualize the protein bands using the ECL western blotting detection kit. The endogenous GGCX protein should not be detected in the GGCX knockout cells, as shown in Fig. 3C.
- Finally, confirm GGCX knockout by sequencing the target region in the genomic DNA.
- Seed the GGCX-deficient HEK293-FIX reporter cells in a 96-well plate. When the cells reach ~90% confluent, aspirate the medium and add 50 μL QuickExtract™ DNA extraction solution.
- Incubate at room temperature for 2–3 min. Detach the cells with a pipette tip, mix thoroughly by pipetting up and down, and transfer the mixture to a PCR tube.
- Extract the genomic DNA by running the following cycling conditions in a PCR machine: 65 °C for 15 min, 68 °C for 15 min, and 98 °C for 10 min.

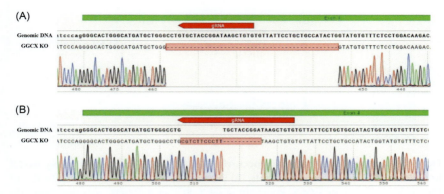

Fig. 4 Genomic DNA sequencing of the gRNA targeted region from one GGCX knockout colony. A 45-bp deletion (A) and 10-bp deletion with 11-bp insertion (B) were identified in the gRNA targeted region from one of the GGCX knockout colonies. Guide RNA sequence was indicated by red arrows.

- Amplify the gRNA target region using the Advantage™ 2 PCR enzyme system and the genomic DNA extract as the template with the following primers.
  Forward primer: 5'-GACTGGATGTATCTTGTCTACACCATC-3'
  Reverse primer: 5'-GGAAAGACTGAAGGGGTGGGATGGAG-3'
- Gel recovery the PCR fragment for sequencing. Fig. 4 shows the sequencing results of one of the GGCX knockout colonies.

3.2.4 Cell-based study of naturally occurring GGCX mutation

- The GGCX S300F mutation is used as an example for the cell-based study of GGCX mutation. First, create the S300F mutation by site-directed mutagenesis using pcDNA3.1-GGCX (Section 3.2.3) as the template.
- To examine the proper expression of the S300F mutant, transiently express the wild-type GGCX and the S300F mutant in the GGCX knockout HEK293-FIX reporter cells in a 12-well plate, as described in Section 2.2.3. Transfected cells are incubated at 37 °C with 5 % CO_2 for forty-eight hours. Cells are harvested with trypsin digestion to examine GGCX expression by western blot, as described in Section 3.2.3. Results in Fig. 5A show that the S300F mutant has a similar expression level as the wild-type GGCX.
- To examine the VKD carboxylation activity of the S300F mutant, transiently express the wild-type GGCX and the S300F mutant in the

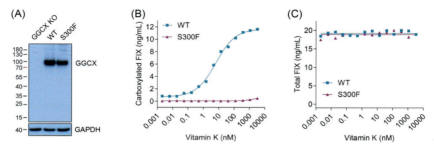

Fig. 5 Assessment the naturally occurring GGCX S300F mutation using GGCX knockout HEK293-FIX reporter cells. (A) Expression of the wild-type GGCX (WT) and the S300F mutant in GGCX knockout (GGCX KO) HEK293-FIX reporter cells. (B) Carboxylation of the reporter protein FIX by the wild-type GGCX and its S300F mutant with increasing concentrations of vitamin K. (C) Total reporter protein FIX determined from samples in Fig. 5B.

GGCX knockout HEK293-FIX reporter cells in a 96-well plate, as described in Section 2.2.5.
- Four hours post-transfection, add 60 μL complete growth medium containing increasing concentrations of vitamin K to each well and gently rock the plate back and forth to mix (see note 8).
- Forty-eight hours later, determine carboxylated and total FIX by ELISA, as described in Section 2.2.5.
- Plot the dose-response curves for carboxylated FIX or total FIX *versus* vitamin K concentrations using the variable slope model by GraphPad Prism. Results in Fig. 5B and C show the effect of the S300F mutant on the carboxylated and total FIX secreted in the cell culture medium.

3.3 Notes

(1) If GGCX gene knockout is an issue for your laboratory, a commercially available GGCX knockout cell line could be used for the proposed studies in this section. Currently available GGCX knockout cell lines are GGCX knockout HEK293T cells from Applied Biological Materials Inc. (Catalog #21469141) and GGCX knockout HeLa cells from Abcam (Catalog #ab265204).

(2) We recommend using the "all-in-one" or one plasmid system for creating the GGCX-knockout reporter cell line. In this system, both Cas9 and the gRNA are encoded on the same vector. We also suggest using transient expression for genome editing, which has a minimum effect on the following function study. To maximize the knockout

efficiency, multi-guide RNA could be introduced into one cell. However, this may result in an increase in off-target.
(3) The detailed protocol for cloning gRNA into the pX330 vector can be found on the Addgene website under Zhang Lab General Cloning Protocol.
(4) Unlike the protocol described in Section 2.2.5, the FIX protein standard is not included in this ELISA screening to ensure the loading of all 96 samples from the cell culture plate to the ELISA plate. Therefore, during the ELISA plate reading, we follow the well with the highest absorbance on each plate and continuously read the plate until it reaches ~0.8. We assume that the well with the highest absorbance in each plate has similar cell numbers (untargeted cells). Alternatively, one can load a FIX protein sample as the control at a certain position in each plate (*e.g.*, H12) and stop the plate reading when this control well reaches the desired value.
(5) Wells with an absorbance less than 0.1 or close to the background most likely do not have reporter cells. We select ~10 colonies per plate with a total of 96 colonies from 10 plates that have the lowest absorbance and visible cell clusters. We determine the total FIX level from these wells by ELISA on one 96-well plate.
(6) We pick up 16 single colonies from each selected well and expand them in the 96-well plate. For 12 selected wells, we have 192 single colonies distributed in two 96-well plates.
(7) If needed, a second round of single colony pickup can be performed to ensure the colony's purity.
(8) If vitamin K concentration titration is not needed, incubate the transfected cells with a fixed concentration of vitamin K for carboxylation activity assay.

4. Assessment of vitamin K epoxidase activity in the cellular milieu

Vitamin K epoxide (KO) is a byproduct of VKD carboxylation (Fig. 1). The epoxide form of vitamin K was first identified as a metabolite of phylloquinone in warfarin-treated rats (Matschiner, Bell, Amelotti, & Knauer, 1970). These authors also found that vitamin K epoxidase activity was inversely proportional to plasma prothrombin levels, thus proposed that the enzymatic interconversion of phylloquinone and its 2,3-epoxide was involved in the VKD production of prothrombin (Willingham & Matschiner, 1974). This hypothesis was further supported by the observations that KH_2 was the active form of VKD carboxylation (Friedman & Shia, 1976; Sadowski, Esmon, & Suttie, 1976) and the metabolite of KO (Fasco & Principe, 1980; Sherman & Sander, 1981). As vitamin K epoxidation occurs during VKD carboxylation, the epoxidase activity of GGCX was determined using the same reaction conditions as for carboxylation activity assay.

Vitamin K epoxidase activity was originally determined by measuring the formation of KO from radioactively labeled vitamin K using thin-layer chromatography separation (Sadowski et al., 1977; Willingham & Matschiner, 1974). As radioactive vitamin K was not commercially available at that time, the assay was replaced by using non-labeled vitamin K in the carboxylation reaction, and the produced KO was determined by measuring UV absorbance at 254 nm after reversed-phase HPLC separation (Elliott et al., 1976; Elliott et al., 1980). Despite the improvements of this HPLC-based assay, all the currently available epoxidase activity was evaluated *in vitro* under artificial conditions. Thus, we developed a cell-based vitamin K epoxidase activity assay to examine the conversion of vitamin K to KO in live HEK293 cells. The following section details the procedure of the cell-based epoxidase activity assay.

4.1 Materials and equipment

- Vitamin K (10 mg/mL, injectable emulsion) (Hospira, Catalog #NDC 0409-9158).
- Vitamin K epoxide (KO, used as the standard) (Sigma–Aldrich, Catalog #I9697)
- Warfarin (Sigma–Aldrich, Catalog #A2250)
- 2-propanol (Fisher Scientific, Catalog #A416).
- Hexane (Fisher Scientific, Catalog #H303).
- Acetonitrile (Fisher Scientific, Catalog #A998).
- Agilent 1100 Series HPLC.
- C18 analytical column (VyDAC, Catalog #201SP54).
- Nitrogen gas.

4.2 Procedure

4.2.1 Evaluation of the endogenous vitamin K epoxidase activity in HEK293 cells

- Seed HEK293 cells into 10 cm cell culture dishes with the complete growth medium so that the cells will reach 90 % confluency after 24-48 h.

- Replace the medium with fresh medium containing 20 µM vitamin K and 5 µM warfarin (see note 1). Incubate the dishes at 37 °C with 5 % CO_2 for 6-12 h, until the time of analysis (see note 2). Cells cultured with medium without vitamin K are used as the negative control.
- Wash the cells 3 times with PBS to remove extracellular vitamin K. Harvest the cells by trypsin digestion.
- Resuspend the cell pellet with 500 µL cell lysis buffer (50 mM Tris-HCl, pH 7.5, 150 mM NaCl, 0.5 % CHAPS) in a 2 mL Eppendorf tube. Mix well to ensure the solubilization and keep samples at room temperature for 5 min.
- Extract K vitamins by adding 500 µL isopropanol and 800 µL hexane sequentially, vortex at high speed for 3 min (see note 3).
- Centrifuge the extraction mixture for 2 min at 8000 × g.
- Carefully transfer the upper phase to 2 mL amber glass autosampler vials.
- Evaporate the solvent under a gentle nitrogen stream.
- Dissolve the dried sample with 300 µL HPLC mobile phase (acetonitrile:isopropanol:water =100:7:2). Vortex briefly to dissolve the sample well.
- Load sample vails to the autosampler and set the injection volume as 100 µL. K vitamins are separated by the Agilent 1100 HPLC system with a C18 column at a flow rate of 2 mL/minute (see note 4).
- K vitamins are detected and quantitated at 248 nm or 254 nm. Fig. 6A shows the chromatography diagrams of the epoxidase activity assay of HEK293 cells and these cells treated with vitamin K or vitamin K and warfarin.

4.2.2 Measurement of vitamin K epoxidase activity of naturally occurring GGCX mutations

- The epoxidase activity of GGCX and its naturally occurring mutations is determined by transiently expressing GGCX or its mutants (the S300F mutant is used as an example) in the GGCX knockout HEK293-FIX reporter cells.
- One day before transfection, seed the reporter cells into 10 cm cell culture dishes with complete growth medium so that the cells will reach 50-70 % confluency at the time of transfection.
- Transiently transfect wild-type GGCX and the S300F mutant in the GGCX knockout cells using the Xfect transfection reagent, as described in Section 2.2.3. Use one cell culture dish with a similar cell density without transfection as the negative control.

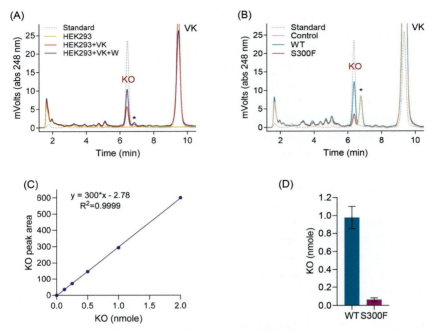

Fig. 6 Cell-based assessment of vitamin K epoxidase activity. (A) Chromatography diagrams of the endogenous epoxidase activity of HEK293 cells and these cells treated with 20 µM vitamin K or with 20 µM vitamin K and 5 µM warfarin for 8 h. HEK293 cells without vitamin K treatment were used as the negative control. The asterisk indicates the unknown metabolite of vitamin K. (B) Chromatography diagrams of GGCX knockout cells (Control), and these cells transiently express wild-type GGCX (WT) or the S300F mutant (S300F). Transfected cells were treated with 20 µM vitamin K and 5 µM warfarin for 12 h. The asterisk indicates the unknown metabolite of vitamin K. (C) The KO standard curve is plotted by GraphPad Prism with linear regression. (D) The production of KO from wild-type GGCX (WT) and the S300F mutant expressing cells in Fig. 6B, calculated from the KO peak area and the standard curve Fig. 6C. Data show the average and standard deviation of three independent experiments.

- Four hours post-transfection, add an equal volume of complete growth medium containing 5 µM warfarin to the transfected cells.
- Incubate the dishes at 37 °C with 5 % CO_2 for forty-eight hours.
- Replace the medium with fresh medium containing 20 µM vitamin K and 5 µM warfarin. Incubate the cell culture dish at 37 °C with 5 % CO_2 for 6–12 h until the time of analysis.
- Collect the cell pellets and wash them with PBS to remove extracellular vitamin K.

- Extract K vitamins and determine KO formation by HPLC, as described in Section 4.2.1. Fig. 6B shows the chromatography diagrams of the epoxidase activity assay of wild-type GGCX, the S300F mutant, and the control.
- Prepare a KO standard curve using pure KO (from Sigma) and plot the curve by GraphPad Prism with linear regression (Fig. 6C)
- KO concentration in each sample can be calculated based on the standard curve (Fig. 6D).

4.3 Notes

1) We recommend using injectable vitamin K, which is miscible with cell culture medium and easily absorbed by the cells. The epoxidase activity of GGCX is evaluated by measuring the formation of KO. To prevent the endogenous VKOR from reducing KO (product) back to vitamin K (reactant), warfarin is included in the cell culture medium to inactivate VKOR.

2) Longer incubation (>12 h) of HEK293 cells with vitamin K produces an unknown metabolite that interferes with the determination of KO (refer to the peaks in Fig. 6A and B indicated by asterisks). Therefore, we use a high concentration of vitamin K with a relatively short incubation time. As higher vitamin K concentrations (>20 µM) show cytotoxicity to HEK293 cells, caution should be taken to ensure minimum cytotoxicity when working with other cell lines.

3) To increase the assay's accuracy and reproducibility, vitamin E acetate or vitamin $K1_{25}$ can be included in the extraction solvent hexane as the internal standard. For the quantitation of KO in the sample, a serial concentration of KO standard is prepared and extracted with the same procedure to obtain the standard curve.

4) A constant HPLC column temperature helps to obtain reproducible results. Various mobile phases have been used for epoxidase activity analysis and the flow rate of each mobile phase depends on its composition.

5. Summary and conclusions

This chapter provides detailed procedures for studying GGCX's carboxylation and vitamin K epoxidation activities in live HEK293 cells. To evaluate the effect of naturally occurring GGCX mutations on the carboxylation and vitamin K epoxidation activities, we knocked out the endogenous GGCX in the HEK293 reporter cells using CRISPR-Cas9-mediated genome editing. The cell-based strategy described in this chapter provides versatile tools for the functional study of vitamin K-related enzymes in their native milieu. However, due to multiple enzymes being involved in VKD carboxylation, the limitation of the assay is that when assessing the function of one enzyme, it assumes that other enzymes are consistent in the assay. Additionally, in these assays, substrate concentration (such as vitamin K) should not reach the concentration that causes cyto-toxicity. Nevertheless, this cell-based strategy is the most successful approach for evaluating VKD carboxylation in the native environment, and it has been successfully extended by other groups using different reporter proteins to study VKD carboxylation. We expect that these novel cell-based assays will improve our understanding of how the various vitamin K cycle enzymes contribute to the complex mechanisms of modifying functional distinct VKD proteins. The detailed understanding of these enzymes and their naturally occurring mutants that can be achieved using these assays will lead to new ways of controlling thrombosis and other vitamin K-related diseases.

Acknowledgments

This work received funding from the National Heart, Lung, and Blood Institute (NHLBI) (R01HL131690). We thank all lab members who were involved in setting up and validating the cell-based assays described in this chapter.

References

Azuma, K., Tsukui, T., Ikeda, K., Shiba, S., Nakagawa, K., Okano, T., ... Inoue, S. (2014). Liver-specific gamma-glutamyl carboxylase-deficient mice display bleeding diathesis and short life span. *PLoS One, 9*(2), e88643. https://doi.org/10.1371/journal.pone.0088643.

Brenner, B., Sanchez-Vega, B., Wu, S. M., Lanir, N., Stafford, D. W., & Solera, J. (1998). A missense mutation in gamma-glutamyl carboxylase gene causes combined deficiency of all vitamin K-dependent blood coagulation factors. *Blood, 92*(12), 4554–4559.

Bristol, J. A., Freedman, S. J., Furie, B. C., & Furie, B. (1994). Profactor IX: The pro-peptide inhibits binding to membrane surfaces and activation by factor XIa. *Biochemistry, 33*(47), 14136–14143. https://doi.org/10.1021/bi00251a024.

Camire, R. M., Larson, P. J., Stafford, D. W., & High, K. A. (2000). Enhanced gamma-carboxylation of recombinant factor X using a chimeric construct containing the prothrombin propeptide. *Biochemistry, 39*(46), 14322–14329. https://doi.org/10.1021/bi001074q.

Carlisle, T. L., & Suttie, J. W. (1980). Vitamin K dependent carboxylase: Subcellular location of the carboxylase and enzymes involved in vitamin K metabolism in rat liver. *Biochemistry, 19*(6), 1161–1167. https://doi.org/10.1021/bi00547a019.

Chatrou, M. L., Reutelingsperger, C. P., & Schurgers, L. J. (2011). Role of vitamin K-dependent proteins in the arterial vessel wall. *Hamostaseologie, 31*(4), 251–257. https://doi.org/10.5482/ha-1157.

Chen, X., Li, C., Jin, D. Y., Ingram, B., Hao, Z., Bai, X., ... Tie, J. K. (2020). A cell-based high-throughput screen identifies drugs that cause bleeding disorders by off-targeting the vitamin K cycle. *Blood, 136*(7), 898–908. https://doi.org/10.1182/blood.2019004234.

Darghouth, D., Hallgren, K. W., Shtofman, R. L., Mrad, A., Gharbi, Y., Maherzi, A., ... Rosa, J. P. (2006). Compound heterozygosity of novel missense mutations in the gamma-glutamyl-carboxylase gene causes hereditary combined vitamin K-dependent coagulation factor deficiency. *Blood, 108*(6), 1925–1931. https://doi.org/10.1182/blood-2005-12-010660.

De Vilder, E. Y., Debacker, J., & Vanakker, O. M. (2017). GGCX-associated phenotypes: An overview in search of genotype-phenotype correlations. *International Journal of Molecular Sciences, 18*(2), https://doi.org/10.3390/ijms18020240.

Dowd, P., Hershline, R., Ham, S. W., & Naganathan, S. (1995). Vitamin K and energy transduction: A base strength amplification mechanism. *Science (New York, N. Y.), 269*(5231), 1684–1691. https://doi.org/10.1126/science.7569894.

Elliott, G. R., Odam, E. M., & Townsend, M. G. (1976). An assay procedure for the vitamin K1 2,3-epoxide-reducing system of rat liver involving high-performance liquid chromatography. *Biochemical Society Transactions, 4*(4), 615–617. https://doi.org/10.1042/bst0040615.

Elliott, G. R., Townsend, M. G., & Odam, E. M. (1980). Assay procedure for the vitamin K1 2,3-epoxide-reducing system. *Methods in Enzymology, 67*, 160–165. https://doi.org/10.1016/s0076-6879(80)67023-2.

Engelke, J. A., Hale, J. E., Suttie, J. W., & Price, P. A. (1991). Vitamin K-dependent carboxylase: Utilization of decarboxylated bone Gla protein and matrix Gla protein as substrates. *Biochimica et Biophysica Acta, 1078*(1), 31–34.

Esmon, C. T., Sadowski, J. A., & Suttie, J. W. (1975). A new carboxylation reaction. The vitamin K-dependent incorporation of H-14-CO$_3$- into prothrombin. *The Journal of Biological Chemistry, 250*(12), 4744–4748.

Fasco, M. J., & Principe, L. M. (1980). Vitamin K1 hydroquinone formation catalyzed by a microsomal reductase system. *Biochemical and Biophysical Research Communications, 97*(4), 1487–1492.

Fregin, A., Czogalla, K. J., Gansler, J., Rost, S., Taverna, M., Watzka, M., ... Oldenburg, J. (2013). A new cell culture-based assay quantifies vitamin K 2,3-epoxide reductase complex subunit 1 function and reveals warfarin resistance phenotypes not shown by the dithiothreitol-driven VKOR assay. *Journal of Thrombosis and Haemostasis: JTH, 11*(5), 872–880. https://doi.org/10.1111/jth.12185.

Friedman, P. A., & Shia, M. (1976). Some characteristics of a vitamin K-dependent carboxylating system from rat liver microsomes. *Biochemical and Biophysical Research Communications, 70*(2), 647–654.

Furie, B. C., Ratcliffe, J. V., Tward, J., Jorgensen, M. J., Blaszkowsky, L. S., DiMichele, D., & Furie, B. (1997). The gamma-carboxylation recognition site is sufficient to direct vitamin K-dependent carboxylation on an adjacent glutamate-rich region of thrombin in a propeptide-thrombin chimera. *The Journal of Biological Chemistry, 272*(45), 28258–28262. https://doi.org/10.1074/jbc.272.45.28258.

Galeffi, P., & Brownlee, G. G. (1987). The propeptide region of clotting factor IX is a signal for a vitamin K dependent carboxylase: Evidence from protein engineering of amino acid-4. *Nucleic Acids Research, 15*(22), 9505–9513. https://doi.org/10.1093/nar/15.22.9505.

Galunska, B., Yotov, Y., Nikolova, M., & Angelov, A. (2024). Extrahepatic vitamin K-dependent gla-proteins-potential cardiometabolic biomarkers. *International Journal of Molecular Sciences, 25*(6), https://doi.org/10.3390/ijms25063517.

Ghosh, S., Kraus, K., Biswas, A., Muller, J., Buhl, A. L., Forin, F., ... Oldenburg, J. (2021). GGCX mutations show different responses to vitamin K thereby determining the severity of the hemorrhagic phenotype in VKCFD1 patients. *Journal of Thrombosis and Haemostasis: JTH, 19*(6), 1412–1424. https://doi.org/10.1111/jth.15238.

Girardot, J. M., Delaney, R., & Johnson, B. C. (1974). Carboxylation, the completion step in prothrombin biosynthesis. *Biochemical and Biophysical Research Communications, 59*(4), 1197–1203.

Hao, Z., Jin, D. Y., Chen, X., Schurgers, L. J., Stafford, D. W., & Tie, J. K. (2021). gamma-Glutamyl carboxylase mutations differentially affect the biological function of vitamin K-dependent proteins. *Blood, 137*(4), 533–543. https://doi.org/10.1182/blood.2020006329.

Haque, J. A., McDonald, M. G., Kulman, J. D., & Rettie, A. E. (2014). A cellular system for quantitation of vitamin K cycle activity: Structure-activity effects on vitamin K antagonism by warfarin metabolites. *Blood, 123*(4), 582–589. https://doi.org/10.1182/blood-2013-05-505123.

Jin, D. Y., Chen, X., Liu, Y., Williams, C. M., Pedersen, L. C., Stafford, D. W., & Tie, J. K. (2023). A genome-wide CRISPR-Cas9 knockout screen identifies FSP1 as the warfarin-resistant vitamin K reductase. *Nature Communications, 14*, 828. https://doi.org/10.1038/s41467-023-36446-8.

Kaesler, N., Schettgen, T., Mutucumarana, V. P., Brandenburg, V., Jahnen-Dechent, W., Schurgers, L. J., & Kruger, T. (2012). A fluorescent method to determine vitamin K-dependent gamma-glutamyl carboxylase activity. *Analytical Biochemistry, 421*(2), 411–416. https://doi.org/10.1016/j.ab.2011.11.036.

Kaesler, N., Schurgers, L. J., & Floege, J. (2021). Vitamin K and cardiovascular complications in chronic kidney disease patients. *Kidney International, 100*(5), 1023–1036. https://doi.org/10.1016/j.kint.2021.06.037.

Knobloch, J. E., & Suttie, J. W. (1987). Vitamin K-dependent carboxylase. Control of enzyme activity by the "propeptide" region of factor X. *The Journal of Biological Chemistry, 262*(32), 15334–15337.

Larson, A. E., Friedman, P. A., & Suttie, J. W. (1981). Vitamin K-dependent carboxylase. Stoichiometry of carboxylation and vitamin K 2,3-epoxide formation. *The Journal of Biological Chemistry, 256*(21), 11032–11035.

Li, Q., Grange, D. K., Armstrong, N. L., Whelan, A. J., Hurley, M. Y., Rishavy, M. A., ... Uitto, J. (2009). Mutations in the GGCX and ABCC6 genes in a family with pseudoxanthoma elasticum-like phenotypes. *The Journal of Investigative Dermatology, 129*(3), 553–563. https://doi.org/10.1038/jid.2008.271.

Lingenfelter, S. E., & Berkner, K. L. (1996). Isolation of the human gamma-carboxylase and a gamma-carboxylase-associated protein from factor IX-expressing mammalian cells. *Biochemistry, 35*(25), 8234–8243. https://doi.org/10.1021/bi9523318.

Matschiner, J. T., Bell, R. G., Amelotti, J. M., & Knauer, T. E. (1970). Isolation and characterization of a new metabolite of phylloquinone in the rat. *Biochimica et Biophysica Acta, 201*(2), 309–315.

Mutucumarana, V. P., Stafford, D. W., Stanley, T. B., Jin, D. Y., Solera, J., Brenner, B., ... Wu, S. M. (2000). Expression and characterization of the naturally occurring mutation L394R in human gamma-glutamyl carboxylase. *The Journal of Biological Chemistry, 275*(42), 32572–32577. https://doi.org/10.1074/jbc.M006808200.

Napolitano, M., Mariani, G., & Lapecorella, M. (2010). Hereditary combined deficiency of the vitamin K-dependent clotting factors. *Orphanet Journal of Rare Diseases, 5*, 21. https://doi.org/10.1186/1750-1172-5-21.

Newcomb, T., Matter, M., Conroy, L., Demarsh, Q. B., & Finch, C. A. (1956). Congenital hemorrhagic diathesis of the prothrombin complex. *The American Journal of Medicine, 20*(5), 798–805. https://doi.org/10.1016/0002-9343(56)90163-2.

Rich, D. H., Lehrman, S. R., Kawai, M., Goodman, H. L., & Suttie, J. W. (1981). Synthesis of peptide analogues of prothrombin precursor sequence 5-9. Substrate specificity of vitamin K dependent carboxylase. *Journal of Medicinal Chemistry, 24*(6), 706–711.

Sadowski, J. A., Esmon, C. T., & Suttie, J. W. (1976). Vitamin K-dependent carboxylase. Requirements of the rat liver microsomal enzyme system. *The Journal of Biological Chemistry, 251*(9), 2770–2776.

Sadowski, J. A., Schnoes, H. K., & Suttie, J. W. (1977). Vitamin K epoxidase: Properties and relationship to prothrombin synthesis. *Biochemistry, 16*(17), 3856–3863.

Shah, D. V., & Suttie, J. W. (1974). The vitamin K dependent, in vitro production of prothrombin. *Biochemical and Biophysical Research Communications, 60*(4), 1397–1402.

Sherman, P. A., & Sander, E. G. (1981). Vitamin K epoxide reductase: Evidence that vitamin K dihydroquinone is a product of vitamin K epoxide reduction. *Biochemical and Biophysical Research Communications, 103*(3), 997–1005.

Stanley, T. B., Jin, D. Y., Lin, P. J., & Stafford, D. W. (1999). The propeptides of the vitamin K-dependent proteins possess different affinities for the vitamin K-dependent carboxylase. *The Journal of Biological Chemistry, 274*(24), 16940–16944.

Suttie, J. W., Geweke, L. O., Martin, S. L., & Willingham, A. K. (1980). Vitamin K epoxidase: Dependence of epoxidase activity on substrates of the vitamin K-dependent carboxylation reaction. *FEBS Letters, 109*(2), 267–270. https://doi.org/10.1016/0014-5793(80)81102-1.

Suttie, J. W., & Hageman, J. M. (1976). Vitamin K-dependent carboxylase. Development of a peptide substrate. *The Journal of Biological Chemistry, 251*(18), 5827–5830.

Suttie, J. W., Lehrman, S. R., Geweke, L. O., Hageman, J. M., & Rich, D. H. (1979). Vitamin K-dependent carboxylase: requirements for carboxylation of soluble peptide and substrate specificity. *Biochemical and Biophysical Research Communications, 86*(3), 500–507.

Tie, J. K., Carneiro, J. D., Jin, D. Y., Martinhago, C. D., Vermeer, C., & Stafford, D. W. (2016). Characterization of vitamin K-dependent carboxylase mutations that cause bleeding and nonbleeding disorders. *Blood, 127*(15), 1847–1855. https://doi.org/10.1182/blood-2015-10-677633.

Tie, J. K., Jin, D. Y., Straight, D. L., & Stafford, D. W. (2011). Functional study of the vitamin K cycle in mammalian cells. *Blood, 117*(10), 2967–2974. https://doi.org/10.1182/blood-2010-08-304303.

Tie, J. K., & Stafford, D. W. (2017). Functional study of the vitamin K cycle enzymes in live cells. *Methods Enzymol, 584*, 349–394. https://doi.org/10.1016/bs.mie.2016.10.015.

Ulrich, M. M., Furie, B., Jacobs, M. R., Vermeer, C., & Furie, B. C. (1988). Vitamin K-dependent carboxylation. A synthetic peptide based upon the gamma-carboxylation recognition site sequence of the prothrombin propeptide is an active substrate for the carboxylase in vitro. *The Journal of Biological Chemistry, 263*(20), 9697–9702.

Vermeer, C., Soute, B. A., De Metz, M., & Hemker, H. C. (1982). A comparison between vitamin K-dependent carboxylase from normal and warfarin-treated cows. *Biochimica et Biophysica Acta, 714*(2), 361–365.

Vermeer, C., Soute, B. A., Hendrix, H., & De Boer-Van Den Berg, M. A. (1984). Decarboxylated bone Gla-protein as a substrate for hepatic vitamin K-dependent carboxylase. *FEBS Letters, 165*(1), 16–20.

Watzka, M., Geisen, C., Scheer, M., Wieland, R., Wiegering, V., Dorner, T., ... Oldenburg, J. (2014). Bleeding and non-bleeding phenotypes in patients with GGCX gene mutations. *Thrombosis Research, 134*(4), 856–865. https://doi.org/10.1016/j.thromres.2014.07.004.

Willingham, A. K., & Matschiner, J. T. (1974). Changes in phylloquinone epoxidase activity related to prothrombin synthesis and microsomal clotting activity in the rat. *The Biochemical Journal, 140*(3), 435–441. https://doi.org/10.1042/bj1400435.

Wood, G. M., & Suttie, J. W. (1988). Vitamin K-dependent carboxylase. Stoichiometry of vitamin K epoxide formation, gamma-carboxyglutamyl formation, and gamma-glutamyl-3H cleavage. *The Journal of Biological Chemistry, 263*(7), 3234–3239.

Wu, G., Ma, Z., Cheng, Y., Hu, W., Deng, C., Jiang, S., ... Yang, Y. (2018). Targeting Gas6/TAM in cancer cells and tumor microenvironment. *Molecular Cancer, 17*(1), 20. https://doi.org/10.1186/s12943-018-0769-1.

Wu, S. M., Cheung, W. F., Frazier, D., & Stafford, D. W. (1991). Cloning and expression of the cDNA for human gamma-glutamyl carboxylase. *Science (New York, N. Y.), 254*(5038), 1634–1636. https://doi.org/10.1126/science.1749935.

Wu, S. M., Morris, D. P., & Stafford, D. W. (1991). Identification and purification to near homogeneity of the vitamin K-dependent carboxylase. *Proceedings of the National Academy of Sciences of the United States of America, 88*(6), 2236–2240. https://doi.org/10.1073/pnas.88.6.2236.

Wu, S. M., Soute, B. A., Vermeer, C., & Stafford, D. W. (1990). In vitro gamma-carboxylation of a 59-residue recombinant peptide including the propeptide and the gamma-carboxyglutamic acid domain of coagulation factor IX. Effect of mutations near the propeptide cleavage site. *The Journal of Biological Chemistry, 265*(22), 13124–13129.

Zhu, A., Sun, H., Raymond, R. M., Jr., Furie, B. C., Furie, B., Bronstein, M., ... Ginsburg, D. (2007). Fatal hemorrhage in mice lacking gamma-glutamyl carboxylase. *Blood, 109*(12), 5270–5275. https://doi.org/10.1182/blood-2006-12-064188.

CHAPTER TEN

Expression, purification, and activation of one key enzyme in anaerobic CO$_2$ fixation: Carbon monoxide dehydrogenase II from *Carboxydothermus hydrogenoformans*

Kareem Aboulhosn[a,b] and Stephen Wiley Ragsdale[a,*]
[a]Department of Biological Chemistry, University of Michigan, Ann Arbor, Michigan, United States
[b]Program in Chemical Biology, Department of Biological Chemistry, University of Michigan, Ann Arbor, Michigan, United States
*Corresponding author. e-mail address: sragsdal@umich.edu

Contents

1. Introduction	238
1.1 Wood-Ljungdahl pathway	238
1.2 Different types of CODH architecture	239
1.3 Relevance for producing CODH-II	240
2. Recombinant expression of CODH-II from *C. hydrogenoformans*	241
2.1 Cloning of CODH-II from *C. hydrogenoformans*	241
2.2 Use of fermenter to grow *E. coli* BL21 cells harboring pCooSII	242
2.3 Anaerobic harvest techniques	243
2.4 Lysis and purification of *Ch*CODH-II	244
2.5 Storage of *Ch*CODH-II	245
2.6 Use of anaerobic techniques in CODH characterization	245
3. Assay for CODH-catalyzed CO oxidation	246
3.1 Standard viologen-based CO oxidation assay	246
3.2 Reductive activation of CODH	247
3.3 EPR-based assay and metal quantification	248
4. Optimization of growth conditions and lysis protocol	249
4.1 Lysis without the use of detergents leads to active *Ch*CODH-II	250
4.2 [Ni] during growth is critical for C-cluster loading	250
4.3 CooC1 and CooC3 do not mature *Ch*CODH-II	251
5. Summary	253
Acknowledgments	254
References	254

Methods in Enzymology, Volume 708
ISSN 0076-6879, https://doi.org/10.1016/bs.mie.2024.10.016
Copyright © 2024 Elsevier Inc. All rights are reserved, including those for text and data mining, AI training, and similar technologies.

Abstract

Climate change due to anthropomorphic emissions will increase global temperature by at least 1.5 °C by the year 2030. One strategy to reduce the severity of the effects of climate change is to sequester carbon dioxide via natural biochemical cycles. Carbon monoxide dehydrogenase (CODH) has the remarkable ability to catalyze the reversible reduction of CO_2 to CO without an overpotential and without reducing protons. It also is a key enzyme in the Wood-Ljungdahl pathway (WLP), which is the only known anaerobic carbon fixation pathway and fixes 10 % of carbon on earth every year. Characterization of this pathway is crucial because it may enable tools to mitigate climate change by using CO_2 to produce biofuels, chemical feedstocks, and polymers. In the WLP, CODH associates with Acetyl-Coenzyme A synthase (ACS), which catalyzes the condensation of CO from CODH, a methyl group from a B_{12}-dependent methyltransferase, and CoA to form acetyl-CoA. In this complex, CO is shuttled through a 138 Å gas tunnel between the two enzymes. One valuable model for studying the CODH component of CODH/ACS is CODH-II from *Carboxydothermus hydrogenoformans* because it is stand-alone and is conducive to recombinant expression. Here we describe a detailed protocol for producing high-activity CODH-II in *E. coli*.

1. Introduction
1.1 Wood-Ljungdahl pathway

The global carbon cycle is arguably the most important natural cycle on Earth because it generates all organic matter available for consumption by heterotrophic organisms and dictates crucial aspects of life such as climate, ocean currents, extreme weather, and agriculture. The Wood-Ljungdahl pathway (WLP) plays a role in the carbon cycle by sequestering CO_2 into acetyl-CoA (Fig. 1), which serves as a source of energy and cell biomass for the microbes that employ this metabolic cycle (Ragsdale & Pierce, 2008). Although the WLP was first discovered in 1965 in *M. thermoacetica* by Harland Wood and Lars Ljungdahl, the key enzyme at the focus of this article was not purified and characterized until 1983 (Ragsdale et al., 1983a). This key enzyme is carbon monoxide dehydrogenase (CODH), which exists in a stand-alone form or in complex with acetyl-CoA synthase (ACS).

The first step of the WLP is the reversible conversion of CO_2 into CO without any overpotential– accomplished by a CODH. The CO produced by CODH is subsequently shuttled through a 138 Å gas tunnel to ACS which catalyzes the formation of acetyl-CoA from CO, a methyl group from B_{12}-dependent-methyl-transferase system, and CoA (Fig. 1) (Cohen et al., 2020). Acetyl-CoA is a central metabolic substrate that other

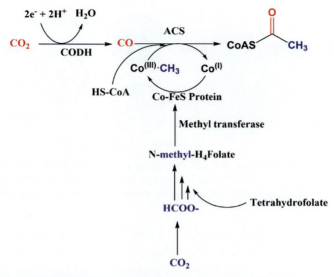

Fig. 1 Wood-Ljungdahl pathway. The WLP is the only known anaerobic carbon fixation pathway that converts CO_2 into acetyl-CoA. Carbon monoxide dehydrogenase (CODH) enzyme first converts CO_2 into CO which is delivered to acetyl-CoA synthase (ACS) via a 138 Å gas tunnel.

pathways convert into cellular biomass and alternatively is converted into acetate and ATP. Both reactions fix CO_2. Organisms using the WLP such as *Clostridium autoethanogenum* have been engineered to yield carbon-negative products such as biofuels and chemical feedstocks (Köpke & Simpson, 2020).

1.2 Different types of CODH architecture

There are two major classes of CODH: the monofunctional Cu-Mo CODH and the bifunctional Ni-Fe CODH (Dobbek et al., 2002). The Cu-Mo CODHs only catalyze CO oxidation and are aerobic whereas Ni-Fe CODHs tend to be highly oxygen sensitive and are capable of catalyzing both CO oxidation and CO_2 reduction (Adam et al., 2018; Techtmann et al., 2009). There are two subtypes within the class of Ni-Fe CODHs, *cooS*-type and *cdh*-type. *CooS*-CODHs are symmetric dimers and are broadly distributed throughout the eubacteria where the *cdh*-CODHs are almost exclusively found in archaea (Techtmann et al., 2012).

Contained within a given *cooS*-CODH monomer is the active site known as the C-cluster, which is a $NiFe_4S_4$, where CO oxidation and CO_2 reduction occur (Fig. 2) (Dobbek et al., 2001; Drennan et al., 2001).

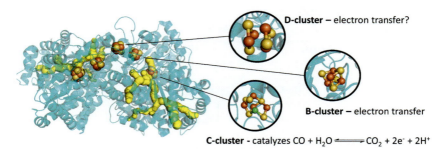

Fig. 2 CODH-II architecture. The ChCODH-II dimer is shown with gas tunnels generated with Pymol Caver. The dotted line on the D-cluster shows the dimer interface whereas the B-cluster and C-cluster are buried deeper in the enzyme. From PDB 1SU7.

Within a given CODH monomer and ~30 Å away from the C-cluster is the B-cluster, which is a Fe_4S_4. When comparing this distance across dimers however, the C-cluster in one monomer and the B-cluster in the other monomer are only 13 Å center to center (8 Å edge to edge) (Drennan et al., 2001) thereby enabling efficient electron transfer. At the interface of the dimer near the protein surface there is the D-cluster whose proposed role is electron transfer. Both the D-cluster and B-clusters are iron–sulfur clusters. In most organisms, the D-cluster is Fe_4S_4 though in some it is a Fe_2S_2 center. Although *cdh*-CODHs have two additional Fe_4S_4 clusters compared to *cooS*-CODHs, the C-cluster $NiFe_3S_4$-(Fe) cluster and the residues that surround it are highly conserved between these two subtypes (Gong et al., 2008).

1.3 Relevance for producing CODH-II

One particularly interesting *cooS*-CODH comes from *C. hydrogenoformans* Z-2901, which possesses five different CODHs (Wu et al., 2005). It is hypothesized that ChCODH-I has a role in energy conservation due to its ability to efficiently reduce CO_2, while ChCODH-II is thought to play a role in regenerating NADPH (Wu et al., 2005; Svetlitchnyi et al., 2001). A bifunctional complex is comprised of ChCODH-III and ChACS, which can perform WLP functions such as fixing CO_2 into acetyl-CoA (Ruickoldt et al., 2022; Svetlitchnyi et al., 2004). ChCODH-IV is more oxygen tolerant than other *cooS*-CODHs and is involved in oxidative stress mitigation (Domnik et al., 2017). The fifth ChCODH is part of a *cooS* operon, however it is not capable of catalyzing CO oxidation (Jeoung et al., 2022).

Of these ChCODHs, CODH-I and CODH-II possess the highest known rates for CO oxidation of 16,400 s^{-1} and 14,825 s^{-1}, respectively, at pH 8.0 and 70 °C when methyl viologen (MV) is used as an electron carrier (Svetlitchnyi et al., 2001). The fast rate makes ChCODH-II an attractive candidate for research and industrial application. For an example, it was recently demonstrated that through tunnel redesign, ChCODH-II could be used to scrub CO out of industrial waste gas and mitigate pollution (Kim et al., 2022). A thorough method for recombinantly expressing active ChCODH-II in *E. coli* would enable mechanistic studies because mutations would be easier to make than in the native organism. Additionally, purification would be more streamlined by adding an affinity tag and a single affinity column compared to the typical multiple column approach with native proteins where proteins are separated based on size/charge/hydrophobicity/etc. While methods have been described for production of recombinant ChCODH-II that supposedly lead to specific activities of 11,000–13,500 U/mg at 70 °C, (Jeoung & Dobbek, 2007) these conditions have not successfully led to the reported activity in our hands. Here we describe a thorough and reproducible ChCODH-II recombinant expression and purification method, which has key differences compared to the published method.

2. Recombinant expression of CODH-II from *C. hydrogenoformans*

2.1 Cloning of CODH-II from *C. hydrogenoformans*

In order to generate an expression construct for ChCODH-II, the *cooSII* gene was amplified from genomic DNA extracted from *C. hydrogenoformans* using forward (GTCGACGGAGCTCGAATTCGGATCCATGGCTAGGCAAAATTTAAAGTC) and reverse (CTTTAAGAAGGAGATATACATATGCCATGGTAATCCCAGGCC) primers. Next, we used the Gibson assembly (NEB HiFi) method (NEBuilder HiFi DNA Assembly Reaction Protocol) to insert *cooSII* into pET28a that had been digested with *Bam*HI and *Nde*I thereby yielding pCooSII, which also contains a thrombin cleavage site and a 6x His tag that can be used for Ni-affinity chromatography. The pCooSII plasmid was transformed into Rosetta II (DE3) *E. coli* BL21-competent cells (Novagen, Madison, WI) alongside an iron-sulfur cluster assembly plasmid - pRKISC (Nakamura et al., 1999) and plated on agar selection plates containing kanamycin (pCooSII), tetracycline (pRKISC), and chloramphenicol (pRARE2, harbored in Rosetta II (DE3)). The pRARE2 plasmid supplies seven codons that are rare in *E. coli*–forgoing this plasmid leads to substantial decreases in protein yield, even if Rosetta I is used (supplies six rare codons).

2.2 Use of fermenter to grow *E. coli* BL21 cells harboring pCooSII

Here we describe our current protocol to produce *Ch*CODH-II. First, a starter culture containing 100 mL of Terrific Broth (TB) is inoculated with Rosetta II (DE3) (Novagen, Madison, WI) *E. coli* containing pRARE2 that supplies tRNAs for 7 rare codons, pCooSII harboring the *Ch*CODH-II-coding gene, and pRKISC with the Fe-S cluster assembly genes. No codon optimization is performed on pCooSII because of the presence of the pRARE2 plasmid. This expression strain is cultured in the presence of 25 µg/mL chloramphenicol, 50 µg/mL kanamycin, and 15 µg/mL tetracycline to maintain pRARE2, pCooSII, and pRKISC in the starter culture. After growing this culture overnight at 37 °C with shaking, it is inoculated into a 5 L growth medium containing TB and 1.5 % glycerol in an autoclaved fermenter containing antibiotics at the same concentrations as in the starter culture (Fig. 3). It is, of course, possible to grow cells without a fermenter (i.e., using large flasks or septum-sealed bottles); however, the fermenter facilitates large scale anaerobic growth and cell harvesting.

Immediately following inoculation, the fermenter is spiked to final concentrations of 40 µM $NiCl_2$, 134 µM $FeSO_4$, 200 µM Cys-HCl, and 100 µM Na_2S through a 0.22 µm filter via a sterile syringe. The fermenter culture is grown at 30 °C aerobically (air supplied through sterile filter) until an optical density at 600 nm (OD_{600}) of 1–1.5 was reached (Fig. 3). An anaerobic purge is then started by bubbling the culture with N_2 gas instead of air; after 10 min, 50 mM KNO_3 (as electron acceptor) was added via sterile syringe. The culture was purged under strong N_2 flow for 45 min, at which point the N_2 flow was turned down to just maintain positive pressure and keep the culture anaerobic. Sterile syringes are used to add 1.8 mM Cys-HCl, 0.9 mM Na_2S, 1.2 mM $FeSO_4$, 2 mM $NiCl_2$, and 200 µM isopropyl thiogalactoside (IPTG) (all final concentrations in the fermenter growth medium). To avoid oxygen contamination during CODH-II induction, all solutions are prepared in the anaerobic chamber by cycling them in as powders and reconstituting with anaerobic water.

Expression, purification, and activation of one key enzyme in anaerobic CO_2 fixation   243

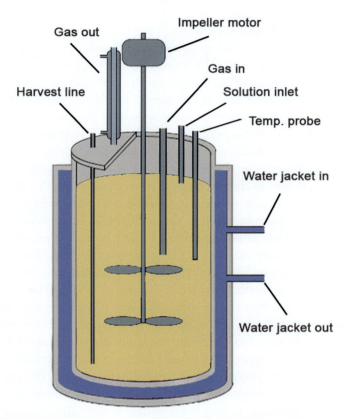

Fig. 3 Fermenter setup for anaerobic growth. All lines remain pinched during growth except water jacket, gas in, and gas out. Water jacket lines are connected to a water bath to maintain temperature of the growth medium and the impeller motor keeps the medium mixed. The gas-in line is connected in series with a sterile filter and a gas source (air or N_2). The gas-out line is connected to a condenser to ensure media is not lost during growth/bubbling. The solution inlet is used to add solutions anaerobically such as those required for induction. Once cell growth is complete, the medium is harvested through the harvest line by pinching the gas-out line and opening the harvest line which extends to the bottom of the fermenter allowing for removal of almost all media when growth is completed.

The positive flow of N_2 gas is maintained during overnight growth at a rate of 0.1–0.5 liters per minute until harvest.

2.3 Anaerobic harvest techniques

It is crucial to harvest cells without introducing oxygen, because although the solution contains reductant (Na_2S), it can be exhausted quickly. A 5 L glass harvest bottle (Kimax Kimble No. 14395) is degassed in the anaerobic chamber overnight before being capped and stoppered with a stopper that has an inlet and outlet tube. The culture is harvested by connecting the harvest line to the inlet of the bottle; the harvest and inlet lines are un-pinched while the outlet line of the harvest bottle remains pinched. The flow of media into the harvest bottle is started by pinching the fermenter gas-out line (Fig. 3) for approx. 30 s to slightly over pressurize the fermenter, forcing the media out of the harvest line. Once the flow has started, the outlet line on the harvest bottle is opened slightly to relieve pressure built up due to displaced air. It is very important that the harvest bottle outlet line is not opened too much otherwise oxygen may be introduced to the anaerobic harvest. Once the media has all been collected, both the inlet and outlet tubing on the harvest bottle are pinched with a clamp. After harvesting the media, the bottle is cycled into the anaerobic chamber (lines still pinched) and the media is loaded into 70 mL poly-carbonate centrifuge bottles (Beckman Coulter 355655). The bottles are balanced in the chamber, sealed with a cap and gasket, and cycled out of the chamber through the antechamber. A Thermo F12–6 × 500 LEX carbon fiber rotor is used to centrifuge the anaerobic bottles at 8000 × g for 20 min. After centrifugation, the harvest bottles are brought back into the anaerobic chamber to remove excess media and to collect the cell pellet. Often, multiple centrifugation cycles are required to harvest all the cell material – the centrifuge bottles must remain sealed when outside of the chamber.

2.4 Lysis and purification of *Ch*CODH-II

The purification steps are performed at room temperature in a Vacuum Atmospheres anaerobic chamber maintained at below 1 ppm O_2 (measured using a calibrated Teledyne O_2 detector). A 50 mL lysis solution is made fresh on the day of sonication by combining the cell pellet (~25 g) with buffer (pH 7.6, 20 mM sodium phosphate, 400 mM NaCl), 4.4 mg of phenylmethylsulfonyl fluoride (PMSF), 12.5 mg of lysozyme, 2 mg of DNase, and 5 mM β-mercaptoethanol (βME). The slurry is sonicated at 80 amps for 10 min total (10 s on, 15 s off) at 4 °C in the anaerobic chamber at 1 ppm O_2. Following sonication, the lysate is heated to 80 °C in a water bath for 15 min with vigorous stirring. The lysate is centrifuged for 1 h at 30,000 RPM (100,000 × g) with a Ti45 rotor at 4 °C. Clarified lysate is purified in the anaerobic chamber over a 5–7 mL Ni-sepharose column that is first equilibrated with 5 column volumes (CV) of purification buffer

(20 mM sodium phosphate, 400 mM NaCl, 5 mM βME) at room temperature. The clarified lysate is loaded onto the column at 2 mL/min (rate maintained throughout purification) via a peristaltic pump and is washed with 3 CV of purification buffer with no imidazole. Increasing amounts of imidazole are added via a step gradient (5 CV of 20 mM, 5 CV of 50 mM, 5 CV of 100 mM, 5 CV of 250 mM) to the purification buffer until *Ch*CODH-II elutes (200–250 mM imidazole). All purification buffers are degassed the night before use and are left in the anaerobic chamber overnight at ~1 ppm to ensure all oxygen is removed. The eluent is buffer exchanged into the storage buffer (50 mM Tris-HCl, 20% glycerol, 1 mM dithiothreitol, 2 mM dithionite, pH 7.6) through a 30 kDa Amicon stirred-cell filter.

2.5 Storage of *Ch*CODH-II

The purified *Ch*CODH-II should be stored in a crimp-top 25 mL vial with a septum at room temperature, ideally in the anaerobic chamber. Additionally, the protein solution should always contain a strong reductant such as DTT or dithionite (1–2 mM) because *Ch*CODH-II is susceptible to oxygen damage (Basak et al., 2023). Although the elution purification buffer contains βME, care should be taken to quickly buffer exchange into the storage buffer (50 mM Tris-HCl, 20% glycerol, 1 mM dithiothreitol, 2 mM dithionite, pH 7.6) after purification. All vials that are used to store the *Ch*CODH-II should be degassed in the chamber at least overnight before use.

2.6 Use of anaerobic techniques in CODH characterization

Performing anaerobic microbiology, biochemistry, and biophysics experiments requires novel design of and investment in specialized equipment, such as large anaerobic chambers that can house equipment, such as chromatographic columns, FPLC columns, spectrometers, and stopped-flow instruments. There are significant challenges to working at < 1 ppm O_2 instead of the 21% found in the air (where the $[O_2]$ is 0.25 mM and the reduction potential is +0.8 V). O_2 must be removed and excluded from every solution. The simplest way to remove oxygen from a heat-stable solution is to boil it vigorously for about one minute in an Erlenmeyer flask and then sparge with oxygen-free gas (N2) as the container cools. For solutions containing a heat-labile substance, the solution must be extensively (~1 h) sparged with an inert gas. Before use, glassware must be heated and degassed in the chamber. Furthermore, intricate work must be performed through thick rubber gloves reaching into the chamber.

However, even after accommodating these extreme methods, it is not possible to remove every molecule of oxygen. As pointed out by Hungate (Hungate, 1969), because of its high reactivity and oxidation potential, it also is impossible to obtain low potentials simply by removing oxygen. To achieve a relatively mild reduction potential of -0.33 V, the solution must contain no more than 10^{-55} molecule of O_2/L of water; this is a statistical function (not even a finite number of O_2 molecules). Thus, a reducing agent must be added to the solution. We typically use freshly made solutions of dithiothreitol (DTT), mercaptoethanol, sodium dithionite, or titanium (III) citrate. It is important to choose a reductant that does not negatively affect the enzyme. For these reasons, it is important to maintain CODH-II and other oxygen-sensitive enzymes like methyl-CoM reductase under reducing conditions at every step of manipulation - from cell growth to protein purification, and during characterization by kinetic, spectroscopic and structural biology methods. 7.

Our work has also taught us that it is just as important to rigorously characterize the redox state of such proteins during each kinetic or crystallographic analysis. For example, metal centers typically have chromophores that can be exploited. For CODH-II, EPR or X-ray absorption spectroscopy are rigorous ways to determine the redox state of the protein. If it is not possible to make such measurements during the analysis, it is crucial to perform such measurements in parallel experiments with the same protein samples. After spending the time preparing highly active enzymes, it also is crucial to use (and, in some cases, develop) rigorously anaerobic instrumentation and methodology to analyze these samples. Following these principles and the protocols described in this article, one can be more confident that the results haven't been compromised by oxygen contamination.

3. Assay for CODH-catalyzed CO oxidation
3.1 Standard viologen-based CO oxidation assay

The specific activities reported here for *Ch*CODH-II were determined by measuring the CO-dependent reduction of MV catalyzed by CODH. To measure this activity, a cuvette is sealed with a septum stopper and the gas phase is exchanged with pure CO (Metro Welding Supply, Detroit, MI) to make the cuvette anaerobic. For this, an inlet needle is placed near the bottom of the cuvette and an outlet needle near the top and gas exchange takes place for at least 5 min. Then a solution containing 20 mM of MV and 5 μM of dithionite is added to 1 mL of CO saturated buffer (50 mM HEPES, pH 8.0). Dithionite is present to provide sufficient reduced MV (shown by a very slight blue color), which scrubs any remaining O_2 from the cuvette. After the cuvette is brought to 25 °C, the assay is started by injecting ChCODH-II (1 μL of a 1–10 μM stock) and the Abs_{578} (the ε_{578} of MV is $9700\,M^{-1}\,cm^{-1}$) of the cuvette is measured over time to determine the initial velocity. Protein concentration is determined by the Rose-Bengal assay (Elliott & Brewer, 1978).

It is important to note that there are differences in the observed rate of CODH-catalyzed CO oxidation when using different electron acceptors. For an example, when benzyl viologen (BV, $\varepsilon_{578} = 7780\,M^{-1}cm^{-1}$) (Spencer & Guest, 1973) is used as an electron acceptor in CO oxidation catalyzed by $M.\ thermoacetica$ CODH/ACS, the rate is more than 2-fold higher compared to MV (Ragsdale et al., 1983b). This difference could be owed to differences in reduction potentials; the MV_{ox}/MV_{red} has a reduction potential of -446 mV whereas BV_{ox}/BV_{red} has a reduction potential of -340 mV. Great care should be taken to ensure rates are being compared against the same electron acceptors.

3.2 Reductive activation of CODH

Because of its extreme oxygen sensitivity, solutions of ChCODH-II must contain reducing agents to react with any trace oxygen. Furthermore, it has been shown that ChCODH-II is subject to reductive activation (Jeoung & Dobbek, 2007); however, a systematically optimized method has not been described. When the C–cluster is treated with CO, the C_{red2} state is formed from C_{red1}, (Kumar & Ragsdale, 1992) thus we hypothesized that poising the C–cluster in the C_{red1} state, defined to be the state that binds CO (Wang et al., 2013), would lead to optimal CO oxidation activity. Dithiothreitol and BV seemed to be the appropriate choice for poising ChCODH-II because the potentials (~-350 mV) are more positive than the coupling potential of CO and CO_2 (-520 mV) (Wang et al., 2013) but still more negative than the C_{ox}/C_{red1} couple (~-100 mV).

UV/Vis was used to determine that 5 mM dithiothreitol in 2 mM BV would poise ChCODH-II at -350 mV because half of the BV is reduced based on the extinction coefficient described above. The ChCODH-II dilution to 2 μM was made in a solution of 5 mM dithiothreitol and 2 mM BV in buffer (50 mM Tris-HCl, 20 % glycerol, pH 8, 25 °C) and CO

Table 1 Reductive activation of *Ch*CODH-II. A dilution of *Ch*CODH-II (2 μM) was made in 5 mM dithiothreitol, 2 mM BV, and was incubated at 25 °C for various amounts of time. The activity was then measured by CO oxidation assay against MV (Abs$_{578}$) in HEPES pH 8 at 25 °C.

Reductive activation time (hours)	k_{cat} (s^{-1})
No activation (as-isolated)	722
1	920
3	1083
6	1324
10	1363

oxidation measurements were run at 1-, 3-, 6-, and 10-hour timepoints (in 50 mM HEPES pH 8, 25 °C). It is clear that *Ch*CODH-II can be reductively activated using BV and dithiothreitol – the peak activity between 6–10 h of approximately ~1300 s^{-1} (Table 1) is higher compared to what has been published in the recent literature (~1000 s^{-1}) (Basak et al., 2022).

3.3 EPR-based assay and metal quantification

Electron paramagnetic resonance (EPR) can be used to determine occupancy of the C-cluster and B-cluster because these clusters have EPR-distinguishable signals when reduced. Incomplete assembly of either of these clusters could decrease activity due to either inhibiting electron transfer (if B-cluster is disturbed) or catalysis (if the C-cluster is disturbed). Damage to these clusters is also possible due to oxygen damage (Basak et al., 2023) hence it is important to have a quantitative method to determine cluster occupancy. It has been determined in previous literature that the CO-reduced state of *Ch*CODH-II which should be C$_{red2}$ (Kumar & Ragsdale, 1992) has g-values of 2.04, 1.99, 1.94, 1.90, 1.80, 1.74 (Fig. 4, **spectrum c**) (Svetlitchnyi et al., 2001). Of these signals, g = 1.99, 1.87, and 1.75 match the closest to C-cluster signals in the *M. thermoacetica* C$_{red2}$ state of the C-cluster (g = 1.97, 1.87, 1.75) (Lindahl et al., 1990). Spectra taken at higher temperatures (Fig. 4, **spectrum g**) show only B-cluster (g = 2.04, 1.94, 1.90); hence quantification of this cluster is also possible.

Using spin quantification on recombinantly produced *Ch*CODH-II by following the previously published protocol (Jeoung & Dobbek, 2007), we found that our recombinant *Ch*CODH-II has fully occupied B-cluster

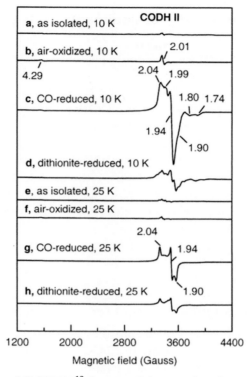

Fig. 4 EPR spectra of ChCODH-II[13] from Svetlichnyi et al. with permission.

(95–99% spin). The C-cluster spin quantification only resulted in 5–10% spin which is consistent with low occupancy of Ni and low observed activity. The method we describe here increases occupancy and spin of the C-cluster (to 30–40%) and activity relative to the activity obtained by following the method in the published protocol (~300 s^{-1} to ~800 s^{-1}, Table 4), while maintaining the nearly complete spin and occupancy of the B-cluster. Inductively coupled plasma optical emission spectroscopy (ICP-OES) was used to determine that contaminating metals such as Zn and Cu were at concentrations below the quantification limit.

4. Optimization of growth conditions and lysis protocol

Below is a description of how the previously published protocol was optimized to generate the protocol that we report here.

4.1 Lysis without the use of detergents leads to active *Ch*CODH-II

We started by following the previously published protocol for producing CODH-II, (Jeoung & Dobbek, 2007) but unfortunately could not produce active protein. Though it was clear that the *E. coli* expression cells did in fact have active CODH-II because there was CO oxidation activity of 1.8 U/mL in whole cells via a BV assay (HEPES pH 8, 25 °C). Thus, we worked on optimization of the cell lysis and protein purification protocols. The published protocol had demonstrated that CODH-II could be solubilized via incubation of the lysate with deoxycholate – this method did not successfully lead to active protein in our hands. We hypothesized that because *C. hydrogenoformans* is a thermophile and CODH-II is a thermophilic protein, heating the solution would increase its solubility. After sonication, the lysate was heated to near the growth temperature of *C. hydrogenoformans* (78 °C) (Wu et al., 2005) in the anerobic chamber. Heating the lysate for 15 min at 80 °C seemed to lead to optimal activity (Table 2).

4.2 [Ni] during growth is critical for C-cluster loading

To assess whether Ni was loaded into the C-cluster co- or post-translationally, we attempted to reconstitute *Ch*CODH-II lysate with $NiCl_2$ and reductant during sonication and heat treatment. However, none of these $NiCl_2$ reconstitution treatments led to a notable increase in activity

Table 2 **Effect of various lysate heating times on specific activity.** Lysate was heated immediately after sonication (10 min total, 10 s on, 15 s off, 80 Amp) and CO oxidation activity was measured by exposing the lysate to CO and measure the increase in MV absorbance at 578 nm (pH 8, 25 °C). Total protein was determined via Rose-Bengal assay (Elliott & Brewer, 1978).

Time heated at 80 °C	Specific activity of lysate (U/mg)
No heat	0.3
5 min	2.1
10 min	10
15 min	14.1
20 min	4.6

Table 3 NiCl$_2$ supplementation during lysis does not increase activity of ChCODH-II. Lysate was exposed to NiCl$_2$, various reductants, and was incubated for varying amounts of time (see Table 1 for sonication/heating details). Specific activity of the lysate was measured against MV in HEPES pH 8 at 25 ºC. Ni supplementation during lysis does not seem to improve the activity of ChCODH-II.

Condition	Specific activity of lysate (U/mg)
No Ni, 5 mM βME during sonication, heat treatment at 80 °C for 15 min	13.7
2 mM NiCl$_2$ + 5 mM βME during sonication, heat treatment at 80 °C for 15 min	14.9
2 mM NiCl$_2$ + 2 mM dithionite during sonication, heat treatment at 80 °C for 15 min	NA, precipitation
No Ni during lysis. Lysate is sonicated, heated for 15 min at 80 °C then incubated for 24 h with 2 mM NiCl$_2$ and 5 mM βME	10.2
No Ni during lysis. Lysate is sonicated, heated for 15 min at 80 °C then incubated for 48 h with 2 mM NiCl$_2$ and 5 mM βME	NA, precipitation
No Ni during lysis. Lysate is sonicated, heated for 15 min at 80 °C then incubated for 72 h with 2 mM NiCl$_2$ and 5 mM βME	NA, precipitation

(Table 3). These results suggest that Ni is incorporated co-translationally and that the C-cluster is not accessible for reconstitution after ChCODH-II has folded.

To test our hypothesis of co-translational Ni incorporation, we increased [NiCl$_2$] during growth. This led to a drastic increase in ChCODH-II activity with the optimal concentration being 2 mM; however, higher Ni^{2+} levels inhibited cell growth and led to inactive protein (Table 4).

4.3 CooC1 and CooC3 do not mature ChCODH-II

While CODH-I and CODH-III from *C. hydrogenoformans* possess maturases within their operons (Fig. 5) that appear to load Ni into the C-cluster, *cooSII* (ChCODH-II) lacks a CooC in its operon. For example, Inoue et al. heterologously expressed CODH-I using CooC3, the complementary maturation factor, and obtained activity amounting to about

Table 4 Activity of purified *Ch*CODH-II when induced with different [NiCl$_2$]. Different concentrations of NiCl$_2$ were used to induce *Ch*CODH-II expression. Specific activity of the purified ChCODH-II was measured using our standard MV-based assay (above) in HEPES pH 8 at 25 °C. Yield is reported as a range over at least 3 protein preparations. NA indicates the cells did not survive and could not yield protein.

[NiCl$_2$] during induction	k_{cat} of purified CODH-II (s^{-1})	CODH-II Yield (mg CODH/L growth culture)
0.5 mM	313	8–9
1 mM	773	15–20
2 mM	865	20–25
3 mM	NA	NA

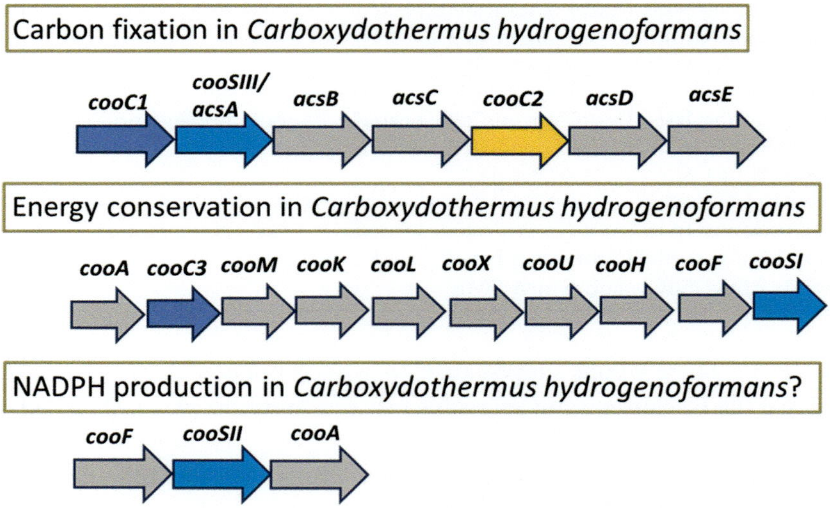

Fig. 5 **Operons related to CODH in *C. hydrogenoformans* (Wu et al., 2005).** Light blue – *cooS* genes, dark blue – *cooC* genes associated with CODH maturation, yellow – *cooC* (also known as *acsF*) gene involved in ACS maturation (Gregg et al., 2016).

half that of the natively expressed protein (Inoue et al., 2014). Thus we hypothesized that CooC1 (for *cooSIII*, aka CODH-III) or CooC3 (for cooSI) might supply Ni to *Ch*CODH-II as well.

However, when CooC1 or CooC3 were co-expressed independently with CODH-II the specific activities of the purified *Ch*CODH-II were 100.5 s^{-1} and 20.1 s^{-1}, respectively (Table 5). These values are significantly lower than expression without CooC. Furthermore, attempts to in–vitro

Table 5 CooC1/CooC3 do not mature *Ch*CODH-II. ChCODH-II was co-expressed with either CooC1 or CooC3 and was then isolated for CO oxidation activity against MV (Abs$_{578}$) in HEPES pH 8 at 25 °C. Yield is reported as a range over at least 3 protein preparations.

Maturase	[Ni] mM	kcat (s-1)	Yield (mg CODH-II/L growth culture)
None	0.5	313	8–9
CooC1	0.5	101	5–8
CooC3	0.5	20	3–4

activate using co-purified *Ch*CODH-II, CooC1/CooC3, ATP, MgCl$_2$ and NiCl$_2$ led to precipitation and this precipitate could not be redissolved by 1 % deoxycholate detergent or heating for 15 min at 80 °C.

Abbreviation	Full name
βME	β-mercaptoethanol
MV	Methyl viologen
BV	Benzyl viologen
CO	Carbon monoxide
CO$_2$	Carbon dioxide
CODH	Carbon monoxide dehydro-genase
ACS	Acetyl-CoA synthase
WLP	Wood-Ljungdahl Pathway
EPR	Electron paramagnetic reso-nance
TB	Terrific broth

5. Summary

The protocols described here allow for the consistent and reproducible preparation of highly active samples of purified CODH, which catalyzes CO$_2$ reduction without an overpotential and without reducing protons. These are remarkable properties because non–enzymatic CO$_2$ reduction catalysts are limited both by competition from protons and by their requirement for a significant overpotential (up to a volt). The ability to prepare large amounts of active CODH will encourage a resurgence of rigorous studies of its kinetic, spectroscopic, and structural properties, which will lead to a deep understanding how CODH catalyzes such specific, efficient and rapid rates of CO_2 reduction. We foresee the translation of these principles into small, inexpensive, earth–abundant metal catalysts that rapidly bind and reduce CO_2 without an overpotential or being derailed by competition with protons.

Acknowledgments

We are grateful to Professor Yasuhiro Takahashi (Department of Biology, Graduate School of Science, Osaka University, Japan) for supplying the pRKisc plasmid used in our CODH expression and growth experiments. We acknowledge NIH (R35-GM141758) for support of this work.

References

Adam, P. S., Borrel, G., & Gribaldo, S. (2018). Evolutionary history of carbon monoxide dehydrogenase/acetyl–CoA synthase, one of the oldest enzymatic complexes. *Proceedings of the National Academy of Sciences of the United States of America, 115*(6), E1166–E1173. https://doi.org/10.1073/pnas.1716667115.

Basak, Y., Jeoung, J. H., Dobbek, H., & Domnik, L. (2023). *Stepwise O2-induced rearrangement and disassembly of the [NiFe4(OH)(M3-S)4] active site cluster of CO dehydrogenase. Angewandte Chemie International Edition, 62.* https://doi.org/10.1002/anie.202305341.

Basak, Y., Jeoung, J.-H., Domnik, L., Ruickoldt, J., & Dobbek, H. (2022). Substrate activation at the Ni,Fe cluster of CO dehydrogenases: The influence of the protein matrix. *ACS Catalysis, 12711–12719.* https://doi.org/10.1021/acscatal.2c02922.

Cohen, S. E., Can, M., Wittenborn, E. C., Hendrickson, R. A., Ragsdale, S. W., & Drennan, C. L. (2020). Crystallographic characterization of the carbonylated A–cluster in carbon monoxide dehydrogenase/acetyl–CoA synthase. *ACS Catalysis, 10*(17), 9741–9746. https://doi.org/10.1021/acscatal.0c03033.

Dobbek, H., Gremer, L., Kiefersauer, R., Huber, R., & Meyer, O. (2002). Catalysis at a dinuclear [CuSMo(==O)OH] cluster in a CO dehydrogenase resolved at 1.1-A resolution. *Proceedings of the National Academy of Sciences of the United States of America, 99*(25), 15971–15976.

Dobbek, H., Svetlitchnyi, V., Gremer, L., Huber, R., & Meyer, O. (2001). Crystal structure of a carbon monoxide dehydrogenase reveals a [Ni-4Fe-5S] cluster. *Science (New York, N. Y.), 293*(5533), 1281–1285.

Domnik, L., Merrouch, M., Goetzl, S., Jeoung, J. H., Leger, C., Dementin, S., ... Dobbek, H. (2017). CODH-IV: A high-efficiency CO-scavenging CO dehydrogenase with resistance to O2. *Angewandte Chemie (International Ed. in English), 56*(48), 15466–15469. https://doi.org/10.1002/anie.201709261.

Drennan, C. L., Heo, J., Sintchak, M. D., Schreiter, E., & Ludden, P. W. (2001). Life on carbon monoxide: X-ray structure of rhodospirillum rubrum Ni-Fe-S carbon monoxide dehydrogenase. *Proceedings of the National Academy of Sciences of the United States of America, 98*(21), 11973–11978. https://doi.org/10.1073/pnas.211429998.

Elliott, J. I., & Brewer, J. M. (1978). The inactivation of yeast enolase by 2,3-butanedione. *Archives of Biochemistry and Biophysics, 190*(1), 351–357. https://doi.org/10.1016/0003-9861(78)90285-0.

Gong, W., Hao, B., Wei, Z., Ferguson, D. J., Tallant, T., Krzycki, J. A., & Chan, M. K. (2008). Structure of the alpha2epsilon2 Ni-dependent CO dehydrogenase component of the methanosarcina barkeri acetyl-CoA decarbonylase/synthase complex. *Proceedings of the National Academy of Sciences of the United States of America, 105*(28), 9558–9563. https://doi.org/0800415105[pii]10.1073/pnas.0800415105.

Gregg, C. M., Goetzl, S., Jeoung, J.-H., & Dobbek, H. (2016). AcsF catalyzes the ATP-dependent insertion of nickel into the Ni, Ni-[4Fe4S] cluster of acetyl-CoA synthase. *The Journal of Biological Chemistry, 291*(35), 18129–18138. https://doi.org/10.1074/jbc.M116.731638.

Hungate, R. E. (1969). A roll tube method for cultivation of strict anaerobes. In J. R. Norris, & D. W. Ribbons (Vol. Eds.), *Methods in microbiology: 3B,* (pp. 117–132). New York: Academic Press, Inc.

Inoue, T., Takao, K., Fukuyama, Y., Yoshida, T., & Sako, Y. (2014). Over-expression of carbon monoxide dehydrogenase-I with an accessory protein co-expression: A key enzyme for carbon dioxide reduction. *Bioscience, Biotechnology, and Biochemistry, 78*(4), 582–587. https://doi.org/10.1080/09168451.2014.890027.

Jeoung, J. H., & Dobbek, H. (2007). Carbon dioxide activation at the Ni,Fe-cluster of anaerobic carbon monoxide dehydrogenase. *Science (New York, N. Y.), 318*(5855), 1461–1464. https://doi.org/318/5855/1461[pii]10.1126/science.1148481.

Jeoung, J.-H., Fesseler, J., Domnik, L., Klemke, F., Sinnreich, M., Teutloff, C., & Dobbek, H. (2022). A morphing [4Fe-3S-nO]-cluster within a carbon monoxide dehydrogenase scaffold. *Angewandte Chemie International Edition, 61*(18), e202117000.

Kim, S. M., Lee, J., Kang, S. H., Heo, Y., Yoon, H.-J., Hahn, J.-S., ... O2-Tolerant, C. O. (2022). Dehydrogenase via tunnel redesign for the removal of CO from industrial flue gas. *Nature Catalysis, 5*(9), 807–817.

Köpke, M., & Simpson, S. D. (2020). Pollution to products: Recycling of 'above ground' carbon by gas fermentation. *Current Opinion in Biotechnology, 65*, 180–189. https://doi.org/10.1016/j.copbio.2020.02.017.

Kumar, M., & Ragsdale, S. W. (1992). Characterization of the carbon monoxide binding-site of carbon monoxide dehydrogenase from clostridium thermoaceticum by infrared spectroscopy. *Journal of the American Chemical Society, 114*(22), 8713–8715. https://doi.org/10.1021/ja00048a062.

Lindahl, P. A., Munck, E., & Ragsdale, S. W. (1990). CO dehydrogenase from clostridium thermoaceticum. EPR and electrochemical studies in CO_2 and argon atmospheres. *The Journal of Biological Chemistry, 265*(7), 3873–3879.

Nakamura, M., Saeki, K., & Takahashi, Y. (1999). Hyperproduction of recombinant ferredoxins in Escherichia coli by coexpression of the ORF1-ORF2-iscS-iscU-iscA-hscB-Hs cA-Fdx-ORF3 gene cluster. *Journal of Biochemistry (Tokyo), 126*(1), 10–18.

NEBuilder HiFi DNA Assembly Reaction Protocol. https://www.neb.com/en-us/protocols/2014/11/26/nebuilder-hifi-dna-assembly-reaction-protocol#:~:text=Use%20NEBuilder%C2%AE%20Protocol%20Calculator,6%20fragments%20are%20being%20assembled.

Ragsdale, S. W., Clark, J. E., Ljungdahl, L. G., Lundie, L. L., & Drake, H. L. (1983a). Properties of purified carbon monoxide dehydrogenase from clostridium thermoaceticum, a nickel, iron-sulfur protein. *The Journal of Biological Chemistry, 258*(4), 2364–2369.

Ragsdale, S. W., Ljungdahl, L. G., & DerVartanian, D. V. (1983b). Isolation of the carbon monoxide dehydrogenase from *Acetobacterium Woodii* and comparison of its properties with those of the *Clostridium Thermoaceticum* enzyme. *Journal of Bacteriology, 155*, 1224–1237.

Ragsdale, S. W., & Pierce, E. (2008). Acetogenesis and the wood-ljungdahl pathway of CO (2) fixation. *Biochimica et Biophysica Acta, 1784*(12), 1873–1898. https://doi.org/10.1016/j.bbapap.2008.08.012.

Ruickoldt, J., Basak, Y., Domnik, L., Jeoung, J.-H., & Dobbek, H. (2022). On the kinetics of CO_2 reduction by Ni, Fe-CO dehydrogenases. *ACS Catalysis, 12*(20), 13131–13142. https://doi.org/10.1021/acscatal.2c02221.

Spencer, M. E., & Guest, J. R. (1973). Isolation and properties of fumarate reductase mutants of Escherichia coli. *Journal of Bacteriology, 114*(2), 563–570.

Svetlitchnyi, V., Dobbek, H., Meyer-Klaucke, W., Meins, T., Thiele, B., Romer, P., Huber, R., & Meyer, O. (2004). A functional Ni-Ni-[4Fe-4S] cluster in the monomeric acetyl-coa synthase from carboxydothermus hydrogenoformans. *Proceedings of the National Academy of Sciences of the United States of America, 101*(2), 446–451.

Svetlitchnyi, V., Peschel, C., Acker, G., & Meyer, O. (2001). Two membrane-associated NiFeS-carbon monoxide dehydrogenases from the anaerobic carbon-monoxide-utilizing eubacterium carboxydothermus hydrogenoformans. *Journal of Bacteriology, 183*(17), 5134–5144. https://doi.org/10.1128/JB.183.17.5134-5144.2001.

Techtmann, S. M., Colman, A. S., & Robb, F. T. (2009). That which does not kill us only makes us stronger: The role of carbon monoxide in thermophilic microbial consortia. *Environmental Microbiology, 11*(5), 1027–1037. https://doi.org/10.1111/j.1462-2920. 2009.01865.x.

Techtmann, S. M., Lebedinsky, A. V., Coleman, A. S., Sokolova, T. G., Woyke, T., & Goodwin, L. (2012). Evidence for horizontal gene transfer of anaerobic carbon monoxide dehydrogenases. *Frontiers in Microbiology, 3*, 132.

Wang, V. C., Can, M., Pierce, E., Ragsdale, S. W., & Armstrong, F. A. (2013). A unified electrocatalytic description of the action of inhibitors of nickel carbon monoxide dehydrogenase. *Journal of the American Chemical Society, 135*(6), 2198–2206. https://doi. org/10.1021/ja308493k.

Wu, M., Ren, Q. H., Durkin, A. S., Daugherty, S. C., Brinkac, L. M., Dodson, R. J., ... Eisen, J. A. (2005). Life in hot carbon monoxide: The complete genome sequence of carboxydothermus hydrogenoformans Z-2901. *PLoS Genetics, 1*(5), 563–574. https:// doi.org/10.1371/journal.pgen.0010065.

CHAPTER ELEVEN

Molybdenum-containing CO dehydrogenase and formate dehydrogenases

Russ Hille[*]

Department of Biochemistry, University of California, Riverside, CA, United States
*Corresponding author. e-mail address: russ.hille@ucr.edu

Contents

1. Overview	257
2. CO Dehydrogenase	258
2.1 Introduction	258
2.2 Genetic context and protein structure	259
2.3 Mechanistic studies of the binuclear Mo/Cu center	261
2.4 Reaction with quinones	263
2.5 Experimental considerations	264
3. Formate dehydrogenases	265
3.1 Introduction	265
3.2 Systematics	265
3.3 Reaction mechanism	269
3.4 Experimental considerations	270
4. Summary	271
Acknowledgment	271
References	271

Abstract

The molybdenum-containing CO dehydrogenase and the formate dehydrogenases catalyze important interconversions of one-carbon compounds, the former oxidizing CO to CO_2, and the latter the reversible interconversion of CO_2 and formate. Methodologies to study these two enzymes are discussed.

1. Overview

Bacterial interconversion of CO, CO_2 and formate plays an important role in the global carbon cycle. Oxidation of CO to CO_2 is catalyzed by two quite different types of CO dehydrogenase. Obligate anaerobes such as *Moorella thermoacetica* (formerly *Clostridium thermoaceticum*) possess a

Methods in Enzymology, Volume 708
ISSN 0076-6879, https://doi.org/10.1016/bs.mie.2024.10.014
Copyright © 2024 Elsevier Inc. All rights are reserved, including those for text and data mining, AI training, and similar technologies.

nickel- and iron-containing CO dehydrogenase that reversibly reduces CO_2 to CO, which is subsequently condensed with methylcobalamin from acetyl-CoA synthase to form acetyl-CoA in the course of acetogenesis. The enzyme is extremely air-sensitive. The reverse process catalyzes the final step in methanogenesis. On the other hand, aerobes such as *Oligotropha carboxidovorans* possess a molybdenum- and copper-containing CO dehydrogenase that oxidizes CO to CO_2. In contrast to the clostridial enzyme, the *O. carboxidovorans* enzyme is quite tolerant of O_2. It is the air-stable CO dehydrogenase that is considered here.

Similarly, there are two broad types of formate dehydrogenase. A variety of organisms, including humans, possess an enzyme that is devoid of redox-active cofactors and catalyzes the oxidation of formate to CO_2. These enzymes use NAD^+ as the electron acceptor in the course of the reaction, which proceeds via direct hydride transfer within a ternary enzyme•formate•NAD^+ ternary complex. On the other hand, a number of bacteria and archaea possess one or more molybdenum- (or occasionally tungsten) containing formate dehydrogenases that catalyzes the reversible interconversion of formate and CO_2. It is these latter enzymes that again constitute the focus of the present account.

The interconversions of CO, CO_2 and formate are also important as feedstock compounds for chemical synthesis and, in the case of the interconversion of CO_2 and formate, have enormous potential for the reversible storage of renewable energy (Appel et al., 2013). What follows is a discussion of considerations that bear in studies of the molybdenum-containing CO dehydrogenase and formate dehydrogenases.

2. CO Dehydrogenase
2.1 Introduction

Oligotropha carboxidovorans and related organisms are able to grow aerobically with CO as sole source of both carbon and energy, and in the presence of O_2 (Meyer et al., 1993). The overall stoichiometry involves the oxidation of approximately six equivalents of CO to CO_2 generating the reducing equivalents that provide the energy necessary to fix one equivalent of CO_2 via the pentose phosphate pathway (Jacobitz and Meyer, 1989). The CO dehydrogenase responsible for this oxidation possesses a binuclear molybdenum- and copper-containing active site, the reducing equivalents thus obtained ultimately being passed into the intramembrane

Molybdenum-containing CO dehydrogenase and formate dehydrogenases 259

quinone pool and on to a CO-insensitive terminal oxidase (Cypionka and Meyer, 1983). It has been estimated that this process remediates as much as 2×108 metric tons of environmental CO each year (Moxley and Smith, 1998; Moersdorf et al., 1992; Jaffe, 1970).

2.2 Genetic context and protein structure

The *O. carboxidovorans* CO dehydrogenase is encoded by a gene cluster encoding twelve distinct proteins (Kraut et al., 1989; Kang and Kim, 1999; Santiago et al., 1999). The structural genes are *coxMSL*, which form the native $(\alpha\beta\gamma)_2$ enzyme. Both CoxF and CoxI are homologous to the *E. coli* XdhC protein that inserts a terminal sulfur group into the molybdenum coordination sphere (Matthies et al., 2004; Neumann et al., 2006; Schumann et al., 2008). CoxD exhibits homology to known membrane-integral ATPases and appears to be involved in incorporation of copper into the maturing binuclear center (Pelzmann et al., 2009). CoxB, CoxC, CoxH and CoxK are predicted to possess multiple transmembrane helices and are likely involved in anchoring CO dehydrogenase on the inner side of the cytoplasmic membrane, where quinones become reduced as the enzyme oxidizes CO (Cypionka and Meyer, 1983; Spreitler et al., 2010).

The demand for carbon necessary to maintain cell growth ensures that *O. carboxidovorans* and related organisms accumulate large amounts of intracellular CO dehydrogenase when grown on CO and it is straightforward to purify the wild-type protein in relatively high quantities (Meyer and Schlegel, 1980; Zhang et al., 2010). CO dehydrogenases from both *O. carboxidovorans* (Dobbek et al., 1999) and *Hydrogenophaga pseudoflava* (Hanzelmann et al., 2000) have been crystallographically characterized and have very similar structures. Fig. 1 shows structure of the *O. carboxidovorans* enzyme, with the 18 kDa CoxS subunit in red containing two [2Fe-2S] iron-sulfur clusters, the 30 kDa CoxM subunit with one equivalent of FAD in yellow, and the 89 kDa CoxL subunit containing the active site Mo/Cu binuclear center in gray. Each subunit has considerable homology to the corresponding parts of other members of the xanthine oxidase family of molybdenum-containing enzymes, the unique nature of the binuclear active site of CO dehydrogenase notwithstanding (Hille, 1996a; Hille et al., 2014a). The structure of the active site binuclear center of CO dehydrogenase in its oxidized form is also shown in Fig. 1 (inset) (Dobbek et al., 2002; Meyer-Klaucke et al., 2001). The oxidized Mo(VI) ion has the distorted square-pyramidal coordination geometry seen in other members of the xanthine oxidase family of molybdenum-containing enzymes, with

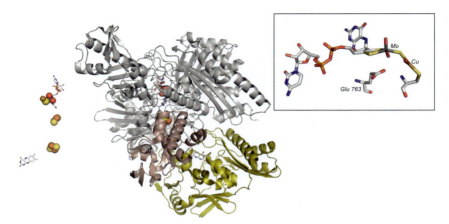

Fig. 1 *The structure of the O. carboxidovorans CO dehydrogenase* (PDB 1N5W). One protomer of the dimeric protein is shown, with the CoxS subunit in *red*, the CoxM subunit in *yellow* and the CoxL subunit in *gray*. In the protomer on the left, the polypeptide has been deleted to show the disposition of the binuclear Mo/Cu center of CoxL, *top*, the two [2Fe-2S] of CoxS, *middle*, and the isoalloxazine of the FAD in CoxM, *bottom*. The inset shows a close-up of the binuclear cluster in the active site.

an apical Mo=O and an equatorial plane consisting of a second Mo=O group rather than the catalytically labile Mo-OH seen in other family members (Zhang et al., 2010) and two sulfurs from the pyranopterin cofactor (present as the dinucleotide of cytosine) that is common to all molybdenum and tungsten enzymes (other than nitrogenase). The final ligand to the molybdenum is a μ-sulfido bridge to the Cu(I) in place of the Mo=S found in other members of this family that is presumably inserted by CoxF and/or CoxI; the Cu(I) ion is also liganded by Cys 388 and a water molecule. Steric access to the deeply buried binuclear center of CO dehydrogenase is quite restricted, and in particular Phe 390 of CO dehydrogenase (equivalent to Phe 914 in the bovine xanthine oxidoreductase which constitutes part of the substrate binding site) is displaced into what would otherwise be the solvent access channel in order to accommodate the copper and Cys 388 that coordinates it. Still, access is sufficient as to allow not only CO but even larger molecules such as n-butylisonitrile (Dobbek et al., 2002), a CO analog.

A heterologously expressed *O. carboxidovorans* CO dehydrogenase in *E. coli* exhibits spectroscopic and kinetic properties comparable to the native enzyme (Kaufmann et al., 2018), and a preliminary study of several variants of active site residues has been undertaken, indicating that mutations at Glu 763,

Phe 390 and Cys 388 all abolished activity. Unfortunately, these variants possessed only low levels of the pyranopterin cofactor, indicating it was likely that loss of activity was trivially due to lack of incorporation of the Mo/Cu center. Several thiol-containing compounds, including cysteine, mercaptoethanol and dithiothreitol, reversibly inhibit CO dehydrogenase with modest K_i's in the mM range (Kress et al., 2014). Extended X-ray absorption fine structure analysis of the copper environment in mercaptoethanol-complexed CO dehydrogenase suggests direct binding of the copper, presumably displacing the water molecule that is otherwise bound to the copper (Kress et al., 2014). CO dehydrogenase is neither inhibited nor inactivated by O_2, but the reduced enzyme is readily reoxidized by O_2 and experiments with the reduced protein are routinely done anaerobically under an atmosphere of N_2 or Ar.

2.3 Mechanistic studies of the binuclear Mo/Cu center

In the course of reaction of CO dehydrogenase with CO, an EPR signal due to the partially reduced Mo/Cu binuclear center (formally Mo^V/Cu^I) accumulates that exhibits extremely strong coupling to the copper ion (Zhang et al., 2010), as shown in Fig. 2. The signal is unchanged on preparation of the sample in D_2O, indicating no strong proton coupling, but modest line-broadening is seen when ^{13}CO is used as substrate, reflecting weak coupling to the substrate carbon (Zhang et al., 2010). A Mo^V/Cu^I model for the EPR-active form of the binuclear active site of CO dehydrogenase having a Mo^VO-μS-Cu^I core exhibits very similar copper hyperfine coupling (Gourlay et al., 2006) that reflects substantial delocalization of unpaired spin density on the copper, despite it being formally Cu^I. ENDOR work with ^{13}CO-reduced CO dehydrogenase shows essentially isotropic ^{13}C hyperfine coupling, indicating that at least two atoms intervene between the molybdenum and carbon (Shanmugam et al., 2013). The signal has been interpreted to arise from a Mo^V/Cu^I species.

having CO bound at the copper of the binuclear center, displacing the water seen in the X-ray crystal structure. Backbonding from the copper CO weakens the C-O bond, thereby activating it for nucleophilic attack upon coordinating to the copper.

A high-resolution (1.1 Å) X-ray crystal structure of CO dehydrogenase with the CO analog n-butylisonitrile shows the inhibitor inserted between the Mo and Cu, rupturing the bridging S-Cu bond (Dobbek et al., 2002). A reaction mechanism has been proposed in which CO similarly inserts into the binuclear center to yield a bridging thiocarbamate concomitant with

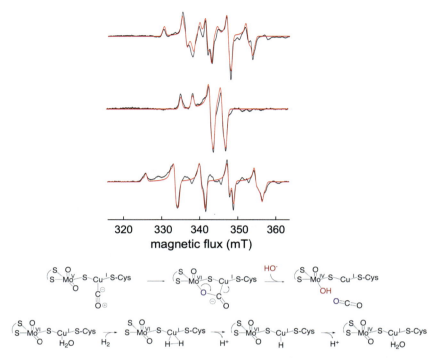

Fig. 2 *EPR signals and reaction mechanisms for CO dehydrogenase.* Top, EPR signals exhibited by the binuclear center of native CO dehydrogenase on reduction with CO, of the silver-substituted enzyme reduced by CO and the native enzyme reduced by H_2. Spectra are in black and simulations in red using the following parameters. Spectra are, from top to bottom: native enzyme reduced with CO, $g_{1,2,3} = 2.0010$, 1.9604, 1.9549 and $A_{1,2,3}(^{63,65}Cu)$ = 117, 164, 132 MHz; silver-substituted enzyme reduced by CO ($g_{1,2,3} = 2.0043$, 1.9595, 1.9540 and $A_{1,2,3}(^{63,65}Cu) = 82.0$, 78.9, 81.9 MHz); and native enzyme reduced with H_2 ($g_{1,2,3} = 2.0127$, 1.9676, 1.9594, $A_{1,2,3}(^{63,65}Cu) = 169$, 200, 170 MHz and $2 \times A_{1,2,3}(^1H) = 80$, 20, 130 MHz). Shown below are the proposed mechanisms for CO oxidation (upper) and H_2 oxidation (lower) as catalyzed by CO dehydrogenase.

reduction of the molybdenum, the thiocarbamate hydrolyzing to regenerate the sulfur bridge to complete the catalytic cycle. The reaction has also been examined computationally (Siegbahn and Shestakov, 2005; Hofmann et al., 2005). One group has concluded that a very stable thiocarbamate intermediate such as that proposed on the basis of the structure of the isonitrile-complexed enzyme indeed forms and passes over a relatively low-lying transition state to release CO_2 as water/hydroxide displaces the thiocarbamate from the binuclear cluster (Siegbahn and Shestakov, 2005). A second group has concluded that such a thiocarbamate intermediate represents a thermodynamic trap and is not likely to form in the course of catalysis; the reaction instead is proposed to proceed through a series of three intermediates that involve the CO carbon bonded to each of the three atoms (Mo, μS and Cu) that contribute to the highly delocalized redox-active orbital of the oxidized binuclear center (Hofmann et al., 2005). It has been suggested these might in fact be better thought of as alternate resonance representations of a single structure, or alternatively discrete, rapidly interconverting intermediates (Shanmugam et al., 2013). Assuming that the equatorial Mo=O of CO dehydrogenase is in fact catalytically labile, as seen in xanthine oxidase (Hille and Sprecher, 1987), the binuclear cluster appears to be constructed so as to provide a substrate binding site for CO adjacent to the molybdenum that positions the CO appropriately for attack by the equatorial Mo=O and spatially extend the formally redox-active Mo d_{xy} orbital such that it can accept an electron pair in the course of the reaction at the more remote (copper) site as shown in Fig. 2.

Like several of the Ni/Fe CO dehydrogenases (Can et al., 2014), the Mo/Cu enzyme is reduced by H_2 (Zhang et al., 2010). A new EPR signal arising from the binuclear center is seen in the course of the reaction that exhibits coupling to two approximately equivalent and solvent-exchangeable protons, in addition to the strong copper hyperfine, as shown in Fig. 2. By analogy to the binding of CO to the Cu of the binuclear center, the signal-giving species is proposed to possess H_2 coordinated (in a side-on η_2 fashion) to the copper of a partially reduced binuclear center (Wilcoxen and Hille, 2013), for which chemical precedent exists (Wang et al., 2013; Frohman et al., 2013). The reaction is thought to proceed by reversible deprotonation of the polarized H_2 to yield a hydride complex (for which chemical precedent also exists (Hulley et al., 2013)), with a second (also reversible) deprotonation resulting in population of the redox-active orbital, as shown in Fig. 2. This mechanism predicts that the enzyme catalyzes exchange of deuterium from D_2O into H_2, but this has not yet been demonstrated.

2.4 Reaction with quinones

Physiologically, CO dehydrogenase passes the reducing equivalents obtained from oxidation of CO into the intramembrane ubiquinone pool via its FAD site, and several quinone species are in fact able to effectively reoxidize dithionite-reduced CO dehydrogenase (Wilcoxen et al., 2011b). Ubiquinone itself reacts with a second-order rate constant of 2.99×10^5 $M^{-1}s^{-1}$, while 1, 4-benzoquinone exhibits hyperbolic dependence on substrate concentration, with a k_{ox}/K_d of $2.13 \times 10^6 \text{ M}^{-1}\text{s}^{-1}$. On the basis of these numbers, it can be concluded that the rate-limiting step for overall turnover resides in the reductive half-reaction.

2.5 Experimental considerations

Growth of any organism on CO as sole carbon source requires attention to safety. In the author's laboratory, a 20 L fermentor is placed in a floor-to-ceiling hood in a small (~120 sq.ft.) room equipped with a CO sensor. After growth to late log phase, the culture is transferred to sealed, 1 L centrifuge bottles within the hood, and the cells then spun down. The supernatant is left in the hood overnight to degas, then disposed of. Harvested cells are handled in a conventional hood on ice, being treated with a solution containing both phenylmethylsulfonylfluoride to inhibit proteases and DNase to minimize viscosity prior to lysis using a French press. CODH is quite air-stable, and no extraordinary measures are required during purification to minimize exposure to O_2. Although membrane-associated in the cell, it is not membrane-integral and is readily soluble up to concentrations of 1 mM or higher. Protein purification (Zhang et al., 2010) involves standard FPLC steps involving Q-Sepharose and Sephacryl S-300 chromatography using a published protocol. Enzyme purity can be gauged by a ratio of absorbance at 450 nm to that at 280 nm of ~5.5.

The bridging sulfur and copper of the binuclear center of CO dehydrogenase can be removed by treatment of the enzyme with cyanide, and a protocol has been developed for reconstituting the binuclear center by successive treatment of the apoenzyme with sulfide and Cu(I)•thiourea (Resch et al., 2005). Interestingly, when Ag(I)•thiourea is used instead, activity is partially recovered (Wilcoxen et al., 2011a) and the enzyme thus reactivated can be fully reduced by CO, although at a fivefold slower rate than seen with native enzyme (8 s^{-1} as opposed to 51 s^{-1}). The EPR signal seen upon partial reduction of the silver-substituted enzyme by CO (Fig. 2) shows the doublets expected for substitution of Ag for Cu. The reduced activity of the silver-substituted enzyme can be rationalized as arising from the lower π-backbonding capacity of silver relative to copper, as weaker backbonding is expected to result in less weakening of the C-O bond of bound substrate and a higher activation barrier to reaction. On the basis of the relative rate constants for reduction of the copper- and silver-containing enzyme, the weaker backbonding in the silver-substituted enzyme reflects ~1 kcal/mol in compromised transition state stabilization upon substitution of silver for copper.

3. Formate dehydrogenases
3.1 Introduction

Enzymes that oxidize formate to CO_2 are broadly distributed and come in two quite different varieties: simple metal-independent enzymes that catalyze a direct hydride transfer from formate to NAD^+ via a ternary E•formate•NAD^+ complex, and metal-dependent enzymes containing molybdenum or tungsten that are frequently components of quite large assemblies. These latter enzymes operate via a ping-pong mechanism in which the molybdenum (or tungsten) center is reduced by formate, with the reducing equivalents thus obtained subsequently transferred to other redox-active centers in the system and eventually to an oxidizing substrate. While most metal-containing formate dehydrogenases function physiologically in the direction of formate oxidation to CO_2, at least some work has focused on the reaction in the opposite direction with the enzyme functioning as a CO_2 reductase, taking reducing equivalents from a source such as H_2 and using them to reduce CO_2 to formate in a reaction of considerable interest for bioremediation of atmospheric CO_2 and the energy economy. Despite a considerable literature devoted to whether a given formate dehydrogenase is or is not able to catalyze the reverse reaction, the reduction of CO_2 to formate, the most recent work strongly suggests that all these metal-containing enzymes are fully reversible when care is taken to use CO_2 rather than bicarbonate as substrate (the bicarbonate/CO_2 equilibrium in aqueous solution being relatively slow compared to the typical rates of enzyme turnover) (Yu et al., 2017). Although not considered further here, it is worth noting that the Mo- and/or W-containing subunits of formate dehydrogenases have considerable structural and functional homology to the corresponding subunits of formylmethanofuran dehydrogenases from methanogenic archaea (Thauer et al., 2008; Wagner et al., 2016). Indeed many organisms encode separate but very similar (in some cases virtually identical) formate dehydrogenases containing tungsten on the one hand or molybdenum on the other; the same is true for formylmethanofuran dehydrogenases.

3.2 Systematics

Whereas the CO dehydrogenase discussed above is a member of the xanthine oxidase family of molybdenum-containing enzymes, the molybdenum-containing formate dehydrogenases are members of the DMSO reductase family (Hille, 1996b; Hille et al., 2014b). Several formate dehydrogenases have been characterized crystallographically (Boyington et al., 1997; Raaijmakers et al., 2002; Jormakka et al., 2002; Oliveira et al., 2020) or by cryoelectron microscopy (Radon et al., 2020), <u>and found to possess very similar overall folds</u>. All contain a metal center coordinated to two equivalents of a pyranopterin-containing cofactor that chelates the metal via an enedithiolate side chain, as shown in Fig. 3; this cofactor is elaborated as the dinucleotide of guanine in all cases. The remainder of the metal coordination sphere is.

made up of a cysteine or selenocysteine residue provided by the polypeptide and a terminal sulfido ligand that is essential for catalytic activity (Thome et al., 2012). The *E. coli* FdhD gene product is a sulfur transferase that, in conjunction with the IscS cysteine desulfurase, catalyzes the insertion of this Mo=S ligand (Arnoux et al., 2015). In addition, most formate dehydrogenases have at least one, and as many as seven, iron–sulfur cluster that lies in close proximity to the molybdenum center.

The metal-containing formate dehydrogenases can be categorized on the basis of the metal present in the active site, molybdenum or tungsten, and whether the protein ligand to the metal is cysteine or selenocysteine. A final level of variation in these enzymes, as discussed further below, is their overall subunit architecture, which ranges from the simple, monomeric FdhF formate dehydrogenases from, e.g., *E. coli* (containing molybdenum and a single [4Fe-4S] cluster) (Boyington et al., 1997) to the cytosolic FdsDABG formate dehydrogenase from *Cupriavidus necator* or *Rhodobacter capsulatus*, containing seven iron-sulfur clusters and FMN in addition to its molybdenum center (Radon et al., 2020).

Mo–containing FdhF formate dehydrogenases from *E. coli* (with selenocysteine coordinated to the metal) (Axley et al., 1990a) or *Pectobacterium atrosepticum* (with cysteine) (Finney et al., 2019) are cytosolic components of membrane-integral formate:hydrogen lyases, fully reversible systems that either generate or consume H_2 depending on the growth conditions (Finney and Sargent, 2019). The W-containing Fdh1AB from *Methylobacterium extorquens* (containing cysteine) and the Mo-containing FdhABC from organisms such as *Desulfovibrio vulgaris* (containing selenocysteine) (Oliveira et al., 2020) are periplasmically localized. The FdnGHI and FdoGHI formate dehydrogenases from, e.g., *E. coli*, are $(\alpha\beta\gamma)_3$ trimers possessing Mo-containing subunits resembling the above FdhF proteins that are attached to membrane-integral, heme-containing subunits via a subunit containing four [4Fe-4S] clusters (Jormakka et al., 2002). FdnGHI is expressed under strictly anaerobic conditions, while FdoGHI is expressed

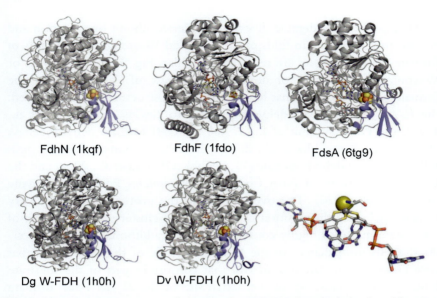

Fig. 3 *Representative structures of the metal-containing domains/subunits of formate dehydrogenases. Top*, from left to right: the molybdenum-containing FdhN formate dehydrogrogenase from *E. coli*, the FdhF formate dehydrogenase from *E. coli*; the C-terminal portion of FdsA from the FdsDABG formate dehydrogenase of *R. capsulatus*. *Bottom*: The tungsten-containing subunits of the formate dehydrogenases from *D. gigas* (left) and *D. vulgaris* (right). In all five cases, the domain containing a [4Fe-4S] cluster is shown in blue. The overall structural homologies among the five structures are evident. The PDB accession numbers used to generate these images are given. At bottom right is a close-up of the active site molybdenum center in the *R. capsulatus* C-terminal domain of FdsA.

in the transition from aerobic to anaerobic growth; both contain selenocysteine. The NAD$^+$-dependent FdsDABG enzymes from *Cupriavidus necator* (Friedebold and Bowien, 1993) and *Rhodobacter capsulatus* (Hartmann and Leimkuehler, 2013) are cytosolic and organized as $(\alpha\beta\gamma)_2$ dimers of trimers. More complicated still is the cytosolic H$_2$-dependent CO$_2$ reductase from, e.g., *Acetobacterium woodii* consisting of an iron-only hydrogenase subunit, two different subunits containing four [4Fe-4S] clusters, and an FdhF-like molybdenum- and [4Fe-4S]-containing subunit (Schuchmann and Mueller, 2013). Interestingly, the *A. woodii* genome encodes tandem genes for this last subunit, one encoding a subunit that has selenocysteine coordinated to the molybdenum and the other cysteine; it is presently not clear under what conditions each can be functionally incorporated.

Despite the considerable diversity in overall subunit organization and architecture, the protein folds of the Mo- and W-containing subunits of formate dehydrogenases are remarkably similar. Fig. 3 shows the Mo- or W-containing subunits (or domain in the case of FdsDABG) for the formate dehydrogenases that have to date been structurally characterized: the *E. coli* FdhF formate dehydrogenase (Mo, SeCys) (Boyington et al., 1997), the FdnG subunit of the *E. coli* FdnGHI formate dehydrogenase (Mo, SeCys) (Jormakka et al., 2002), the FdhA subunit of the *D. gigas* FdhAB formate dehydrogenase (W, SeCys) (Oliveira et al., 2020) and the C-terminal domain of FdsA from the *R. capsulatus* FdsDABG enzyme (Radon et al., 2020). The overall folds of the proteins are clearly very similar, including the position of the [4Fe-4S] cluster present in each of these subunits. Also highly conserved are two additional active site residues, a His immediately to the C-terminal side of the Cys/SeCys that coordinates the metal, and an Arg that constitutes part of the substrate binding site. A close-up of the active site of the *E. coli* FdhF formate dehydrogenase is shown in Fig. 4.

Formate dehydrogenases are thought to be evolutionarily ancient, tracing back to the Last Universal Common Ancestor to all extant life forms. Both the acetogenic Wood-Ljungdahl pathway of bacteria and the version found in methanogenic archaea begin with CO_2 reduction by a Mo- or W-containing enzyme (an H_2-oxidizing CO_2 reductase in the former case, a formylmethanofuran dehydrogenase in the latter) (Ragsdale and Pierce, 2008). The two branches share many common intermediates, including sequential formyl-, methenyl-, methylenyl- and methylpterin species.

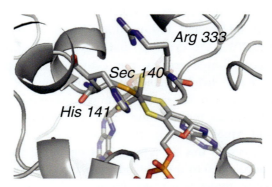

Fig. 4 *The active site of the E. coli FdhF formate dehydrogenase.* Shown are the highly conserved His 141 and Arg 333, as well as the molybdenum-coordinating Sec 140.

3.3 Reaction mechanism

Given the overall similarities in active site structures, it is likely that all molybdenum- (and tungsten)-containing formate dehydrogenases act via fundamentally the same reaction mechanism regardless of whether they possess Cys or SeCys coordinated to the metal, and over the years various mechanisms have been proposed for one or another enzyme (Khangulov et al., 1998; Leopoldini et al., 2008; Mota et al., 2011; Maia et al., 2016). A key mechanistic clue was the demonstration that oxygen from solvent was not incorporated into product CO_2 in the course of the reaction (Khangulov et al., 1998). Although the observation that oxygen from solvent is not incorporated into product CO_2 has recently been called into question (Kumar et al., 2023), the most recent work has clearly demonstrated that oxygen derived from solvent is not incorporated into product CO_2 (D. Niks, R. Hille and coworkers, *J. Am. Chem. Soc.*, in press).

With both the SeCys-containing *E. coli* FdhF (Khangulov et al., 1998) and Cys-containing *C. necator* FdsABG (Niks et al., 2016; Niks and Hille, 2019), the C_α hydrogen of formate is transferred to the molybdenum center in the course of formate oxidation and is strongly coupled in the EPR signal of the Mo(V) oxidation state. It has been proposed that the hydrogen is present as a Mo-SH group (Niks et al., 2016), consistent with the similarly strong hyperfine coupling of Mo-SH protons that is observed in both model complexes (Wilson et al., 1991; Young and Wedd, 1997) and other enzymes (Malthouse et al., 1981). It has been concluded that the reaction proceeds as a simple hydride transfer to give the fully reduced Mo (IV) species, with the EPR-detectable Mo(V) valence state formed upon subsequent transfer of one electron out of the molybdenum to other redox-active centers in the protein, as shown in Fig. 5. In support of this mechanism, formate is known to be a good hydride donor (the reduction of NAD^+ by the metal-free formate dehydrogenases necessarily occurs via hydride transfer) and the Mo=S group of other molybdenum-containing enzymes (notably xanthine oxidase (Xia et al., 1997), 1999) is known to be a good hydride acceptor. An important implication of this mechanism is

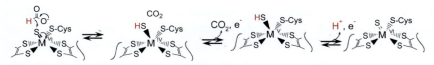

Fig. 5 *A simple, fully reversible hydride transfer mechanism for the metal-containing formate dehydrogenases.* After Niks et al. (2016).

that the molybdenum coordination sphere is coordination stable, remaining six-coordinate throughout the catalytic sequence with no dissociation of either the Mo=S or Mo-(Se)Cys ligands to the metal.

The hydride transfer mechanism has received wide, if not universal, acceptance, but it has recently been suggested that the reaction proceeds via sequential one-electron transfers rather than a single hydride transfer step (Nazemi et al., 2022; Liu et al., 2022). This seems unlikely on the basis of work with xanthine oxidase, another enzyme in which a Mo=S group functions as a hydride acceptor, where it has been shown unambiguously that there is no correlation between the one-electron reduction potential for a wide series of substrates (determined radiolytically) with either $\log(k_{red})$ or $\log(k_{red}/K_d)$ from rapid-reaction studies, indicating that the reaction does proceed in sequential one-electron steps (Stockert et al., 2002).

3.4 Experimental considerations

Selenocysteine- and/or tungsten-containing formate dehydrogenases are generally extremely air-sensitive, but some Mo/cysteine enzymes, e.g., the *C. necator* FdsABG enzyme, are less so when 10 mM nitrate, is added as a protectant (Friedebold and Bowien, 1993). O_2 inactivation of FdsDABG has been shown to involve $O_2{}^{\bullet-}$, and superoxide dismutase has been found to be as effective in retaining activity as nitrate; inhibition appears to involve oxidation of the catalytically essential sulfido group to sulfite (Hakopian et al., 2022). A general assay for the oxidation of formate to CO_2 follows the anaerobic reduction of benzyl or methyl viologen in the presence of formate at pH 7–7.5 (Axley et al., 1990a, 1990b). In addition to the above general activity assay, the NAD^+-dependent FdsABG activity can be conveniently followed using the absorbance change associated with reduction of NAD^+ (Niks et al., 2016). Assays in the reverse reaction should use a solution saturated with CO_2 (29.5 mM at standard temperature and pressure), with NADH or dithionite-reduced benzyl/methyl viologen as source of reducing equivalents (Yu et al., 2017).

Both native (Oliveira et al., 2020; Friedebold and Bowien, 1993; Axley et al., 1990b) and recombinant (Hartmann and Leimkuhler, 2013) formate dehydrogenase preparations have been reported. A common issue in both cases has to do with assessing the degree of functionality of the isolated material. Recombinant systems are prone to incomplete incorporation of molybdenum (or tungsten) to varying degrees, but even most native preparations may still have lower than maximal specific activity, frequently due to loss of (or failure to incorporate) the sulfido ligand to the metal that is known to be catalytically essential (Thome et al., 2012). Maintaining activity can be facilitated by inhibitors such as azide (Axley et al., 1990a), nitrate (Friedebold and Bowien, 1993) or superoxide dismutase (Hakopian et al., 2022). The extent of functionality can be easily estimated by comparing the extent of reduction with formate with that seen by a non-specific reagent such as sodium dithionite, as monitored by UV/visible spectroscopy.

4. Summary

Relatively straightforward procedures for expression and purification of both native and recombinant the CO dehydrogenase from aerobes such as *O. carboxidovorans* and metal-containing formate dehydrogenases from various obligate or facultative anaerobes have been described in the literature. EPR has proven extremely useful in characterizing the molybdenum (or tungsten, in the case of the formate dehydrogenases) centers that constitute their active sites, and their reaction mechanisms are fairly well understood despite the absence of convenient UV–visible spectroscopic probes that directly report on the active site. Still, these enzymes invariably possess one or more additional redox-active centers in addition to the molybdenum center (minimally a [4Fe-4S] cluster adjacent to the active site) and these provide convenient spectral changes that permit the accumulation of reducing equivalents that accumulate in the course of reaction with substrate to be monitored. There is a general, if not universal, agreement as to the reaction mechanism of both types of enzyme, although considerable work remains to elucidate the specific catalytic roles of active site amino acid residues, particularly in the case of the formate dehydrogenases. Not directly related to reaction mechanism, considerable efforts are under way to develop industrially effective systems based on one or another formate dehydrogenase for the interconversion of formate and CO_2, both in remediation of atmospheric CO_2 and the reversible storage of energy from renewable sources.

Acknowledgment

Work in the author's laboratory is supported by the US Department of Energy (grant DE-FG02–13ER16411).

References

Appel, A. M., Bercaw, J. E., Bocarsly, A. B., Dobbek, H., DuBois, D. L., et al. (2013). *Chemical Reviews, 113*, 6621–6658.

Arnoux, P., Ruppelt, C., Oudouhou, F., Lavergne, J., Siponen, M. I., et al. (2015). *Nature Communications, 6.*

Axley, M. J., Grahame, D. A., & Stadtman, T. C. (1990a). *Journal of Biological Chemistry, 265,* 18213–18218.

Axley, M. J., Grahame, D. A., & Stadtman, T. C. (1990b). *FASEB Journal, 4* A1978-A.

Boyington, J. C., Gladyshev, V. N., Khangulov, S. V., Stadtman, T. C., & Sun, P. D. (1997). *Science, 275,* 1305–1308.

Can, M., Armstrong, F. A., & Ragsdale, S. W. (2014). *Chemical Reviews, 114,* 4149–4174.

Cypionka, H., & Meyer, O. (1983). *Journal of Bacteriology, 156,* 1178–1187.

Dobbek, H., Gremer, L., Kiefersauer, R., Huber, R., & Meyer, O. (2002). *Proceedings of the National Academy of Sciences of the United States of America, 99,* 15971–15976.

Dobbek, H., Gremer, L., Meyer, O., & Huber, R. (1999). *Proceedings of the National Academy of Sciences of the United States of America, 96,* 8884–8889.

Finney, A. J., Lowden, R., Fleszar, M., Albareda, M., Coulthurst, S. J., & Sargent, F. (2019). *Molecular Microbiology, 112,* 1440–1452.

Finney, A. J., & Sargent, F. (2019). Chapter Eight - Formate hydrogenlyase: A group 4 [NiFe]-hydrogenase in tandem with a formate dehydrogenase. In R. K. Poole (Vol. Ed.), *Advances in Microbial Physiology. Vol. 74. Advances in Microbial Physiology* (pp. 465–486). Elsevier.

Friedebold, J., & Bowien, B. (1993). *Journal of Bacteriology, 175,* 4719–4728.

Frohman, D. J., Grubbs, G. S., II, Yu, Z., & Novick, S. E. (2013). *Inorganic Chemistry, 52,* 816–822.

Gourlay, C., Nielsen, D. J., White, J. M., Knottenbelt, S. Z., Kirk, M. L., & Young, C. G. (2006). *Journal of the American Chemical Society, 128,* 2164–2165.

Hakopian, S., Niks, D., & Hille, R. (2022). *Journal of Inorganic Biochemistry, 231.*

Hanzelmann, P., Dobbek, H., Gremer, L., Huber, R., & Meyer, O. (2000). *Journal of Molecular Biology, 301,* 1221–1235.

Hartmann, T., & Leimkuehler, S. (2013). *Febs Journal, 280,* 6083–6096.

Hartmann, T., & Leimkuhler, S. (2013). *FEBS Journal, 280,* 6083–6096.

Hille, R., Hall, J., & Basu, P. (2014a). *Chemical Reviews, 114,* 3963–4038.

Hille, R., Hall, J., & Basu, P. (2014b). *Chemical Reviews, 114,* 3963–4038.

Hille, R., & Sprecher, H. (1987). *Journal of Biological Chemistry, 262,* 10914–10917.

Hille, R. (1996a). *Chemical Reviews, 96,* 2757–2816.

Hille, R. (1996b). *Chemical Reviews, 96,* 2757–2816.

Hofmann, M., Kassube, J. K., & Graf, T. (2005). *Journal of Biological Inorganic Chemistry, 10,* 490–495.

Hulley, E. B., Welch, K. D., Appel, A. M., DuBois, D. L., & Bullock, R. M. (2013). *Journal of the American Chemical Society, 135,* 11736–11739.

Jacobitz, S., & Meyer, O. (1989). *Journal of Bacteriology, 171,* 6294–6299.

Jaffe, L. S. (1970). *Annals of the New York Academy of Sciences, 174,* 76.

Jormakka, M., Tornroth, S., Byrne, B., & Iwata, S. (2002). *Science, 295,* 1863–1868.

Kang, B. S., & Kim, Y. M. (1999). *Journal of Bacteriology, 181,* 5581–5590.

Kaufmann, P., Duffus, B. R., Teutloff, C., & Leimkuehler, S. (2018). *Biochemistry, 57,* 2889–2901.

Khangulov, S. V., Gladyshev, V. N., Dismukes, G. C., & Stadtman, T. C. (1998). *Biochemistry, 37,* 3518–3528.

Kraut, M., Hugendieck, I., Herwig, S., & Meyer, O. (1989). *Archives of Microbiology, 152,* 335–341.

Kress, O., Gnida, M., Pelzmann, A. M., Marx, C., Meyer-Klaucke, W., & Meyer, O. (2014). *Biochemical and Biophysical Research Communications, 447,* 413–418.

Kumar, H., Khosraneh, M., Bandaru, S. S. M., Schulzke, C., & Leimkuehler, S. (2023). *Molecules, 28.*

Leopoldini, M., Chiodo, S. G., Toscano, M., & Russo, N. (2008). *Chemistry—A European Journal, 14*, 8674–8681.

Liu, M., Nazemi, A., Taylor, M. G., Nandy, A., Duan, C., et al. (2022). *Acs Catalysis, 12*, 383–396.

Maia, L. B., Fonseca, L., Moura, I., & Moura, J. J. G. (2016). *Journal of the American Chemical Society, 138*, 8834–8846.

Malthouse, J. P. G., George, G. N., Lowe, D. J., & Bray, R. C. (1981). *Biochemical Journal, 199*, 629–637.

Matthies, A., Rajagopalan, K. V., Mendel, R. R., & Leimkuhler, S. (2004). *Proceedings of the National Academy of Sciences of the United States of America, 101*, 5946–5951.

Meyer, O., Frunzke, K., & Mördorf, G. (1993). *Microbial Growth on C1 Compounds, 27*.

Meyer, O., & Schlegel, H. G. (1980). *Journal of Bacteriology, 141*, 74–80.

Meyer-Klaucke, W., Gnida, M., Ferner, R., Gremer, L., & Meyer, O. (2001). *Journal of Inorganic Biochemistry, 86*, 339.

Moersdorf, G., Frunzke, K., Gadkari, D., & Meyer, O. (1992). *Biodegradation, 3*, 61–82.

Mota, C. S., Rivas, M. G., Brondino, C. D., Moura, I., Moura, J. J. G., et al. (2011). *Journal of Biological Inorganic Chemistry, 16*, 1255–1268.

Moxley, J. M., & Smith, K. A. (1998). *Soil Biology & Biochemistry, 30*, 65–79.

Nazemi, A., Steeves, A. H., Kastner, D. W., & Kulik, H. J. (2022). *Journal of Physical Chemistry B, 126*, 4069–4079.

Neumann, M., Schulte, M., Junemann, N., Stocklein, W., & Leimkuhler, S. (2006). *Journal of Biological Chemistry, 281*, 15701–15708.

Niks, D., Duvvuru, J., Escalona, M., & Hille, R. (2016). *Journal of Biological Chemistry, 291*, 1162–1174.

Niks, D., & Hille, R. (2019). *Protein Science, 28*, 111–122.

Oliveira, A. R., Mota, C., Mourato, C., Domingos, R. M., Santos, M. F. A., et al. (2020). *Acs Catalysis, 10*, 3844–3856.

Pelzmann, A., Ferner, M., Gnida, M., Meyer-Klaucke, W., Maisel, T., & Meyer, O. (2009). *Journal of Biological Chemistry, 284*, 9578–9586.

Raaijmakers, H., Macieira, S., Dias, J. M., Teixeira, S., Bursakov, S., et al. (2002). *Structure, 10*, 1261–1272.

Radon, C., Mittelstadt, G., Duffus, B. R., Burger, J., Hartmann, T., et al. (2020). *Nature Communications, 11*, 1912.

Ragsdale, S. W., & Pierce, E. (2008). *Biochimica et Biophysica Acta-Proteins and Proteomics, 1784*, 1873–1898.

Resch, M., Dobbek, H., & Meyer, O. (2005). *Journal of Biological Inorganic Chemistry, 10*, 518–528.

Santiago, B., Schubel, U., Egelseer, C., & Meyer, O. (1999). *Gene, 236*, 115–124.

Schuchmann, K., & Mueller, V. (2013). *Science, 342*, 1382–1385.

Schumann, S., Saggu, M., Moller, N., Anker, S. D., Lendzian, F., et al. (2008). *Journal of Biological Chemistry, 283*, 16602–16611.

Shanmugam, M., Wilcoxen, J., Habel-Rodriguez, D., Cutsail, G. E. I., Kirk, M. L., et al. (2013). *Journal of the American Chemical Society, 135*, 17775–17782.

Siegbahn, P. E. M., & Shestakov, A. F. (2005). *Journal of Computational Chemistry, 26*, 888–898.

Spreitler, F., Brock, C., Pelzmann, A., Meyer, O., & Koehler, J. S. (2010). *Chembiochem, 11*, 2419–2423.

Stockert, A. L., Shinde, S. S., Anderson, R. F., & Hille, R. (2002). *Journal of the American Chemical Society, 124*, 14554–14555.

Thauer, R. K., Kaster, A. K., Seedorf, H., Buckel, W., & Hedderich, R. (2008). *Nature Reviews Microbiology, 6*, 579–591.

Thome, R., Gust, A., Toci, R., Mendel, R., Bittner, F., et al. (2012). *Journal of Biological Chemistry, 287*, 4671–4678.

Wagner, T., Ermler, U., & Shima, S. (2016). *Science, 354*, 114–117.

Wang, N., Wang, M., Chen, L., & Sun, L. (2013). *Dalton Transactions, 42*, 12059–12071.

Wilcoxen, J., & Hille, R. (2013). *Journal of Biological Chemistry, 288*, 36052–36060.

Wilcoxen, J., Snider, S., & Hille, R. (2011a). *Journal of the American Chemical Society, 133*, 12934–12936.

Wilcoxen, J., Zhang, B., & Hille, R. (2011b). *Biochemistry, 50*, 1910–1916.

Wilson, G. L., Greenwood, R. J., Pilbrow, J. R., Spence, J. T., & Wedd, A. G. (1991). *Journal of the American Chemical Society, 113*, 6803–6812.

Xia, M., Dempski, R., & Hille, R. (1999). *Journal of Biological Chemistry, 274*, 3323–3330.

Xia, M., Ilich, P., Dempski, R., & Hille, R. (1997). *Biochemical Society Transactions, 25*, 768–773.

Young, C. G., & Wedd, A. G. (1997). *Chemical Communications,* 1251–1257.

Yu, X. J., Niks, D., Mulchandani, A., & Hille, R. (2017). *Journal of Biological Chemistry, 292*, 16872–16879.

Zhang, B., Hemann, C. F., & Hille, R. (2010). *Journal of Biological Chemistry, 285*, 12571–12578.

> CHAPTER TWELVE

Assessing the role of redox carriers in the reduction of CO_2 by the oxo-acid: ferredoxin oxidoreductase superfamily

Sheila C. Bonitatibus, Mathew Walker, and Sean J. Elliott*

Department of Chemistry, Boston University, Cummington Mall, Boston, MA, United States
*Corresponding author. e-mail address: elliott@bu.edu

Contents

1. Introduction	276
2. General considerations	279
3. Materials and reagents	279
3.1 Molecular biology of the *Chlorobaculum* Fd proteins	279
3.2 Purification and characterization of the recombinant *Chlorobaculum* Fd proteins	280
3.3 Additional reagents	281
4. Assay design and implementation	281
4.1 The basic metronitazole assay	281
4.2 Coupled biochemical assay with WT *Ct* PFOR	282
4.3 Fd-mediated electrocatalytic assay	282
5. Verification of redox potentials of Fd proteins	284
5.1 General method	284
5.2 Electrochemical characterization of Fd proteins	285
6. Comparison of assay results	286
6.1 Comparison of *Ct* Fd reactivity with *Ct* PFOR – *metronidazole assay*	286
6.2 Comparison of *Ct* Fd reactivity with *Ct* PFOR – *reductive assay*	286
6.3 *Ct* Fds as electron acceptors for *Ct* PFOR	288
7. Conclusions and future directions	292
Acknowlegments	293
References	293

Abstract

The oxo-acid:ferredoxin oxidoreductase (OFOR) superfamily of enzymes are responsible for the reversible interconversion of CO_2 and oxo-acids, using CoA-derivatives as co-substrates, and requiring redox equivalents in the form of a soluble redox-carrier protein ferredoxin (Fd). Ultimately, these enzymes are responsible for the reduction of CO_2 to form pyruvate (in the case of PFOR) and oxo-glutarate (in the case of OGOR),

---

Methods in Enzymology, Volume 708
ISSN 0076-6879, https://doi.org/10.1016/bs.mie.2024.10.022
Copyright © 2024 Elsevier Inc. All rights are reserved, including those for text and data mining, AI training, and similar technologies.

by the reductive carboxylation reaction of acetyl-CoA and succinyl-CoA, respectively. The nature and kind of Fd that is the best redox-carrier to support the reductive reaction has been poorly studied to date. Most organisms that possess an OFOR contain multiple Fd redox-carriers (in addition to flavin-based flavodoxins). Here, we provide a guide for the comparison of various, similar, but non-identical Fd proteins that can interact with the PFOR from *Chlorobaculum tepidum*, as a model system. The conventional assay is presented, alongside an electrochemically detected assay, which demonstrates the inequivalence of Fd proteins in supporting either component of catalysis.

1. Introduction

Members of the oxo-acid:ferredoxin oxidoreductase superfamily interconvert oxo-acids and CO_2 through the oxidative/reductive decarboxylation chemistry that is illustrated by Eq. 1 (Ragsdale, 2003; Ragsdale and Pierce, 2008).

$$\underset{R}{\overset{O}{\underset{O}{\bigvee}}}OH + CoASH + 2Fd_{ox} \rightleftharpoons CO_2 + \underset{R}{\overset{O}{\bigvee}}SCoA + 2Fd_{red} + 2H^+ \tag{1}$$

The best studied member of the OFOR family is PFOR, wherein pyruvate can be oxidized to CO_2 and acetyl-CoA, and two electrons are liberated in the form of a soluble redox carrier, ferredoxin (Fd) (Chen et al., 2018; Furdui and Ragsdale, 2000, 2002). OFORs are thought to be reversible enzymes, though the extent to which oxidation and reduction reactions are favored is not well understood, as is the requirement for potentially specific Fd proteins. As OFORs are found in the reverse TCA cycle and other reductive pathways where they must engage in the carboxylation reaction of acetyl-CoA and succinyl-CoA (to synthesize oxoglutarate) (Yoon et al., 1999; Chen et al., 2019; Bock et al., 1996), we have presumed that Fd proteins that are low in redox potential will favor CO_2 reduction.

To date, little has been shown to indicate which Fd may be the "best" or "native" redox partner for a given redox enzyme, though it has been established that Fd proteins can have specific pairings (Burkhart et al., 2019), including OFORs. In terms of OFOR based CO_2 reduction it is similarly unclear what the redox potentials of various Fd proteins are, or what types of interactions (electrostatic or hydrophobic) promote Fd recognition and binding by an OFOR partner protein. An underexplored question is whether these redox or structural properties of a Fd determine its "fitness" to support higher rates of oxidative or reductive chemistry. Here, we provide foundational data for the *Chlorobaculum tepidum* (*Ct*) system. *Ct* possess three similar, but not identical, Fd proteins that vary in sequence, but all bind two [4Fe−4S] clusters. Here we present a comparison of the solution-based assays that might be typically used to assess this question, as well as an electrochemical assay that provides the simultaneous read-out of oxidative and reductive activity.

There are good reasons to hypothesize that there may be some specificity with respect to Fd:OFOR interactions. Bioinformatic studies have indicated that the homologs of small electron transfer (ET) proteins (such as Fds) have evolved to increase specificity towards a particular subset of oxidoreductase partners by tuning both protein structure (partner binding) and reduction potentials of redox cofactors (Campbell et al., 2019). Recent expression and pull-down studies from the Santangelo lab demonstrated that each of the three Fds encoded by *Thermococcus kodakarensis* has a unique expression pattern. These differences in transcriptional regulation suggest that the Fds play unique roles within the cell and are partnered with specific oxidoreductase complexes (Burkhart et al., 2019). Together, these results support the idea that organisms evolve multiple Fd isoforms and use Fd:oxidoreductase pairs which may be specific, where the specificity could be governed by protein-protein interaction on/off rates, ET rates, and favorable or less favorable redox potentials.

We have made significant advances toward understanding the influence of the redox potential of a Fd on the catalysis of an OFOR partner protein (Chen et al., 2019; Li et al., 2021; Li and Elliott, 2016). Specifically, studies of the KorAB OGOR enzyme demonstrated that the reduction potential of the interacting Fd determined OFOR directionality and the magnitude of reactivity (turnover frequency), suggesting that redox potential may serve as one mechanism for promoting redox partner specificity (Li et al., 2021). Yet there are few experimental studies that help establish the ground-rules for what determines how or why some Fd proteins may be required for a given reaction, and whether distinct pairings of Fds with a given enzyme make a difference in the enzyme's reactivity. We believe that the *Ct* Fd:PFOR pairing may be a good model system to explore the biochemical aspects that govern Fd binding and reactivity.

In terms of the specific *Ct* system, Tabita and co-workers have demonstrated that a high-potential rubredoxin (Rd, −87 mV vs. standard hydrogen electrode (SHE)) was able to accept electrons from PFOR. Thus, the Rd was proposed to be the physiological electron acceptor for the oxidative reaction (Yoon et al., 1999), and subsequently, a low-potential ferredoxin was designated as the electron donor for the reductive reaction (Yoon et al., 2001). Here, we compare the three possible Fd proteins of *Chlorobaculum* side-by-side, in terms of reactivity and biochemical activity of *Ct* PFOR, in both the oxidative and reductive directions. We aim to identify whether the PFOR enzyme displays a difference in reactivity when using various low-potential Fds and/or to determine if the enzyme shows a preference towards any specific Fd.

We have selected to study the *Ct* system, in part because this organism only encodes three ferredoxins, which is a manageable number, and each of these ferredoxins are of the 2x[4Fe–4S] cluster type (Eisen et al., 2002). *Ct* Fd1 and Fd2 have been hypothesized to serve as electron donors to the *Ct* PFOR, however, their redox potentials, and rate constants for electron transfer, as well as their ability to bind to PFOR and other enzymes are not well established (Eisen et al., 2002). *Ct* Fd3 had not been identified previous to our work, and therefore all aspects of its biochemical function and characteristics are unknown.

From a physiological point of view, we hypothesize that the preferred Fd redox partner should exhibit better kinetic properties (faster turnover rates) in its interaction with *Ct* PFOR during catalysis in comparison to the other available Fds. Yet the lack of previous data (reduction potentials, ET rates, K_m values, or even measured catalytic rates) makes it unclear how large a kinetic effect can be anticipated. To address this knowledge gap in this Chapter, we study all three *Ct* Fds with *Ct* PFOR by first determining the redox potentials of each Fd by direct protein film voltammetry, and then examining the ability of each Fd to participate in oxidative or reductive PFOR chemistry via kinetic assays. We also utilize an electrochemical assay developed by the Elliott lab (Li and Elliott, 2016) to gain insight into the thermodynamic properties of the Fd:PFOR system as a whole by determining the catalytic potential associated with oxidative vs. reductive catalysis and its dependence on the partnering protein. The biochemical and electrochemical data will help to identify the native redox partner for *Ct* PFOR and highlight the importance of identifying the correct cognate redox partner for accurately studying catalytic bias for reversible enzymes.

2. General considerations

The methods described here require typical biochemical laboratory equipment for the purification of redox-sensitive ferredoxin proteins:
- sterile environments
- temperature controlled shakers
- temperature controlled incubators
- temperature controlled centrifuges
- gravity-type columns or an FPLC for protein purification purposes
- an optical spectrometer for time-dependent and wavelength dependent experiments
- protein purification of Fd proteins can require anaerobic conditions, so purification steps are performed in a Coy-style anaerobic chamber
- electrochemical assays require anaerobic conditions, meaning that electrochemical analyzers must be interfaced with an anaerobic glove box, such as an MBraun or Vac-Atmospheres hard-style glove box.
- electrochemical analyzers to empower the voltammetric and amperometric assays. Such equipment has been largely described previously by Maiocco, et al (Maiocco et al., 2018), with the following additional points of consideration:
- In order to engage in the steady-state evaluation of Fd-mediated catalytic current, one requires rotating disk electrodes, as described by Armstrong et al. (1997), Blandford and Armstong (2006), Jeuken (2016). Here, the material of choice is pyrolytic graphite, as the potentials are sufficiently low that typical alkane-thiol modified gold, and indium–doped tin oxide materials may suffer.
- The use of a rotating electrode requires the use of a rotating system (Ametek Scientific Instruments), which can rotate a disk electrode in a cell solution, to provide for the flux required to bring molecules to the electrode surface, and minimize the impact of diffusionally controlled events at the electrode surface.

3. Materials and reagents
3.1 Molecular biology of the *Chlorobaculum* Fd proteins

The DNA encoding *Ct* Fd1 (gene locus: CT1261), *Ct* Fd2 (gene locus: CT1260), and *Ct* Fd3 (gene locus: CT1736) were codon optimized for expression in *E. coli*, and the corresponding oligonucleotides were synthesized (IDT) and inserted into pCDFduet-1 vector (Novagen/Millipore) between the *Nco*I and *Xho*I restriction sites. All plasmid constructs were confirmed by Sanger DNA sequencing (Genewiz). In this strategy a second cloning site is available if needed for co-expression of a second gene, such as an OFOR. We have also opted for a tag-free purification strategy, described below, to prevent the potential complication from a typical His_6-tag that might bind metal ions adventitiously, and then compromise electrochemical/biophysical characterization.

3.2 Purification and characterization of the recombinant *Chlorobaculum* Fd proteins

The plasmids for expression of *Ct* Fds were separately transformed into *E. coli* BL21(DE3) $\Delta iscR$ cells and selected on LB-agar plates with 100 µg/mL streptomycin. For the starter culture, a single colony was inoculated into 5 mL of LB, supplemented with the same concentration of streptomycin, and grown overnight at 37°C. The starter culture was then inoculated into 1 L of 2x Yeast Extract Tryptone (YT) media and grown at 37°C with shaking at 200 rpm until an OD_{600} of 0.4–0.6. At this point, the cells were cooled at 4°C for 30 min and then supplemented with 1 mM ferrous ammonium sulfate and 500 µM IPTG. Cells were grown overnight at 20°C with a shaking rate of 100 rpm. The following morning, cells were harvested by centrifugation and were stored at -20°C until purification. A typical growth yielded 3 g/L (wet weight).

The tag-less purification of each ferredoxin was accomplished using tandem anion exchange chromatography under anaerobic conditions. All buffers were made anaerobic before use. For each Fd, the cell pellet (~3 g, from a 1 L growth) was allowed to thaw before being resuspended with 30 mL of lysis buffer (50 mM phosphate (pH 7.0), 1 mM PMSF, 1 mM DTT, 5 U/mL DNase, 1 mg/mL lysozyme). The suspension was incubated on cold beads for 30 min before sonication. The crude cell lysate was transferred to a centrifuge bottle and secured with a sealed cap (O-ring) and vinyl tape to minimize oxygen exposure and clarified by centrifugation (15,000 g, 30 min, 4°C).

The clarified cell lysate was loaded onto a DEAE-Sepharose column (25 mL column bed) that was preequilibrated with 5 CV (125 mL) of wash buffer (20 mM Tris-HCl, 1 mM DTT, pH 7.8). The column was washed with 5 CV of the same buffer supplemented with 150 mM NaCl, followed by elution with a step gradient from 150 to 350 mM NaCl (increasing in steps of 25 mM NaCl/step, 2 CV per step, 450 mL total). Fractions containing ferredoxin were visually identified by their dark brown color.

Ct Fd1 eluted at 325 mM NaCl, *Ct* Fd2 eluted at 300 mM NaCl, and *Ct* Fd3 eluted at 275 mM NaCl.

The pooled, protein containing fractions were concentrated from ~ 50 mL to 1 mL using an Amicon ultra centrifugal filter (3 kDa MWCO), diluted 30-fold (30 mL) in a salt-free buffer (20 mM Tris-HCl (pH 7.8), 1 mM DTT) and applied to a Q-Sepharose high-performance column (25 mL column volume) preequilibrated with 5 CV of the wash buffer supplemented with 250 mM NaCl, and each Fd bound to the column as a brown band. *Ct* Fd1 and *Ct* Fd2 were eluted using a step gradient from 300 to 500 mM NaCl (increasing by 25 mM NaCl/step, 2 CV per step). *Ct* Fd 3 was eluted in the same manner, except the step gradient was between 250 and 450 mM NaCl. Ferredoxin-containing fractions were pooled, exchanged into a storage buffer containing 50 mM HEPES (pH 8.0), 100 mM NaCl, 1 mM $MgCl_2$, and 10 % glycerol (w/v) and concentrated by Amicon ultra centrifugal filters (3 kDa MWCO, EMD-Millipore). Protein purity was established by SDS-PAGE analysis, however, the Fds often migrated with the dye-front due to their small size. Thus protein purity was estimated by comparing the ratio between the ferredoxin concentration determined by experimentally determined molar extinction coefficients of 31.6 $mM^{-1} \cdot cm^{-1}$ (*Ct* Fd1 and Fd3) and 30.9 $mM^{-1} \cdot cm^{-1}$ (*Ct* Fd2) at 395 nm (Yoon et al., 2001) and the protein concentration determined by the standard Bradford assay. Typical protein purity was ~ 90 %. From a 2 L growth, 6 g of cell pellet, a purification would yield a 1 mL stock of 600 μM Fd.

3.3 Additional reagents

Purifications of WT *Mm* OGOR was achieved as described in reference (Chen et al., 2019), and the *Ct* PFOR was achieved as described (Bonitatibus, 2024, *submitted*). Lactate dehydrogenase (isolated from rabbit muscle, 1005 U/mg, 9.3 mg/mL) was purchased from Sigma.

4. Assay design and implementation
4.1 The basic metronitazole assay

UV–vis absorption spectra of the *Ct* Fds were recorded under anaerobic conditions in quartz cuvettes (1 cm optic path) with an Olis spectrophotometer. To assess the electron transfer efficiency between the WT *Ct* PFOR and three *Ct* Fds, a time-dependent reoxidation of reduced Fds (by PFOR) was coupled to metronidazole (MNZ) reduction (Fig. 1). The

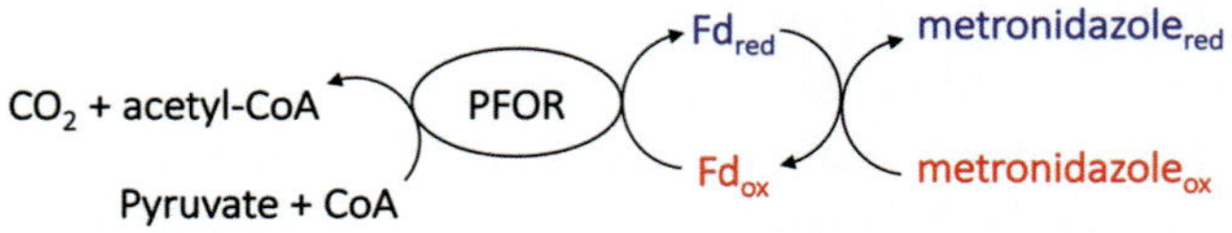

Fig. 1 Schematic representation of the coupled enzyme assay between Fd and MNZ. Fd acts as an electron acceptor for *Ct* PFOR during pyruvate oxidation. Metronidazole is utilized as an electron acceptor ($\varepsilon 320 = 9.3\ \text{mM}^{-1}\text{•cm}^{-1}$).

absorbance change was monitored by collecting UV–Vis spectra every 5 s at 320 nm. The assay for the coupled metronidazole reduction contained 10 mM pyruvate, 200 μM CoA, 0 or 1 μM *Ct* Fds, 100 μM metronidazole, and 100 nM *Ct* PFOR and was carried out in a 50 mM HEPES buffer (pH 7.5). A control reaction without Fd was used to measure the background rate of reaction between *Ct* PFOR and metronidazole.

4.2 Coupled biochemical assay with WT *Ct* PFOR

The coupled assay for CO_2 reduction of *Ct* PFOR using *Ct* Fds included a reduced Fd-regeneration system (upstream reaction), by *Mm* OGOR (Chen et al., 2019), shown in Fig. 2 and a pyruvate detection system (downstream, by lactate dehydrogenase (LDH)) such that the overall readout arises from $NADH/NAD^+$ detection via the LDH enzyme. The reaction was performed at 30°C in a multi-component buffer (10 mM MOPS, CAPS, TAPS, CHES, and MES at pH 7.0) under anaerobic conditions. The assay mixture consisted of 10 mM 2-oxoglutarate, 0.2 mM CoA, 500 nM of *Mm* OGOR, 5 mM DTT, 0.1 mM NADH, 20 mM sodium bicarbonate, 10 μM Fd, 100 nM *Ct* PFOR, and 25 nM LDH. Reaction components were allowed to sit for 5 min to allow the Fds to reduce from the upstream reaction, and pyruvate synthesis (CO_2 reduction) was initiated by the addition of acetyl-CoA (final concentration 1 mM). The change in absorbance at 340 nm was monitored for the oxidation of NADH ($\varepsilon_{340} = 6.22\ \text{mM}^{-1}\text{•cm}^{-1}$) (Furdui and Ragsdale, 2000). The oxidation of one equivalent of NADH corresponded to the synthesis of one equivalent of pyruvate (and, therefore, the reduction of one equivalent amount of CO_2).

4.3 Fd-mediated electrocatalytic assay

Ct PFOR interacts with a given Fd in the presence of its oxidative or reductive substrates (Fig. 3), and the electrochemical readout in the form of current is due to the shuttling of electrons from the redox mediator to the

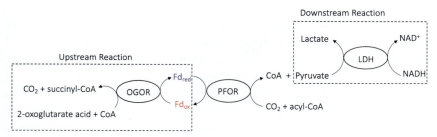

Fig. 2 Schematic representation of the CO$_2$ reduction coupled assay. Fd as the electron donor for *Ct* PFOR during pyruvate synthesis (CO$_2$) reduction. Fd is kept reduced by the upstream reaction (*Mm* OGOR) and pyruvate as the product is detected by LDH (downstream).

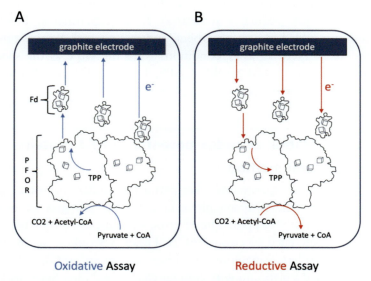

Fig. 3 The schematic presentation of the electrocatalytic assay. Flow of electrons in (A) the oxidative assay and (B) the reductive assay. OFOR enzyme, Fds and substrate were present in the electrochemical cell. Both 2-oxoacid oxidation (blue arrow) and CO$_2$ reduction (red arrow) are depicted. Fds mediate ET between the PFOR and electrode; therefore, the expected catalytic current is reflective of either the continuous re-oxidation of Fd (A, anodic current) or the continuous re-reduction of Fd (B, cathodic current). The arrowheads indicate the direction of electron flow.

electrode surface (Li et al., 2021; Li and Elliott, 2016; Firer-Sherwood et al., 2011). Here, through control experiments it was established that at rotation rates used and over the range of redox potential studied, PFOR on its own would give rise to no current beyond the baseline, and that all catalytic features can be attributed to redox shuttling. The experiments were carried out over the potential range of −0.3 to −1.05 V (vs. standard calomel electrode (SCE)). Cyclic voltammetry was conducted with a scan rate of 1 mV/s with a 0.15 mV step potential and using a rotation rate of 200 rpm at the working electrode. All experiments must be conducted in an anaerobic glove-box, due to the non-specific reduction of molecular oxygen that might occur under laboratory atmosphere. The rotation that is appropriate must be determined empirically by increasing the rotation rate through multiple experiments, until there is no increase in the maximal current. PFOR was added to a final concentration of 1 μM, in a 1:5 (PFOR:Fd) ratio with ferredoxins, in the presence of 10 mM pyruvate and 0.2 mM CoA (oxidative substrates), or 20 mM sodium bicarbonate and 1 mM acetyl-CoA (reductive substrates). In all experiments, 1 mM of neomycin was added as a co-absorbent, and as a control, Fd was omitted from the assay. Here all assay conditions were the same as the spectrophotometric assays (10 mM MOPS, CAPS, TAPS, CHES, and MES at pH 7.0, at room temperature).

5. Verification of redox potentials of Fd proteins
5.1 General method

All electrochemical experiments were performed under anaerobic conditions in a Mbraun LAB-master glove box under a nitrogen atmosphere with an O_2 level of < 1 ppm. The electrochemical cell consists of a standard calomel reference electrode (SCE, Fisher Scientific), an in-house, edge-plane pyrolytic graphite working electrode (PGE, Minerals Technology), and a platinum wire counter electrode (Fisher Scientific). For the Fd redox potential measurement a polished PGE electrode was further modified by the addition of multi-walled carbon nanotubes (MWCNTs): to the polished electrode a MWCNTsolution (14 μL of a 3 mg/mL solution of multi-walled carbon nanotubes in dimethylformamide (DMF)) was added, and incubated overnight at room temperature. Voltammetry experiments were conducted with a PGSTAT 12 potentiostat operated by GPES (Metrohm/ Eco Chemie Autolab); experiments were conducted at 8°C in a multi-component buffer (5 mM acetate, MOPS, CAPS, TAPS, CHES, and MES at pH 7.0). Voltammograms were collected at 50 or 100 mV/s with a step potential of 0.3 mV.

5.2 Electrochemical characterization of Fd proteins

We expected the clusters of the *Ct* Fds to possess low reduction potentials (< −450 mV) and hypothesized that their values may be important factors in supporting the reductive catalysis for CO_2 fixation. Therefore, cyclic voltammetry experiments of the *Ct* Fds were performed to determine the cluster potentials of each Fd to help rationalize which one is most or least likely to be a physiologically relevant partner to PFOR. Each Fd was electrochemically active and gave rise to broad features spanning the potential range of −600 mV to −500 mV vs. SHE (Fig. 4). The broad envelop signal for each Fd was fit to two, one-electron transfers that correspond to the two $[4Fe-4S]^{2+/1+}$ redox couples. The reduction potentials for each ET event were extracted using QSOAS and are tabulated in Table 1.

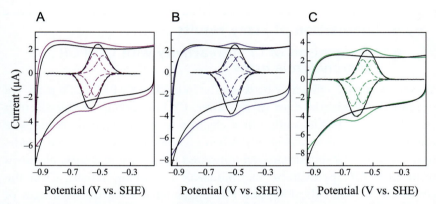

Fig. 4 Cyclic voltammetry of *Ct* Fds. (A) *Ct* Fd1, (B) *Ct* Fd2, and (C) *Ct* Fd3. Experimental conditions: temperature: 8°C; pH 8.0; scan rate: 100 mV/s; step potential: 0.15 mV. For each of the figures the raw data is shown in color, compared to the background response of the unmodified electrode. The inset data are the data (black-solid) corrected for background, and deconvoluted into two one-electron redox components (dashed lines).

Table 1 Reduction potentials of *Ct* Fds.

Ferredoxin	E_{M1}	E_{M2}
Fd 1	−510	−560
Fd 2	−500	−550
Fd 3	−540	−600

E_M (mV vs. SHE)

For all three *Ct* Fds, cluster I is ligated by a clostridial-type cluster motif and cluster II is ligated by a chromatium-type cluster motif. Electrochemistry on its own does not establish which potential corresponds to which cluster, yet it has been established that the [4Fe–4S] clusters of the chromatium-type Fds show very negative (< −500 mV) reduction potentials in comparison to clostridial-type motifs (Giastas et al., 2006), suggesting that the lower potential features (E_{M2}) of the *Ct* Fds are due to the chromatium-type cluster environment. The reduction potentials for *Ct* Fd1 and *Ct* Fd2 compared well with the previously reported values of −584 and −515 mV vs. SHE at pH 7.5 and 25 °C, measured by diffusional voltammetry (a method which would not allow for the deconvolution of two, relatively closely spaced redox potentials) (Yoon et al., 2001).

6. Comparison of assay results

6.1 Comparison of *Ct* Fd reactivity with *Ct* PFOR – metronidazole assay

To survey the ET efficiency between the individual *Ct* PFOR:Fd pairs, we examined the kinetic traits of the three *Ct* Fds serving as the electron acceptor for *Ct* PFOR with a final electron acceptor, metronidazole, as a reporting chromophore (Fig. 5). When compared under the same conditions, all three *Ct* ferredoxins demonstrated the ability to shuttle electrons from *Ct* PFOR/pyruvate to metronidazole to varying extents. Over the course of 30 min, Fd1, Fd2, and Fd3 engendered absorbance changes of 0.211, 0.307, and 0.068 at 320 nm, corresponding to 23, 33, and 7 μM reduced metronidazole (ε_{320} = 9.3 mM^{-1}•cm^{-1}), indicating the Fd2 was most successful redoxrelay, in this platform. Already, through this assay, Fd3 gave rise to the least amount of electron-shuttling, both in terms of overall amount of an optical change, but also in terms of the apparent rate.

6.2 Comparison of *Ct* Fd reactivity with *Ct* PFOR – *reductive assay*

Where metronidazole is a convenient indicator of the oxidative chemistry of Fd:PFOR pairs, we also used to a coupled enzymatic assay to compare each of the three *Ct* Fds. In these assays, either *Ct* Fd1, Fd2, or Fd3 (previously reduced with *Mm* OGOR) served as the electron donor for *Ct* PFOR. The biochemical assay was designed to ensure that the activity of the enzymes performing the upstream (*Mm* OGOR) and downstream

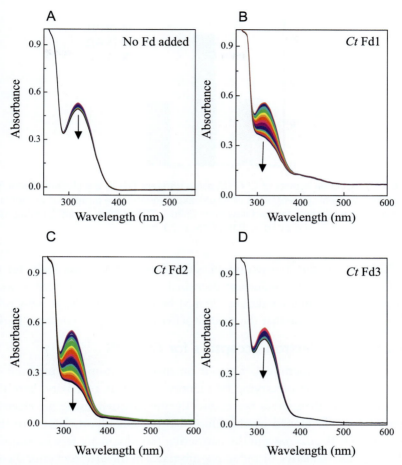

Fig. 5 Reduction of metronidazole by *Ct* PFOR, mediated by *Ct* Fds. 50 nM *Ct* PFOR was added to a mixture of 10 mM pyruvate, 200 μM CoA, 0 or 1 μM *Ct* Fds, and ~60 μM metronidazole. (A) Control: no Fd added. (B) *Ct* Fd1, (C) *Ct* Fd2, and (D) *Ct* Fd3. The spectra were collected every 5 s for 30 min. The reduction of metronidazole was monitored at 320 nm, and its background reaction with *Ct* PFOR was found to be negligible.

(LDH) reactions exceeded that of the PFOR. Under these experimental conditions, PFOR catalysis was rate-limiting, and the detection of NADH oxidation represented the rate of pyruvate synthesis/CO_2 consumption.

Measured under the same conditions (10 μM Fd), *Ct* PFOR achieved the highest CO_2 reduction activity at 156 ± 1.3 min^{-1} as a turnover frequency (TOF), which corresponds to a specific activity of ~1160 $nmol^{-1}$ min^{-1} mg^{-1}, with *Ct* Fd2 as the electron donor. When *Ct* Fd1 (TOF of 80 min^{-1}) and *Ct*

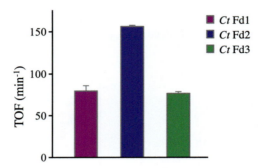

Fig. 6 CO_2 reduction activity of Ct PFOR using different Fds as the electron donor. Fd was added to a final concentration of 10 μM. Experiments were performed in triplicate and error bars represent standard deviation. Ct Fd 1: TOF of 80 ± 6 min^{-1}, Ct Fd2: TOF of 156 ± 1.3 min^{-1}, and Ct Fd 3: TOF of 76 ± 2.1 min^{-1}.

Fd3 (TOF of 75 min^{-1}) were utilized as the electron donor, the ability for Ct PFOR to catalyze CO_2 reduction decreased by 50% (Fig. 6). Studies were conducted at saturating conditions (20 mM bicarbonate, 1 mM acetyl-CoA), and as such, we assume that the specific activity values approximate a TOF.

6.3 Ct Fds as electron acceptors for Ct PFOR

While the prior experiments focused on spectrophotometric methods to report on the impact of varying Fd identity on PFOR reactivity, here the redox reactions themselves were directly interrogated using an electrochemical assay developed by the Elliott lab (Li and Elliott, 2016). In this assay, the PFOR enzyme and the partnering ferredoxin are free in solution with either the oxidative or reductive substrates in the electrochemical cell. The scan rate in the cyclic voltammetry experiment is kept low (1 or 2 mV/s), so that the diffusional voltammetry of mediated electron transfer from the Fd dominates the electrochemical shape, and the direct interaction between the Ct PFOR and the electrode does not contribute significantly. As shown previously, this approach has been useful to illustrate how different Fd mediators impact catalytic behavior (Li and Elliott, 2016) (Pieulle et al., 1999). The catalytic current observed is reflective of the Fd either being oxidized (positive current, when acting as an electron acceptor,) or being reduced (negative current, when acting as an electron donor), as it participates in PFOR chemistry. This relationship divides the electrocatalytic assay into two components: **(1)** the enzymatic step performed by the PFOR and **(2)** the electrochemical step accomplished by the Fd interacting with the electrode.

(1) 2-oxoacid + 2Fd$_{ox}$ + CoA → CO$_2$ + 2Fd$_{red}$ + acetyl-CoA *(enzymatic step)*

(2) 2Fd$_{red}$ → 2Fd$_{ox}$ *(electrochemical step)*

Accordingly, the catalytic currents are interpreted as a coupled reaction between the PFOR and the corresponding Fd, and the total current output (which is a surrogate for PFOR turnover, mediated by the Fd) can be compared to determine which Fd communicates the most effectively with the PFOR enzyme.

6.3.1 Current dependence on Ct PFOR specific activity

As a control experiment and acting as the "baseline" cyclic voltammogram, the Fd was omitted from the electrochemical cell and no catalytic wave was observed (Fig. 7 black trace). This result indicates that the PFOR does not communicate electrochemically with the electrode, even when substrates are present. Upon the addition of ferredoxin, the enzymatic reaction was coupled to the electrochemical step, effectively wiring PFOR activity to the electrode. Upon these conditions a large, pseudo-sigmodal anodic wave was observed, due to the continuous re-oxidation of the interacting Fd (Fig. 7, purple trace).

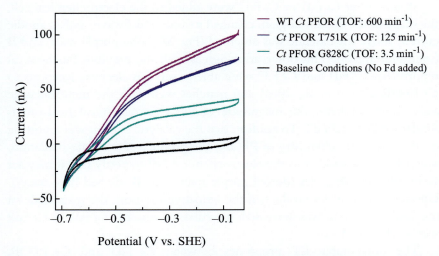

Fig. 7 Current output dependency on *Ct* PFOR activity. Experiments were performed under oxidative conditions and using *Ct* Fd1 as an ET mediator. The activity values listed in the legend of the various PFOR enzymes were calculated from solution-based activities.

As a secondary control to confirm that the magnitude of the catalytic current reflected the turnover of substrate by the PFOR enzyme, we compared the current output of the WT enzyme (purple trace) with two variant *Ct* PFOR enzymes, T751K and G828C, whose activities are lower than that of the WT *Ct* PFOR (Fig. 7). As expected, the magnitude of the catalytic current scaled with the ability of the *Ct* PFOR enzymes to turnover substrate, indicating that if the *Ct* PFOR is kept constant, any changes to current output can be attributed to the partnering Fds. However, the scaling of catalytic current output was non-linear with the two mutants tested thus far in terms of the TOF measurements made by spectrophotometric assays (e.g., we would expect that G828C PFOR would have a minimal current of barely 1 nA, but the teal trace of Fig. 7 is about 50 % of the current of the trace for the WT enzyme). The non-linearity may reveal that there are other complexities in the solution-based and/or electrochemical assays that we do not yet appreciate.

6.3.2 Comparison of Ct Fds ability to accept electrons from Ct PFOR

Next, we compared each Fd in the electrocatalytic assay with the WT *Ct* PFOR (TOF: $600 \, min^{-1}$, determined by the MV-coupled biochemical assay) to determine how each individual Fd influenced PFOR catalysis.

When either *Ct* Fd1 or *Ct* Fd2 were added to the electrochemical cell, the shape of the voltammogram changed to a sigmodal wave, and an anodic current above the baseline was observed (Fig. 8A (solid purple trace) and **B** (solid blue trace)). This observation agrees with the previous biochemical reports that *Ct* Fd1 and *Ct* Fd2 were able to serve as electron acceptors to *Ct* PFOR (Yoon et al., 2001) and matches well with the metronidazole assays reported here: the magnitude of the electrochemical read-out is similar for Fd1 and Fd2. To confirm that the catalytic current was reporting on the active site chemistry of PFOR, the product of pyruvate oxidation, acetyl-CoA, was added to the electrochemical cell to verify that it may act as an inhibitor (Fig. 8A (dotted purple trace) and **B** (dashed blue trace)). Upon addition of the product of the oxidative reaction, the magnitude of the catalytic current was decreased, regardless of whether *Ct* Fd1 or *Ct* Fd2 was the ET mediator.

The unfavorable ET properties between *Ct* Fd3 and *Ct* PFOR observed in the metronidazole coupled assay were further corroborated by the oxidative electrocatalytic experiments. The overall current output (Fig. 9A) indicates that Fd3 is not as efficient a redox mediator as Fd1 or Fd2. Upon multiple scans, the voltammograms of Fd3-mediated

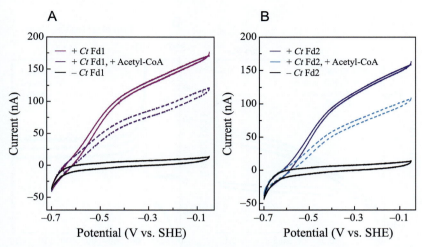

Fig. 8 Electrocatalytic assay in the 2-oxoacid oxidation direction using (A) Ct Fd1 and (B) Ct Fd2 as the Fd mediator. In both panels, the solid black traces correspond to WT Ct PFOR with oxidative substrates in the absence of Fd, the solid colored traces correspond to WT Ct PFOR with oxidative substrates with the indicated Fd, and the dotted/dashed traces correspond to WT Ct PFOR with oxidative substrates, Fd, and acetyl-CoA (1 mM) added. Oxidative substrates: pyruvate (10 mM), CoA (0.2 mM).

electrocatalysis decayed, as indicated by the decreasing traces and crossing over of voltammograms in Fig. 9A. The decay in catalytic output can also be examined by chronoamperometry (Fig. 9B). In these experiments, the same reaction components were present in the cell as the oxidative electrocatalytic experiments, but the potential was poised first at a reductive potential (no oxidative current), then the potential is jumped to a more positive value (here −550 mV), where positive current was observed following the initial current spike due to the shift in voltage. Repeating the cycle of voltage changes in succession three times illustrates that current overall diminished, and that with each chronoamperometry experiment (Fig. 9B), the current profile was not stable, but declined slowly on the timescale of the electrocatalytic experiment (Fig. 9A). The basis for this finding is not currently clear, but given the overall poorer reactivity displayed by Fd3 in the metronidazole assay (Fig. 5), it may be that this Fd is not stable upon redox-cycling, or, as time progresses, is so very oxygen sensitive that it becomes inactivated by trace molecular oxygen, even in a glovebox operating at < 1 ppm oxygen.

Finally, we sought to directly compare the oxidative results described above with comparable reductive electrochemical assays. Unfortunately,

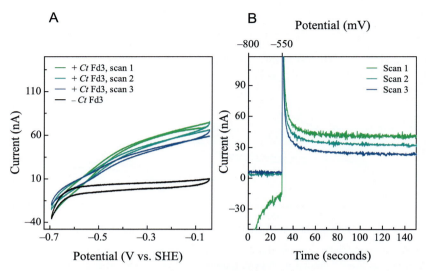

Fig. 9 Electrocatalytic assay in the 2-oxoacid oxidation direction using Ct Fd3 as the Fd mediator. (A) Comparison of the catalytic wave over three scans, showing a decrease in detectable Ct PFOR activity. (B) Chronoamperometric traces of Ct PFOR oxidative activity mediated by Ct Fd3, observing the decay of activity over time at a set potential of −550 mV.

we were not able to determine the potential differential activity supported by Ct Fds, due to the nonspecific reduction of the substrate pyruvate at the low potentials of these experiments, as discussed below.

7. Conclusions and future directions

Here we lay out a set of tools that will be of use to study of redox-linked CO_2 reductions, where one can consider the nature and characteristics of the redox-carrier protein, in addition to the actual decarboxylase/carboxylase. Notably, in the case of the *Chlorobaculum* PFOR enzyme system, three candidate 2x[4Fe−4S]-bearing Fds were identified, and direct electrochemistry indicates that their redox potentials are quite close to one another. With the redox potential for the CO_2/pyruvate couple centered at −520 mV (Fuchs, 2011), all three Fds appeared to possess a sufficiently low redox potential to act as reductants for Ct PFOR. Yet, when Ct Fd2 was the redox partner, Ct PFOR was able to perform CO_2 reduction at a rate twice as fast (TOF of ~150 min^{-1}) in comparison to reactions using Ct Fd1 or Ct Fd3 (TOF of ~80 and 75 min^{-1}, respectively).

In the oxidative electrocatalytic experiments, a different Fd trend emerged, where *Ct* Fd1 and *Ct* Fd2 displayed similar electron acceptor properties, while the *Ct* Fd3 appeared to struggle. In the assay, the Fds acted as the electron carriers (electron acceptors) by wiring the oxidative enzymatic reaction of the PFOR to the electrode surface (Li and Elliott, 2016). This interaction allowed us to probe the ET properties between the Fd partner and the *Ct* PFOR by comparing the total current output, which was attributed to PFOR turnover and verified by the mutant PFOR study. As a side note, the output of current produced by the various mutant PFORs did not scale with their measured solution-based activity. This observation suggests that elements of electrochemical assay may have created an environment that promoted better interactions between the Fd and the PFOR or that the detection method was more sensitive. Clearly, experiments such as these must be paired with concentration dependencies of the Fd, as well as determination of the Fd:PFOR binding constants.

Finally, the mediated electrochemical comparison of CO_2 reduction chemistry was not determinable. In comparison to the prior work studying OGOR and oxo-glutarate reduction (Chen et al., 2019), here pyruvate can be non-specifically reduced at electrodes at a potential that makes the reductive electrochemistry experiment challenging (see the reductive "tail" in the background traces of Figs. 7–9). Prior studies on OGOR did not suffer from this dilemma as oxo-glutarate is only non-specifically reduced at yet-lower potentials. Similarly, when compared to the prior work using the *Da* PFOR and the library of single [4Fe–4S] cluster ferredoxins prepared by Li and co-workers (Li et al., 2021), the potentials of the *Ct* Fd are all so low that the electrochemical reactions of interest competed with non-specific pyruvate reactivity. Thus, a major experimental condition yet to be overcome is the optimization of an appropriate potential window, or making use of an electrode material that will report on the low-potential redox cycling of Fds, but not give rise to artifactual pyruvate reduction.

Acknowlegments

This work was supported by the Division of Chemical Sciences, Geosciences, and Biosciences, Office of Basic Energy Sciences of the U.S. Department of Energy, (DE-SC0012598) and National Institutes of Health (R35 GM136294).

References

Amstrong, F. A., Heering, H. A., & Hirst, J. (1997). Reaction of complex metalloproteins studied by protein-film voltammetry. *Chemical Society Reviews, 26*, 169–179.

Blanford, C. F., & Armstrong, F. A. (2006). The pyrolytic graphite surface as an enzyme substrate: microscopic and spectroscopic studies. *Journal of Solid State, Electrochemistry, 10,* 826–832.

Bock, A. K., Kunow, J., Glasemacher, J., & Schönheit, P. (1996). Catalytic properties, molecular composition and sequence alignments of pyruvate: Ferredoxin oxidoreductase from the methanogenic archaeon Methanosarcina barkeri (strain Fusaro). *European Journal of Biochemistry, 237*(1), 35–44.

Burkhart, B. W., Febvre, H. P., & Santangelo, T. J. (2019b). Distinct physiological roles of the three ferredoxins encoded in the hyperthermophilic archaeon Thermococcus kodakarensis. *mBio, 10*(2), https://doi.org/10.1128/mbio.02807-18.

Campbell, I. J., Bennett, G. N., & Silberg, J. J. (2019). Evolutionary relationships between low potential ferredoxin and flavodoxin electron carriers. *Frontiers in Energy Research, 7,* 79.

Chen, P. Y.-T., Aman, H., Can, M., Ragsdale, S. W., & Drennan, C. L. (2018). Binding site for coenzyme A revealed in the structure of pyruvate: Ferredoxin oxidoreductase from Moorella thermoacetica. *Proceedings of the National Academy of Sciences, 115*(15), 3846–3851.

Chen, P. Y.-T., Li, B., Drennan, C. L., & Elliott, S. J. (2019). A reverse TCA cycle 2-oxoacid: Ferredoxin oxidoreductase that makes CC bonds from CO_2. *Joule, 3*(2), 595–611.

Eisen, J. A., Nelson, K. E., Paulsen, I. T., Heidelberg, J. F., Wu, M., Dodson, R. J., ... Haft, D. H. (2002). The complete genome sequence of Chlorobium tepidum TLS, a photosynthetic, anaerobic, green-sulfur bacterium. *Proceedings of the National Academy of Sciences, 99*(14), 9509–9514.

Firer-Sherwood, M. A., Bewley, K. D., Mock, J. Y., & Elliott, S. J. (2011). Tools for resolving complexity in the electron transfer networks of multiheme cytochromes c. *Metallomics, 3*(4), 344–348.

Fuchs, G. (2011). Alternative pathways of carbon dioxide fixation: Insights into the early evolution of life? *Annual Review of Microbiology, 65,* 631–658.

Furdui, C., & Ragsdale, S. W. (2000). The role of pyruvate ferredoxin oxidoreductase in pyruvate synthesis during autotrophic growth by the Wood-Ljungdahl pathway. *Journal of Biological Chemistry, 275*(37), 28494–28499.

Furdui, C., & Ragsdale, S. W. (2002). The roles of coenzyme A in the pyruvate: ferredoxin oxidoreductase reaction mechanism: Rate enhancement of electron transfer from a radical intermediate to an iron− sulfur cluster. *Biochemistry, 41*(31), 9921–9937.

Giastas, P., Pinotsis, N., Efthymiou, G., Wilmanns, M., Kyritsis, P., Moulis, J.-M., & Mavridis, I. M. (2006). The structure of the 2 [4Fe−4S] ferredoxin from Pseudomonas aeruginosa at 1.32-Å resolution: Comparison with other high-resolution structures of ferredoxins and contributing structural features to reduction potential values, JBIC. *Journal of Biological Inorganic Chemistry, 11,* 445–458.

Jeuken, L. J. C. (2016). Biophotoelectrochemistry: from bioelectrochemistry to biophotovoltaics. In L. Jeuken (Vol. Ed.), *Advances in Biochemical Engineering/Biotechnology. 158.* Cham: Springer.

Li, B., & Elliott, S. J. (2016). The catalytic bias of 2-oxoacid: Ferredoxin oxidoreductase in CO_2: Evolution and reduction through a ferredoxin-mediated electrocatalytic assay. *Electrochimica Acta, 199,* 349–356.

Li, B., Steindel, P., Haddad, N., & Elliott, S. J. (2021). Maximizing (electro) catalytic CO_2 reduction with a ferredoxin-based reduction potential gradient. *ACS Catalysis, 11*(7), 4009–4023.

Maiocco, S. J., Walker, L. M., & Elliott, S. J. (2018). Determining redox potentials of the iron-sulfur clusters of the AdoMet radical enzyme superfamily. *Methods in Enzymology, 606,* 319–339.

Pieulle, L., Charon, M. H., Bianco, P., Bonicel, J., Pétillot, Y., & Hatchikian, E. C. (1999). Structural and kinetic studies of the pyruvate–ferredoxin oxidoreductase/ferredoxin complex from Desulfovibrio africanus. *European Journal of Biochemistry, 264*(2), 500–508.

Ragsdale, S. W. (2003). Pyruvate ferredoxin oxidoreductase and its radical intermediate. *Chemical Reviews, 103*(6), 2333–2346.

Ragsdale, S. W., & Pierce, E. (2008). Acetogenesis and the Wood–Ljungdahl pathway of CO_2 fixation. *Biochimica et Biophysica Acta (BBA)-Proteins and Proteomics, 1784*(12), 1873–1898.

Yoon, K.-S., Bobst, C., Hemann, C. F., Hille, R., & Tabita, F. R. (2001). Spectroscopic and functional properties of novel 2 [4Fe-4S] cluster-containing ferredoxins from the green sulfur bacterium Chlorobium tepidum. *Journal of Biological Chemistry, 276*(47), 44027–44036.

Yoon, K.-S., Hille, R., Hemann, C., & Tabita, F. R. (1999). Rubredoxin from the green sulfur bacterium Chlorobium tepidum functions as an electron acceptor for pyruvate ferredoxin oxidoreductase. *Journal of Biological Chemistry, 274*(42), 29772–29778.

CHAPTER THIRTEEN

Measuring carbonic anhydrase activity in alpha-carboxysomes using stopped-flow

Nikoleta Vogiatzi and Cecilia Blikstad*

Department of Chemistry—Ångström Laboratory, Uppsala University, Uppsala, Sweden
*Corresponding author. e-mail address: cecilia.blikstad@kemi.uu.se

Contents

1. Introduction	298
2. Cultivation of *H. neapolitanus*	301
2.1 Equipment and materials	303
2.2 Buffer and reagents	303
2.3 Chemostat growth of *H. neapolitanus*	303
3. Purification of carboxysomes	304
3.1 Equipment and materials	305
3.2 Buffers and reagents	306
3.3 Purification of α-carboxysomes	306
4. Measuring carbonic anhydrase activity using stopped-flow	309
4.1 Equipment	311
4.2 Reagents	311
4.3 SX20 stopped-flow start-up and setting up the equipment	311
4.4 Cleaning and preparing the SX20 stopped-flow before starting measurements	312
4.5 Preparing assay buffer and making and diluting the CO_2 solution	313
4.6 Making a standard measurement	313
4.7 Measuring a saturation curve	315
4.8 Measuring at different pH values	316
5. Data analysis	317
5.1 Calculate the experimental buffer factor	317
5.2 Calculate the kinetic constants k_{cat}, K_M and k_{cat}/K_M	317
6. Summary	319
Acknowledgment	319
References	319

Abstract

Carboxysomes are protein-based organelles that serve as the centerpiece of the bacterial CO_2 concentration mechanism (CCM). They are present in all cyanobacteria and many chemoautotrophic proteobacteria and encapsulate the key enzymes for CO_2 fixation, carbonic anhydrase and the carboxylase Rubisco, within a protein shell. The CCM actively accumulates bicarbonate in the cytosol, which diffuses into the carboxysome where carbonic anhydrase rapidly equilibrates it to CO_2. This creates a high CO_2 concentration around Rubisco, ensuring efficient carboxylation. In this chapter, we present a general method for purifying α-carboxysomes and measuring carbonic anhydrase activity within these purified compartments. We exemplify this with α-carboxysomes purified from the chemoautotroph *Halothiobacillus neapolitanus c2*, a model organism for the α-carboxysome based CCM. However, this purification protocol can be adapted for other species, such as carboxysomes from α-cyanobacteria or carboxysomes expressed in heterologous hosts. Further, we describe the Khalifah/pH indicator assay for measuring steady-state kinetics of carbonic anhydrase catalyzed CO_2 hydration. This method allows us to determine the kinetic parameters k_{cat}, K_M and k_{cat}/K_M for the purified α-carboxysomes. It uses a stopped-flow spectrometer for rapid mixing and detection, crucial for capturing the fast equilibrium between CO_2 and bicarbonate. The reaction progress is monitored by absorbance via a pH indicator that changes color due to the proton release. While the method specifically focuses on measuring carbonic anhydrase activity on carboxysomes, it can be used to measure activity on carbonic anhydrases from other contexts as well.

1. Introduction

In plants, algae, and some autotrophic bacteria multiple types of carbon dioxide (CO_2) concentration mechanisms (CCMs) have arisen independently over the last 2 billion years (Flamholz & Shih, 2020; Raven, Cockell, & De La Rocha, 2008). By concentrating CO_2 around Ribulose-1,5-bisphosphate carboxylase-oxygenase (Rubisco), the carboxylase of the Calvin–Benson–Bassham cycle, CCMs compensate for Rubisco's relatively slow kinetics and its inability to distinguish between CO_2 and the off-target substrate oxygen (O_2) (Andersson, 2008; Prywes, Phillips, Tuck, Valentin-Alvarado, & Savage, 2022). Carbonic anhydrases (CAs), enzymes which catalyze the rapid interconversion between CO_2 and bicarbonate (HCO_3^-), play diverse but critical roles in these mechanisms (Badger, 2003). In a multitude of ways, CAs ensure efficient supply of inorganic carbon by e.g. directly supplying carboxylases with CO_2, aiding the entry of CO_2 into the cell and accumulating CO_2/HCO_3^- intermediate pools (Badger, 2003; DiMario, Machingura, Waldrop, & Moroney, 2018). In a larger perspective, rapid CO_2/HCO_3^- equilibration is central in metabolism and CAs are essential in all organisms where it has been tested (Badger & Price, 1994; Merlin, Masters, McAteer, & Coulson, 2003; Supuran, 2023). CAs are a large superfamily, and known as one of the fastest enzymes, often acting near diffusion limited rates (Khalifah, 1971, 1973). To further understand their roles in nature, it is essential to understand and characterize these enzymes' kinetic behavior.

Bacteria employ a biophysical CCM which is found in all cyanobacteria and many chemoautotrophic proteobacteria (Fig. 1A). It consists of two main components: (I) energy-coupled inorganic carbon transporters that accumulate bicarbonate in the cytosol, and (II) carboxysomes, specialized protein organelles co-encapsulating the key enzymes for CO_2 fixation, Rubisco and CA, within a protein shell (Desmarais et al., 2019; Flamholz et al., 2020; Kerfeld & Melnicki, 2016; Rae, Long, Badger, & Price, 2013). The accumulated HCO_3^- diffuses into the carboxysome through pores in the shell. Inside, it rapidly equilibrates to CO_2, catalyzed by the CA, and is then fixed by Rubisco. This mechanism saturates Rubisco's active sites with CO_2, and thereby accelerates carboxylation while competitively inhibiting oxygenation. As such, the overall carboxylation rate in the organism is enhanced. Carboxysomes are self-assembling icosahedral structures which consist entirely of proteins and normally range from 100–200 nm in diameter. They are built up by hexameric shell proteins forming the facets, shell pentamers capping the corners, one to two scaffolding proteins directing assembly and organizing the interior, and the encapsulated enzymes Rubisco and CA (Fig. 1B). Two types of carboxysomes have evolved convergently, the α-type found in oceanic cyanobacteria and proteobacteria and the β-type found in freshwater cyanobacteria (Kerfeld & Melnicki, 2016; Melnicki, Sutter, & Kerfeld, 2021).

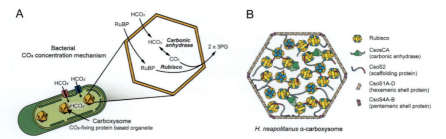

Fig. 1 Schematic of the bacterial CO_2 concentration mechanism (CCM) and an α-carboxyosome from *H. neapolitanus*. (A) The bacterial CCM consists of inorganic carbon transporters and carboxysomes, proteinaceous organelles that co-encapsulate Rubisco and carbonic anhydrase (CA). (B) Architecture of α-carboxysomes from *H. neapolitanus*: the shell is composed of hexameric proteins (CsoS1A-D), capped in the corners by pentameric proteins (CsoS4A-B). The scaffolding protein (CsoS2) directs the assembly and organization of the organelle by interacting with both the shell and Rubisco, which in turn interacts with the carbonic anhydrase (CsoSCA). B is based on Fig. 5 in Blikstad et al., 2023.

Reversible hydration of CO_2 is central in metabolism and CAs are found throughout the tree of life. CAs are an evolutionarily diverse class of enzymes consisting of at least 8 different protein families (α, β, γ, δ, ζ, η, θ and ι) which share no sequence or structural homology (Akocak & Supuran, 2019; Khalifah, 1971; Supuran, 2016; Tripp, Smith, & Ferry, 2001). Typical turnover numbers (k_{cat}) for CAs range between 10^4–10^6 s^{-1}. Many of these enzymes approach the diffusion-limit, with catalytic efficiencies (k_{cat}/K_M) reaching up to 10^9 s^{-1} M^{-1}, making them some of the fastest enzymes in nature. All α-carboxysomes contain a β-CA known as CsoSCA (Baker, Williams, Aldrich, Gambrell, & Shively, 2000; Blikstad et al., 2023; Heinhorst et al., 2006; Pulsford et al., 2024; Sawaya et al., 2006; So et al., 2004), while β-carboxysomes have either an active γ-CA domain on the scaffolding protein CcmM (de Araujo et al., 2014; Peña, Castel, de Araujo, Espie, & Kimber, 2010) or a β-CA named CcaA (McGurn et al., 2016; Zang, Wang, Hartl, & Hayer-Hartl, 2021). For the bacterial CCM to function, there is an absolute requirement for a CA to be localized in the carboxysome and efficient regulation of carboxysomal CA activity is crucial (Dou et al., 2008; Fukuzawa, Suzuki, Komukai, & Miyachi, 1992; Kimber, 2014; Price & Badger, 1989). For instance, kinetic measurements have demonstrated that CcmM from *Thermosynechococcus vestitus BP-1* (formerly known as *Thermosynechococcus elongatus BP1*) and CsoSCA from the proteobacterium *Halothiobacillus neapolitanus c2 (H. neapolitanus)* are regulated by their redox state, suggesting that these enzymes are inactive in the reducing cytosol and active in the oxidizing environment of the carboxysomes (Heinhorst et al., 2006; Peña et al., 2010). Further, in contrast to CsoSCA from *H. neapolitanus*, the orthologue from the α-cyanobacteria *Cyanobium PCC 7001 (Cyanobium)* is activated by the Rubisco substrate ribulose-1,5-bisphosphate (RuBP), suggesting that CsoSCAs from photosynthetic bacteria have a specific regulation linked to large fluctuations in cellular RuBP concentration (Pulsford et al., 2024). The carboxysomal CAs so far characterized follow Michaelis-Menten kinetics and have k_{cat} values in the range of 3×10^3–9×10^4 s^{-1} and K_M values for CO_2 in the low mM range (de Araujo et al., 2014; Heinhorst et al., 2006; McGurn et al., 2016), making them fairly modest CAs compared to e.g. human CA II ($k_{cat} = 1 \times 10^6$ s^{-1}, $K_M = 8$ mM) (Khalifah, 1971). To further underpin the function and regulation of carboxysomal carbonic anhydrases, measuring CA kinetics on purified enzymes as well as on carboxysomes is of the essence. This information is also crucial to aid in engineering the carboxysome-based CCM into crops and industrially relevant microbes (Borden & Savage, 2021; Flamholz et al., 2020; Nguyen et al., 2024), which has the potential to enhance CO_2 uptake and increase yields.

In this chapter, we describe an experimental procedure to purify α-carboxysomes and subsequently measure carbonic anhydrase enzymatic activity on the purified compartments. First, we describe how to grow the chemoautotroph *H. neapolitanus*, the model organism for the α-carboxysome based CCM, and how to purify its carboxysomes (Metskas et al., 2022). Thereafter, we describe how to measure steady-state kinetic parameters (k_{cat}, K_M and k_{cat}/K_M) for CA catalyzed CO_2 hydration using the Khalifah/pH indicator assay (Blikstad et al., 2023; Khalifah, 1971). This assay uses stopped-flow spectrometry for rapid mixing and detection, which is needed to capture the fast initial rates of CA catalyzed reactions. The reaction is followed by absorbance using a pH indicator that monitors proton release. The outlined purification protocol can be adapted to purify α-carboxysomes from other species, for example from different α-cyanobacteria or carboxysomes engineered into heterologous hosts. Likewise, the kinetic measurements can be used to measure in vitro enzyme activity with any carbonic anhydrase, making this a versatile protocol that extends beyond carboxysomes.

2. Cultivation of *H. neapolitanus*

Carboxysomes were isolated for the first time in 1973 by Shively et al., (Shively, Ball, Brown, & Saunders, 1973). They were isolated from the sulfur oxidizing chemoautotroph *H. neapolitanus*, and this bacterium has since served as a model organism for studying α-carboxysomes. *H. neapolitanus* is a γ-proteobacteria that relies solely on CO_2 and sulfur to satisfy its carbon and energy needs. In nature it plays important roles in the global biogeochemical carbon and sulfur cycles. However, in the scientific community it has gained most attention as a model for the α-carboxysome based CCM. To date, α-carboxysomes from different origins have been purified using similar protocols as we describe below. Despite not being photosynthetic, *H. neapolitanus* has several advantages over cyanobacteria when investigating carboxysomes. Firstly, it harbors α-carboxysomes which can readily be purified as opposed to β-carboxysomes which have proven challenging to isolate in the high yields and purity needed for kinetic measurements (however, using a precipitation protocol, β-carboxysomes have been isolated in sufficient quantity for electron microscopy (Kong et al., 2024; Price, Coleman, & Badger, 1992)). Secondly, to our knowledge, to date no α-cyanobacteria can be genetically manipulated. *H. neapolitanus* is genetically tractable (Blikstad et al., 2023; Menon, Heinhorst, Shively, & Cannon, 2010) and thus constitutes the only organism in which we can probe carboxysome function with both in vivo experiments and in vitro studies on purified carboxysomes.

Compared to e.g. *E. coli*, *H. neapolitanus* grows only to low cell densities and large volumes of cell culture are therefore needed to purify carboxysomes. The following protocol describes how to grow *H. neapolitanus* in a 10 L bioreactor setup to function as a chemostat (Fig. 2) (Metskas et al., 2022). In a chemostat, fresh media is continuously added (feed) at the same rate that the growing culture is removed (effluent), resulting in a steady-state growth. We have found that using this chemostat-mode protocol, instead of conventional growth in a bioreactor, it is possible to reach higher cell densities and thereby significantly increase the final yield of purified carboxysomes. However, if a 10 L chemostat is not available, this protocol

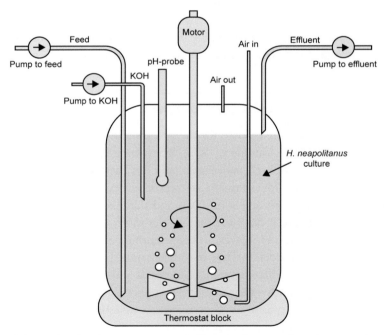

Fig. 2 Schematic of a chemostat setup. A *H. neapolitanus* culture is grown in a 10 L bioreactor set up as a chemostat. The culture is kept at 30 °C, stirred and sparged with air and adjusted to a constant pH of 6.4 using KOH. Fresh media (feed) is pumped in at the same rate as the growing culture (effluent) is pumped out, at a rate which keeps the culture in a steady state. A 20 L flask with fresh media is connected to the feed, and a 20 L empty flask is connected to the effluent (not shown in figure).

can be adapted and it is possible to grow *H. neapolitanus* in other ways. For instance, in a smaller bioreactor/chemostat (Cannon & Shively, 1983; So et al., 2004), in a bioreactor without chemostat mode (Shively et al., 1973; Sun et al., 2022) or without a bioreactor in a large vessel and manually adjusting the pH every day.

2.1 Equipment and materials
- Orbital shaker at 30 °C
- BioFlo©/CelliGen© 115 10 L Bioreactor (New Brunswick)
- BE2100 Sensor biomass monitor (Buglab)
- Large scale centrifuge and 1 L centrifuge bottles
- 50 mL falcon tube
- 250 mL E-flask
- 2–3 L E-flask
- 2 × 25 L autoclavable flasks
- Gas tube or pipe with ambient air

2.2 Buffer and reagents
- *Halothiobacillus neapolitanus c2* cells
- DSMZ 68 media. For 1 L: 5.09 g KH_2PO_4, 2.60 g K_2HPO_4, 0.8 g $MgSO_4$ × 7 H_2O, 0.4 g NH_4Cl, 5 mL of Vishniac and Santer trace element solution, 0.5 mL saturated bromocresol purple solution and 100 mL of 10 % (w/v) sterile filtered $Na_2S_2O_3$ (Koblitz et al., 2023; Vishniac & Santer, 1957).
- Vishniac and Santer trace element solution. For 1 L: 50.0 g Na_2-EDTA, 22.0 g $ZnSO_4$ × 7 H_2O, 5.54 g $CaCl_2$, 5.06 g $MnCl_2$ × 4 H_2O, 5 g $FeSO_4$ × 7 H_2O, 1.10 g $(NH_4)_6Mo_7O_{24}$ × 4 H_2O, 1.57 g $CuSO_4$ × 5 H_2O, and 1.61 g $CoCl_2$ × 6 H_2O, pH 6.0 adjusted with KOH
- 5 M sterile filtered KOH

2.3 Chemostat growth of *H. neapolitanus*
1. Prepare trace element solution: Dissolve EDTA in MQ water and adjust the pH to 7.0 using KOH. Thereafter, add all the trace elements and adjust the pH to 6.0 by slow addition of KOH (slow addition prevents precipitation).
2. Prepare DSMZ 68 media: For 1 L media, dissolve KH_2PO_4 and K_2HPO_4 in 450 mL dH_2O. In a second bottle, dissolve $MgSO_4$, NH_4Cl, trace element solution and the bromocresol purple solution in 450 mL dH_2O. Autoclave the two bottles separately. Once cooled down, mix the two solutions and add $Na_2S_2O_3$.
3. From a glycerol stock of *H. neapolitanus* cells, start a 10 mL liquid culture in a 50 mL falcon tube in DSMZ 68 media. Let the culture

grow at 30 °C under continuous shaking until the culture is turbid and the pH indicator bromocresol purple has turned from purple to yellow. This normally takes 3 days.

4. Dilute the culture into 100 mL DSMZ 68 in a 250 mL E-flask and grow at 30 °C under continuous shaking until turbid and the pH indicator turns yellow, which takes approximately 2 days.

5. Dilute the 100 mL culture into 500 mL DSMZ 68 in a 2–3 L E-flask and grow at 30 °C under continuous shaking until turbid and the pH indicator turns yellow, which takes approximately 2 days.

6. Autoclave all parts of the bioreactor, including tubings and the 25 L feed and effluent vessels needed to connect as a chemostat. Autoclave 9.5 L of DSMZ 68 media for the bioreactor and 20 L to connect the feed. When new feed is needed, autoclave 20 L more.

7. Add the 500 mL *H. neapolitanus* culture from step 5 to a 10 L bioreactor containing 9.5 L of DSMZ 68 media. In this example a BioFlo©/ CelliGen© 115 10 L Bioreactor (New Brunswick) bioreactor is used, with the following settings: Temperature: 30 °C, pH: 6.4 adjusted with 5 M KOH, Agitation: 200, Sparging: 100 % air and Gasflow: 2.0. Attach the biomass monitor around the bioreactor.

8. When the 10 L culture has reached mid-log phase, set up the bioreactor in chemostat mode. This is done using two of the peristaltic pumps on the bioreactor. Connect the first pump to a 25 L bottle containing 20 L DSMZ 68 media so that it pumps in fresh media (feed) into the bioreactor. Connect the second pump to an empty 25 L bottle so that it pumps out culture (effluent) from the bioreactor. A dilution rate of 2.5 %/h is normally suitable. If the OD of the culture decreases, decrease the dilution rate. We have been running the chemostat for up to 16 days, producing ~100 L of culture without any problem.

9. Harvest cells from the effluent flask every 2–3 days by 6000 × *g* centrifugation.

10. Scrape up the pellet from 10–15 L cells and transfer to a 50 mL falcon tube. Freeze the cells at −80 °C for later use. Cell pellets from 10 L normally range from 6–7 g.

3. Purification of carboxysomes

The purification protocol of α-carboxysomes from *H. neapolitanus* that we describe below consists of chemical lysis and a series of consecutive centrifugation steps including sucrose gradient ultracentrifugation fractionation (Fig. 3) (Cannon & Shively, 1983; Metskas et al., 2022; Ni et al., 2022; So et al., 2004). Finally, the quality of the final sample is determined by SDS-PAGE analysis and negative stain transmission electron microscopy (TEM). With modifications in the lysis step, the protocol can be adjusted for purifying α-carboxysomes from different source organisms, e.g. the cyanobacteria *Cyanobium marinum* PCC 7001 (Evans et al., 2023; Long et al., 2018) and *Prochlorococcus* MED4 (Roberts, Cai, Kerfeld, Cannon, & Heinhorst, 2012; Zhou et al., 2024), as well as recombinant α-carboxysomes expressed in *E. coli* (Blikstad et al., 2023; Bonacci et al., 2012; Flamholz et al., 2020; Nguyen et al., 2023) or in tobacco chloroplasts (Chen et al., 2023; Long et al., 2018).

3.1 Equipment and materials
- Magnetic stirrer
- Avanti J-25 centrifuge (Beckman Coulter)
- Optima L-90K ultracentrifuge (Beckman Coulter)
- Microcentrifuge 5427 R (Eppendorf)

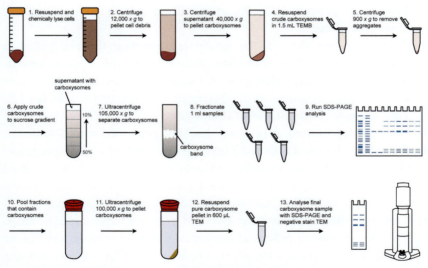

Fig. 3 Workflow of α-carboxysome purification. Chemical lysis of the cell pellet is followed by a series of centrifugation steps (1–5). The crude carboxysome sample is thereafter applied to a sucrose gradient for further purification (6–8). Samples from all steps are analyzed by SDS-PAGE (9) and the fractions containing carboxysomes are pooled together and ultracentrifuged to pellet carboxysomes (10–11). The pellet is resuspended in TEM buffer and the final sample is analyzed by SDS-PAGE and TEM (12–13).

- Rotor JA 25.50 (Beckman Coulter),
- Rotor 70Ti (Beckman Coulter)
- Rotor SW32 Ti Swinging bucket (Beckman Coulter)
- 32 mL, Open-Top Thickwall Polycarbonate Tube (Beckman Coulter)
- 26.3 mL, Polycarbonate Bottle with Cap Assembly (Beckman Coulter)
- Protein electrophoresis equipment (Bio-Rad)
- Mini-PROTEAN precast gels (Bio-Rad)
- Nanodrop spectrophotometer (Nanodrop 2000, Thermo Fisher)

3.2 Buffers and reagents

- TEMB buffer: 10 mM Tris-HCl, 1 mM EDTA, 10 mM $MgCl_2$, 20 mM $NaHCO_3$, pH 8.0
- TEM buffer: 10 mM Tris-HCl, 1 mM EDTA, 10 mM $MgCl_2$, pH 8.0
- Lysis buffer: B-PER II (Thermo Fisher) diluted to 1 × with TEMB buffer, 0.1 mg/mL lysozyme, 1 mM PMSF, 0.1 μL/mL benzonase (Sigma-Aldrich).
- Sterile filtered 50 % w/v sucrose solution dissolved in TEMB buffer

3.3 Purification of α-carboxysomes

1. Thaw a cell pellet from ~10 L of chemostat culture (6–7 g of cells) at room temperature.
2. Chemically lyse the cells by gently resuspending the cell pellet in 40 mL of lysis buffer by pipetting up and down with a plastic pasteur pipette. Once resuspended, pour the suspension into a 100 mL beaker and stir with a magnetic stirrer (500 rpm) at room temperature for 1 h. *Take a sample (L) for SDS-PAGE analysis.*
3. From this step forward, all processes should be performed at 4 °C or on ice.
4. Centrifuge the lysed cells at 12,000 × g (Avanti J-25 centrifuge, rotor: JA 25.50) for 15 min to pellet cell debris. Gently decant the clarified lysate (supernatant) and save it; this contains the carboxysomes. *Take a sample (CL) for SDS-PAGE analysis.*
5. Centrifuge the clarified lysate at 40,000 × g (Avanti J-25 centrifuge, rotor: 70Ti) for 30 min to pellet carboxysomes. Carefully remove and discard the supernatant and save the pellet. The resulting pellet is brown-red and translucent. *Take a sample (supernatant 40K-S) for SDS-PAGE analysis.*
6. Resuspend the pellet in 1.5 mL of TEMB buffer. This is a crucial step for keeping the carboxysomes intact. Pipette gently up and down and avoid creating foam or touching the pellet with the pipette tip (as any mechanical damage can break the compartments). Incubation on ice

for 30-60 min with slow shaking makes the pellet easier to resuspend and can speed up the process.

7. Transfer the resuspended pellet to a 2 mL microcentrifuge tube and centrifuge at $900 \times g$ for 3 min to pellet any aggregated carboxysomes. Remove and save the supernatant that contains intact carboxysomes. *Take two samples (supernatant 0.9K–S and pellet 0.9K-P) for SDS-PAGE analysis.*

8. Prepare a 25 mL stepwise 50–10 % w/v sucrose gradient, with each step being 5 mL. Make a stock of 50 % w/v sucrose solution in TEMB and sterilize it by filtration. Dilute the stock to make 5 mL solutions of 40 %, 30 %, 20 % and 10 % w/v sucrose. Using a pasteur pipette, apply 5 mL of the 50 % solution to the bottom of the open-top 32 mL thickwall ultra-centrifuge tubes. Continue with the next solution (40 % w/v) by carefully applying it on the surface of the previous one. Avoid mixing the two densities by pipetting slowly and against the wall of the tube. Apply the rest of the sucrose solutions with descending order of densities. Be careful to not introduce air bubbles when creating the gradient. It is recommended to prepare the gradient on the same day but at an earlier stage of the process (e.g. step 4 while centrifuging) and store it at 4 °C until use.

9. To further purify the carboxysomes, load the supernatant on top of the sucrose gradient and centrifuge at $105,000 \times g$ (Optima L-90K ultra-centrifuge, rotor: SW 32Ti Swinging bucket) for 35 min at 4 °C. Ultracentrifugation with sucrose gradient separates molecules based on their sedimentation rates, which is dependent on the shape, mass and size of the particles. The sample travels through the different sucrose densities, and after the spin the carboxysomes can be seen as a white band towards the middle-bottom part of the gradient (30–40 %).

10. Fractionate the gradient by carefully collecting 1 mL samples with a pipette, starting from the surface of the gradient and slowly moving down through the entire 25 mL. *Take samples for SDS-PAGE analysis, label them by number of fraction (1–26). Resuspend the pellet and take a sample of the pellet from the gradient (PG).*

11. Determine the purity by SDS-PAGE analysis (Fig. 4A). Load the samples indicated in each step of the protocol. For the sucrose gradient, it is enough to run every other or every third fraction (1, 3, 5, etc.). For L, CL, 40K-S and 0.9K-S/P dilute the sample 1:4 in dH_2O, for sucrose gradient fractions run undiluted.

12. Pool the fractions that contain carboxysomes, i.e. the ones in which all carboxysomal proteins can be identified on the SDS-PAGE gel (typically fractions ~12–22).

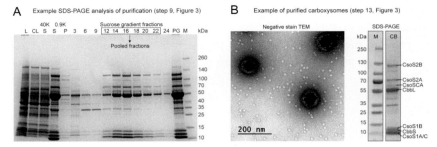

Fig. 4 (A) SDS-PAGE analysis of all purification steps (step 9 in Fig. 3) showing lysate (L), clarified lysate (CL), supernatant (S) and pellet (P) of respective centrifugation step (40,000 and 900 $x\ g$), sucrose gradient fractions (3–24), pellet from gradient (PG) and molecular weight marker (M). Pooled sucrose gradient fractions (12–22) are marked with a box. (B) Analysis of final purified carboxysomes (step 13 in Fig. 3) Left: negative stain TEM micrograph of final carboxysome sample. Right: SDS-PAGE analysis, where the scaffolding protein (CsoS2A/B), carbonic anhydrase (CsoSCA), both Rubisco large (CbbL) and small (CbbS) subunits, and the shell proteins (CsoS1A/B/C) can be identified. The shell proteins CsoS1D and CsoS4A-B are not abundant enough to be identified by SDS-PAGE.

13. Pellet the pooled fractions by ultracentrifugation at $100,000\ x\ g$ (Optima L-90K ultracentrifuge, rotor: 70Ti, tubes: polycarbonate bottle with cap assembly) for 90 min at 4 °C. This will remove sucrose and concentrate the final sample. The resulting pellet is light red and translucent.
14. Gently resuspend the pellet in 600 μL TEM buffer by pipetting up and down to obtain the final purified carboxysome sample. (To avoid carrying over HCO_3^- in the subsequent activity measurements, we exclude HCO_3^- from the TEMB buffer in this final step. If the carboxysomes will be used for another purpose, the standard TEMB buffer can be used here).
15. Determine the concentration of the final sample by measuring absorbance at 280 nm, using a nanodrop spectrophotometer. Assume 1 Absorbance unit = 1 mg/mL. Average yields are around 10 mg/mL.
16. Determine the purity of the final carboxysome sample by SDS-PAGE analysis and visualize the integrity and quality of the compartments by TEM (Fig. 4B). For negative stain TEM, 0.2 mg/mL sample applied to formvar/carbon coated copper grids and stained with 2% aqueous uranyl acetate should yield good results.
17. Store purified carboxysomes at 4 °C until further use. Under these storage conditions, carboxysomes are typically stable for 2–3 weeks.

4. Measuring carbonic anhydrase activity using stopped-flow

In the context of measuring kinetics, the uncatalyzed rate of CO_2 to HCO_3^- conversion is rapid and reaches equilibrium after 1–2 min at 25 °C (Gibbons & Edsall, 1963; Tripp & Ferry, 2000). Consequently, CA catalyzed CO_2 hydration reaches equilibrium in an even shorter time frame. To measure steady-state kinetics, we need to determine initial rates, that is conditions under which the readout is linear and the change in substrate concentration can be assumed to be negligible (Fersht, 2017). Due to the fast equilibrium for CA catalyzed reactions, this requires measurements within a time-frame of 0.5 s. To measure these fast events, rapid mixing and detection is necessary and a stopped-flow instrument is therefore required (Blikstad et al., 2023; Khalifah, 1971; McGurn et al., 2016; Rowlett, Gargiulo, Santoli, Jackson, & Corbett, 1991; Supuran, 2008; Tripp & Ferry, 2000). In a stopped-flow experiment, one syringe is filled with enzyme and one syringe with substrate. A drive ram then rapidly shoots the two solutions into a mixing chamber and the mixed solution thereafter enters an optical cell. The flow stops with the help of a stop syringe and the reaction in the optical cell is recorded using a suitable detector (typically absorbance or fluorescence) (Fig. 5A). The dead time (i.e. the time between mixing and observation) is typically less than 1 ms, making it ideal for detection of fast kinetics.

In water, inorganic carbon exists in equilibrium between CO_2, HCO_3^- and CO_3^{2-}, depending on the pH of the solution. At low pH, the dominant form is CO_2, with a relative concentration of 100 % at pH 4. As the pH increases, so does the relative concentration of the HCO_3^- form, reaching 100 % at pH ~8 (Fig. 5C). In the following protocol, we will describe how to measure CA kinetics using the Khalifah/pH indicator assay (Khalifah, 1971) and how to measure a saturation curve to determine the kinetic parameters k_{cat} (turnover number), K_M (Michaelis constant/apparent dissociation constant) and k_{cat}/K_M (catalytic efficiency). We will exemplify this with purified carboxysomes, but the protocol can be used for in vitro measurement of any CA catalyzed CO_2 hydration reaction.

The Khalifah/pH indicator assay measures the rate of proton release when CO_2 is hydrated to form HCO_3^- and H^+. The released H^+ protonates a pH indicator, resulting in a measurable change in absorbance (Fig. 5B) (Khalifah, 1971). Briefly, a saturated CO_2 solution is prepared by bubbling CO_2 into unbuffered MQ water, resulting in a ~34 mM CO_2

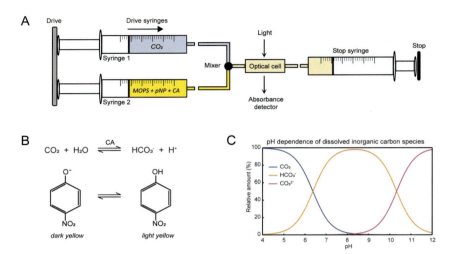

Fig. 5 (A) Schematic representation of a stopped-flow instrument which ensures rapid mixing of CO_2 (drive syringe 1) and buffer, pH indicator and carbonic anhydrase (CA) (drive syringe 2). The mixed solution enters the optical cell, the flow is rapidly stopped by the stop syringe and the reaction is recorded by an absorbance detector. (B) Reaction scheme of CA catalyzed hydration of CO_2 using *para*-nitrophenol (pNP) as pH indicator. When CO_2 hydrates to form HCO_3^- and H^+, pNP gets protonated and shifts color from dark yellow to light yellow. (C) Plot of relative concentrations of dissolved inorganic carbon species as a function of pH.

solution at pH ~4 at 25 °C. For the stopped-flow measurement, one drive syringe will contain the CO_2 solution and the other drive syringe will contain a buffer at the desired pH, a pH indicator with a similar pK_a as the buffer and the CA (in this case carboxysomes) (Fig. 5A). When the unbuffered CO_2 is mixed with the buffer, the CO_2/HCO_3^- equilibrium shifts and the rate at which the new equilibrium is reached is measured. To determine the kinetics constants, the measurement is repeated with varying concentrations of CO_2 and the Michaelis-Menten equation is fitted to the resulting saturation curve. We describe the protocol for measurement at pH 7.5 using the colorimetric pH indicator *para*-nitrophenol (pNP). For measurements at other pH values other buffer/pH indicator pairs are used (for specifics, see Section 4.8). The protocol uses an SX20 Stopped-Flow spectrometer (Applied Photophysics). From the different instruments tested, we found this one to be most suitable for measuring CA activity. However, the assay and the protocol can be adapted for other stopped-flow equipment.

Carboxysomal CA activity

In the following sections we will describe how to prepare the stopped-flow and CO_2 solution for the measurements (4.3–4.5), how to make a standard measurement (4.6), how to measure a saturation curve (4.7), how to measure pH dependence (4.8) and lastly how to analyze the results to determine the kinetic parameters (5.1–5.2).

4.1 Equipment

- SX20 Stopped Flow spectrometer (Applied Photophysics) equipped with a monochromator and absorbance detector, and connected to a circulating water bath.
- 3 mL Luer Lock syringes (HENKE-JECT)
- 2.5 mL gas tight, glass syringe with Luer Lock (Hamilton)
- 100 mL round bottom flask
- Syringe needle and tygon tubing
- CO_2 gas tank

4.2 Reagents

- pH indicator stock solution: *para*-nitrophenol (pNP) 25 mM in MQ water
- Assay buffer: MOPS 100 mM, pH 7.5, ionic strength 100 mM (adjusted with Na_2SO_4)
- Degassed MQ water

4.3 SX20 stopped-flow start-up and setting up the equipment

1. Start up the SX20 stopped-flow spectrometer according to the manufacturer's protocol.
2. Turn on the water bath and set it to 25 °C.
3. Select the 10 mm pathlength of the optical cell. Set the slit width on the monochromator to 2.
4. Open the Pro-Data SX20 software and load Pro-Data Viewer. The Viewer needs to be opened from the Pro-Data software, otherwise no connection is made between the software and the viewer.
5. Set file location and set working directory here. Set seed name - click the Spec(0) icon and put the # to 0. Now all the data will be saved with this name and an increasing number starting from 0.
6. Set up the instrument panel in the Pro-Data software to the following settings:
   Signal: Absorbance
   Monochromator: set wavelength to 400 nm
   Trigger: External.

4.4 Cleaning and preparing the SX20 stopped-flow before starting measurements

1. Fill 2 clean reservoir syringes (3 mL plastic HENKE-JECT syringes) with 2 mL MQ, make sure to remove all air bubbles.
2. Having the drive valves in the Load position (pointing outwards) – connect the reservoir syringe to the drive syringe – push the syringes up and down a couple of times to clean the drive syringe. Perform the cleaning three times with fresh MQ.
3. Load both drive syringes with MQ, make sure to not introduce air bubbles. Level out the drive syringes using the Drive Ram - they should both touch the Ram. Turn the drive valves to Drive position (pointing forward).
4. In Pro-Data SX20 software: Click the "Drive" button. This makes a shot but without recording. To rinse the flow circuit, click the "Drive" button 10–15 times to empty one 2 mL syringe. Perform the cleaning three times.
5. Set the detector high voltage and baseline: fill the drive syringes with baseline solution (Syringe 1: MQ, Syringe 2: MOPS, Table 1) and flush the system by clicking the "Drive" button 10–15 times. Click on "…" and then AutoPM to set the voltage of the detector. Under the Baseline tab, click "Reference". Absorbance should go down to zero. The stopped-flow is now clean and ready for measurements.

Table 1 Content of syringes and final concentrations of reagents after mixing for each step of the stopped-flow protocol.

	Syringe 1	Syringe 2	Final concentrations
Baseline	MQ	100 mM MOPS	50 mM MOPS
Reference	MQ	100 mM MOPS 100 μM pNP	50 mM MOPS 50 μM pNP
Uncatalyzed background reaction	3.4 mM–34 mM CO_2 solution	100 mM MOPS 100 μM pNP	50 mM MOPS 50 μM pNP 1.7 mM–17 mM CO_2 solution
CA catalyzed reaction	3.4 mM–34 mM CO_2 solution	100 mM MOPS 100 μM pNP 0.2–0.4 μM CsoSCA (= 0.6–0.12 mg/mL *H. neapolitanus* carboxysomes)	50 mM MOPS 50 μM pNP 1.7 mM–17 mM CO_2 solution 0.1–0.2 μM CsoSCA (= 0.3–0.6 mg/mL *H. neapolitanus* carboxysomes)

4.5 Preparing assay buffer and making and diluting the CO_2 solution

It is important to remember that all reagents will be mixed in a 1:1 vol ratio in the stopped-flow mixing chamber. The concentrations of the solutions in the syringes therefore need to be 2x the final desired concentrations (Table 1). When handling the CO_2 solution, it is important to use a 2.5 mL gas–tight Hamilton syringe as the reservoir syringe. The dilution of CO_2 is done directly in the syringe (do not use a pipette). This procedure should be done carefully but rather quickly to prevent CO_2 from "escaping". A saturated CO_2 solution at 25 °C is 34 mM. Due to the 1:1 mixing the highest final CO_2 concentration which can be measured is thus 17 mM. We normally perform the measurements in the range of 1.7–17 mM final concentration (3.4–34 mM syringe concentration).

1. Using the 25 mM stock solution, add 100 µM pNP (pH indicator) to the assay buffer to prepare the buffer/indicator solution. Since a small change in the buffer/indicator concentration will affect the absorbance magnitude, prepare a sufficient volume so that all measurements within an experiment (including uncatalyzed background) can be done with the same solution.

2. To prepare a saturated CO_2 solution (34 mM at 25 °C), connect a tygon tubing to a CO_2 gas tank. Attach a needle to the other end and place it in the 100 mL round bottom flask 2/3rd filled with MQ water. Place the flask in the 25 °C water bath. To ensure saturation, let it gently bubble for 30 min before starting measuring and then keep it bubbling continuously throughout the experiment.

3. To prepare 2 mL of CO_2 solution at different concentrations, gently pull up the required volume of saturated CO_2 (e.g. for 1.7 mM final concentration pull up 200 µL and for 17 mM final concentration pull up 2000 µL). This should be done slowly otherwise there is a risk of "pulling out" CO_2 from the solution, resulting in bubbles and a lower concentration of dissolved CO_2.

4. In the same syringe, pull up the required volume of degassed MQ water (e.g. 1800 µL for 1.7 mM final concentration, nothing for 17 mM final concentration).

5. Connect the Hamilton syringe to the drive syringe on the stopped-flow. Gently push the syringe up and down two times to mix.

6. Fill up the drive syringe and turn the valve to drive position.

4.6 Making a standard measurement

To measure CA kinetics two different measurements are needed: (I) 0.5 s traces: this is the initial rate (v_0) which is used to plot the saturation curve (v_0 vs. [CO_2]) to determine k_{cat}, K_M and k_{cat}/K_M (Fig. 6A). (II) 120 s traces:

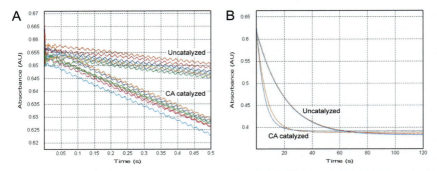

Fig. 6 Screenshot from Pro-Data Viewer of representative experimental traces for CA catalyzed CO_2 hydration using the Khalifah/pH indicator assay, with MOPS/pNP pH 7.5, at 400 nm. (A) 0.5 s traces of uncatalyzed background and CA catalyzed reaction that are used to measure initial rates. (B) 120 s traces of uncatalyzed background and CA catalyzed reaction that show the full progression curve until equilibrium that are used to calculate the experimental buffer factor (Q).

this is the full progression curve from start until equilibrium is reached (Fig. 6B). The total change in absorbance is used to calculate the experimental buffer factor (Q), which in turn is used to convert the initial rates from change in absorbance to change in CO_2 concentration. Below, the standard procedure to do a reference, uncatalyzed background and CA catalyzed measurement is described. In the next Section (4.7) we will describe how to use these to measure a saturation curve.

1. Wash and set up the equipment and software and prepare the reagents as described in (4.3–4.5).
2. Fill the drive syringes with 2 mL of reference, uncatalyzed background or CA catalyzed solution (see Table 1 for details).
3. Click "Drive" 3 times to fill the tubings and cuvette with the correct solution.
4. Run 6–10 × 0.5 shots. Use the following settings:
   Time(s): 0.5 s
   Points: 1000
   Sequencer: 6–10
   Pressure hold: enabled
   Click "Acquire" to start collecting the data.
5. Run 2 × 120 s shots. Use the following settings:
   Time(s): 120 s
   Points: 1000
   Sequencer: 2
   Pressure hold: disabled
   Click "Acquire" to start collecting the data.

4.7 Measuring a saturation curve

To measure steady-state kinetics, the first step is to find an enzyme concentration at which the kinetic trace is linear, so that initial rates are measured. We normally test this out at the highest CO_2 concentration (17 mM), but also confirm the result at one of the lower CO_2 concentrations. When a suitable enzyme concentration has been established, you can begin measuring a full saturation curve. For this, keep the enzyme concentration fixed and vary the concentration of CO_2 (final concentration between 1.7–17 mM). How to prepare and dilute the CO_2 solution is described in Section 4.5 and how to make a standard measurement in Section 4.6. A saturation curve is measured as follows:

1. Run a reference measurement without CO_2: Fill the drive syringes with reference solution (Syringe 1: MQ, Syringe 2: MOPS + pNP, Table 1). Measure 0.5 s and 120 s traces, as described in 4.6. This step only needs to be done once at the beginning to ensure that the buffer/indicator and the instrument behave correctly.
2. Measure the uncatalyzed background reaction at 17 mM CO_2: Fill the drive syringes with background solution (Syringe 1: CO_2, Syringe 2: MOPS + pNP, Table 1). Measure 0.5 s and 120 s traces, as described in 4.6 (Fig. 6A and B).
3. Measure the CA catalyzed reaction at 17 mM CO_2: Fill the drive syringes with CA solution (Syringe 1: CO_2, Syringe 2: MOPS + pNP + carboxysomes/CA, Table 1). Measure 0.5 s and 120 s traces, as described in 4.6 (Fig. 6A and B). Find a CA concentration that results in a linear 0.5 s trace.

   This is to determine a suitable CA concentration for measuring initial rates. If the 0.5 s kinetic trace bends, decrease the enzyme concentration. If the trace is linear, but the signal low, try increasing the enzyme concentration. To verify that the trace is linear, decreasing the enzyme concentration by half should result in half the rate. For *H. neapolitanus* α-carboxysomes a final carboxysome concentration of 0.3–0.6 mg/mL is suitable. This corresponds to 0.1–0.2 μM CsoSCA (CsoSCA is estimated to account for 1.9 % of the total molecular weight of a *H. neapolitanus* carboxysome (Sun et al., 2022)). Note that this concentration can vary from different preps and between species and should therefore always be determined experimentally.
4. Repeat the measurement (background and catalyzed) at one of the lowest CO_2 concentrations to verify the linear relationship. Once a suitable enzyme concentration has been established, continue to step 5 to measure a saturation curve.

5. To measure a full saturation curve, keep the enzyme concentration fixed and vary the final CO_2 concentration from 1.7–17 mM. Repeat the measurement as described in step 2 (background) and 3 (catalyzed) for each CO_2 concentration. Dilute the CO_2 solution as described in 4.5 steps 3–6.

Start with the lowest CO_2 concentration (1.7 mM: 200 µL CO_2 solution + 1800 µL MQ). Thereafter, continue with the next concentration point by increasing the volume of saturated CO_2 solution by steps of 200 µL for each point of the curve (3.4 mM: 400 µL CO_2 solution + 1600 µL MQ, 5.1 mM: 600 µL CO_2 solution + 1400 µL MQ etc.) until reaching 17 mM. This results in 10 points in the range of 1.7–17 mM and is usually suitable. However, depending on the K_M value, additional points may be required to obtain a good saturation curve.

To avoid unnecessary cleaning of the instrument between measurement points, start with the lowest CO_2 concentration and continue with increasing CO_2 concentrations. Measure the full curve for the uncatalyzed background reaction, clean the instrument, and then move on to measure the curve for the CA catalyzed reaction.

4.8 Measuring at different pH values

pH dependence is commonly investigated in carbonic anhydrases. The protocol described in 4.3–4.7 can easily be adapted for measurements at other pHs by using different buffer-indicator pairs. The most commonly used buffer-indicator pairs are listed in Table 2.

Table 2 Buffers and pH-indicator pairs with matching pK_a values for measurement of CA catalyzed CO_2 hydration at different pH values.

pH range	Buffer	pK_a buffer (at 25 °C)	pH indicator	pK_a indicator (at 25 °C)	Wavelength (nm)
5.5–6.7	MES	6.1	Chlorophenol red	6.00	574
6.5–7.9	MOPS	7.2	*para*-Nitrophenol	7.15	400
6.8–8.2	HEPES	7.5	Phenol red	7.90	560
7.7–9.1	TAPS	8.4	*m*-Cresol purple	8.32	578
9.0–9.5	AMPSO	9.0	Thymol blue	8.90	596

Carboxysomal CA activity

5. Data analysis

To convert the change in absorbance (ΔAbs) to the change in CO_2 concentration ($\Delta[CO_2]$) the experimental buffer factor is used. The kinetic constants are thereafter determined by fitting the Michaelis Menten equation to the saturation curve obtained from the initial rates. In the following sections we will describe how to extract the data and perform the calculations to determine k_{cat}, K_M and k_{cat}/K_M.

5.1 Calculate the experimental buffer factor

The experimental buffer factor (Q) is calculated by plotting a standard curve of $\Delta[CO_2]$ vs. ΔAbs for the full progression curve (120 s trace) for all CO_2 concentrations. The slope of the resulting curve is the buffer factor (Khalifah, 1971).

1. Step 1–3 in the current Section (5.1) should be repeated for all CO_2 concentrations measured. Open the experimental file for the 120 s trace in Pro-Data Viewer. Make an average of all traces. Note down the starting and the end point absorbance.
2. Calculate: ΔAbs = Abs$_{start}$ - Abs$_{end\ point}$
3. Calculate: $\Delta[CO_2]$ = $[CO_2]_{start}$ − $[CO_2]_{end\ point}$. At pH 7.5, 93 % of CO_2 is converted to HCO_3^- (Fig. 5C). The fraction of HCO_3^- at a given pH is calculated using the Henderson-Hasselbalch equation (K_1 = 4.3×10^{-7} for H_2CO_3 dissociation, and K_2 = 4.7×10^{-11} for HCO_3^- dissociation). See example calculation for pH 7.5 below.
   a. Calculate the ratio of $[HCO_3^-]/[H_2CO_3]$ = $K_1/[H^+]$ = 13.60 and the ratio of $[CO_3^{2-}]/[HCO_3^-]$ = $K_2/[H^+]$ = 1.49×10^{-3}
   b. Set $[H_2CO_3]$ = x, which gives $[HCO_3^-]$ = 13.60x and $[CO_3^{2-}]$ = 0.020x
   c. Calculate dissolved inorganic carbon (DIC) = $[H_2CO_3]$ + $[HCO_3^-]$ + $[CO_3^{2-}]$ = 14.62x
   d. Calculate the fraction of $[HCO_3^-]$ = $[HCO_3^-]/DIC$ = 0.93 = 93 %
4. Plot: x = ΔAbs vs. y = $\Delta[CO_2]$. The slope of this curve is the experimental buffer factor (Q).

5.2 Calculate the kinetic constants k_{cat}, K_M and k_{cat}/K_M

To determine k_{cat}, K_M and k_{cat}/K_M the initial rates (v_0) obtaned from the 0.5 s traces are used. The raw absorbance data are first converted to $\Delta[CO_2]$ (M/s) using the buffer factor, and the Michaelis Menten equation is then fitted to the saturation curve, v_0 (s^{-1}) vs. $[CO_2]$ (M) (Fig. 7).

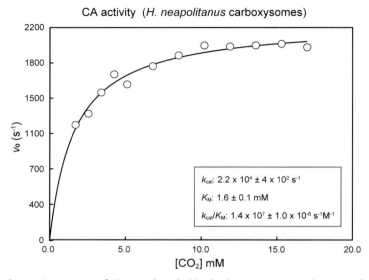

Fig. 7 Saturation curve of CA catalyzed CO_2 hydration measured on purified carboxysomes from *H. neapolitanus*. The Michaelis-Menten equation is fitted to the experimental data to calculate the kinetic constants: $k_{cat} = 2.2 \times 10^4$ s^{-1}, $K_M = 1.6$ mM and $k_{cat}/K_M = 1.4 \times 10^7$ s^{-1} M^{-1} (unpublished data). Measurements were performed in TAPS/*m*-cresol purple pH 8, at 578 nm.

1. Open the experimental file for the 0.5 s traces in Pro-Data Viewer. Remove obvious outliers which can occur from e.g. air bubbles in the system. Make an average of all traces.
2. Open curve fitting, and fit a linear equation to the data between 0.05–0.5 s (due to mixing effects the data can be noisy in the beginning which is why the first 0.05 s are removed from the fit). Repeat for all CO_2 concentrations and for both the catalyzed reaction and the uncatalyzed background reaction. Parameter a is the initial rate, v_0 in ΔAbs/s. It is important to make sure that the trace is linear within the fitted range. If the trace bends, initial rates are not measured and the measurements need to be repeated with a lower enzyme concentration. As a general rule of thumb, the change in absorbance over 0.5 s needs to be smaller than 10 % of the total change over 120 s (or until equilibrium is reached) otherwise steady-state cannot be assumed.
3. Subtract the background rate from the catalyzed rate: $v_0 = v_{0c} - v_{0b}$
4. Convert v_0 in ΔAbs/s to M/s using the experimental buffer factor: v_0 (M/s) = v_0 (ΔAbs/s) * Q

5. Divide by the enzyme concentration to get v_0 in s^{-1}: v_0 (s^{-1}) = v_0 (M/s)/E (M)

6. Plot the saturation curve: x = [CO_2] vs. y = v_0 (s^{-1}) and fit the Michaelis-Menten equation, $v_0 = V_{max}[S]/(K_M + [S])$ to the experimental data using nonlinear regression. If v_0 is converted to s^{-1} (step 5) V_{max} will equal k_{cat} ($k_{cat} = V_{max}/[E]$). Use your favorite program to fit the data, we normally use SimFit where MMFit is used to extract k_{cat} and K_M and RFFit to extract k_{cat}/K_M.

6. Summary

In this chapter, we have provided a detailed description for purifying α-carboxysomes and measuring their carbonic anhydrase activity. As an example we have used α-carboxysomes from the chemoautotroph *H. neapolitanus* and thus also describe how to cultivate this bacterium in a chemostat. The purification protocol described can however be used to purify α-carboxysomes from any source. The kinetic assay uses a stopped-flow spectrometer and relies on a pH indicator which changes color when CO_2 is hydrated to form HCO_3^- and H^+. We have described how to use this method to measure steady-state kinetics to obtain a saturation curve and determine the kinetic constants k_{cat}, K_M and k_{cat}/K_M. Given that CA's role in photosynthesis is to supply the carboxylase Rubisco with CO_2 through various mechanisms, we believe that this method holds significant value for the photosynthesis research community. Additionally, the kinetic protocol can be used to measure in vitro activity for any CA. CA's central role in metabolism across the tree of life, therefore makes it a highly versatile protocol for many research areas.

Acknowledgment

We thank Guillaume Gaullier for helpful advice and comments on the manuscript. This work was funded by grants from the Swedish research council (2019–03700 and 2023–05296) and the Swedish research council for sustainable development (2019–01171) to CB. Drawing of electron microscope was adapted from bioincons.com and is licensed under CC-BY 4.0.

References

Akocak, S., & Supuran, C. T. (2019). Activation of α-, β-, γ- δ-, ζ- and η- class of carbonic anhydrases with amines and amino acids: A review. *Journal of Enzyme Inhibition and Medicinal Chemistry, 34*(1), 1652–1659.

Andersson, I. (2008). Catalysis and regulation in Rubisco. *Journal of Experimental Botany, 59*(7), 1555–1568.

Badger, M. (2003). The roles of carbonic anhydrases in photosynthetic CO_2 concentrating mechanisms. *Photosynthesis Research, 77*(2–3), 83–94.

Badger, M. R., & Price, G. D. (1994). The role of carbonic anhydrase in photosynthesis. *Annual Review of Plant Physiology and Plant Molecular Biology, 45*(1), 369–392.

Baker, S. H., Williams, D. S., Aldrich, H. C., Gambrell, A. C., & Shively, J. M. (2000). Identification and localization of the carboxysome peptide Csos3 and its corresponding gene in Thiobacillus neapolitanus. *Archives of Microbiology, 173*(4), 278–283.

Blikstad, C., Dugan, E. J., Laughlin, T. G., Turnšek, J. B., Liu, M. D., Shoemaker, S. R., Vogiatzi, N., et al. (2023). Identification of a carbonic anhydrase-Rubisco complex within the alpha-carboxysome. *Proceedings of the National Academy of Sciences of the United States of America, 120*(43), e2308600120.

Bonacci, W., Teng, P. K., Afonso, B., Niederholtmeyer, H., Grob, P., Silver, P. A., & Savage, D. F. (2012). Modularity of a carbon-fixing protein organelle. *Proceedings of the National Academy of Sciences of the United States of America, 109*(2), 478–483.

Borden, J. S., & Savage, D. F. (2021). New discoveries expand possibilities for carboxysome engineering. *Current Opinion in Microbiology, 61*, 58–66.

Cannon, G. C., & Shively, J. M. (1983). Characterization of a homogenous preparation of carboxysomes from Thiobacillus neapolitanus. *Archives of Microbiology, 134*(1), 52–59.

Chen, T., Hojka, M., Davey, P., Sun, Y., Dykes, G. F., Zhou, F., Lawson, T., et al. (2023). Engineering α-carboxysomes into plant chloroplasts to support autotrophic photosynthesis. *Nature Communications, 14*(1), 2118.

de Araujo, C., Arefeen, D., Tadesse, Y., Long, B. M., Price, G. D., Rowlett, R. S., Kimber, M. S., et al. (2014). Identification and characterization of a carboxysomal γ-carbonic anhydrase from the cyanobacterium Nostoc sp. PCC 7120. *Photosynthesis Research, 121*(2–3), 135–150.

Desmarais, J. J., Flamholz, A. I., Blikstad, C., Dugan, E. J., Laughlin, T. G., Oltrogge, L. M., Chen, A. W., et al. (2019). DABs are inorganic carbon pumps found throughout prokaryotic phyla. *Nature Microbiology, 4*(12), 2204–2215.

DiMario, R. J., Machingura, M. C., Waldrop, G. L., & Moroney, J. V. (2018). The many types of carbonic anhydrases in photosynthetic organisms. *Plant Science, 268*, 11–17.

Dou, Z., Heinhorst, S., Williams, E. B., Murin, C. D., Shively, J. M., & Cannon, G. C. (2008). CO_2 fixation kinetics of Halothiobacillus neapolitanus mutant carboxysomes lacking carbonic anhydrase suggest the shell acts as a diffusional barrier for CO_2. *The Journal of Biological Chemistry, 283*(16), 10377–10384.

Evans, S. L., Al-Hazeem, M. M. J., Mann, D., Smetacek, N., Beavil, A. J., Sun, Y., Chen, T., et al. (2023). Single-particle cryo-EM analysis of the shell architecture and internal organization of an intact α-carboxysome. *Structure (London, England: 1993), 31*(6), 677–688.e4.

Fersht, A. (2017). *Structure and mechanism in protein science: A guide to enzyme catalysis and protein folding.* World Scientific Publishing.

Flamholz, A., & Shih, P. M. (2020). Cell biology of photosynthesis over geologic time. *Current Biology, 30*(10), R490–R494.

Flamholz, A. I., Dugan, E., Blikstad, C., Gleizer, S., Ben-Nissan, R., Amram, S., Antonovsky, N., et al. (2020). Functional reconstitution of a bacterial CO_2 concentrating mechanism in Escherichia coli. *eLife, 9*.

Fukuzawa, H., Suzuki, E., Komukai, Y., & Miyachi, S. (1992). A gene homologous to chloroplast carbonic anhydrase (icfA) is essential to photosynthetic carbon dioxide fixation by Synechococcus PCC7942. *Proceedings of the National Academy of Sciences of the United States of America, 89*(10), 4437–4441.

Gibbons, B. H., & Edsall, J. T. (1963). Rate of hydration of carbon dioxide and dehydration of carbonic acid at 25 degrees. *The Journal of Biological Chemistry, 238*, 3502–3507.

Heinhorst, S., Williams, E. B., Cai, F., Murin, C. D., Shively, J. M., & Cannon, G. C. (2006). Characterization of the carboxysomal carbonic anhydrase CsoSCA from Halothiobacillus neapolitanus. *Journal of Bacteriology, 188*(23), 8087–8094.

Kerfeld, C. A., & Melnicki, M. R. (2016). Assembly, function and evolution of cyanobacterial carboxysomes. *Current Opinion in Plant Biology, 31*, 66–75.

Khalifah, R. G. (1971). The carbon dioxide hydration activity of carbonic anhydrase. I. Stop-flow kinetic studies on the native human isoenzymes B and C. *The Journal of Biological Chemistry, 246*(8), 2561–2573.

Khalifah, R. G. (1973). Carbon dioxide hydration activity of carbonic anhydrase: Paradoxical consequences of the unusually rapid catalysis. *Proceedings of the National Academy of Sciences of the United States of America, 70*(7), 1986–1989.

Kimber, M. S. (2014). Carboxysomal carbonic anhydrases. *Sub-Cellular Biochemistry, 75*, 89–103.

Koblitz, J., Halama, P., Spring, S., Thiel, V., Baschien, C., Hahnke, R. L., Pester, M., et al. (2023). MediaDive: The expert-curated cultivation media database. *Nucleic Acids Research, 51*(D1), D1531–D1538.

Kong, W.-W., Zhu, Y., Zhao, H.-R., Du, K., Zhou, R.-Q., Li, B., Yang, F., et al. (2024). Cryo-electron tomography reveals the packaging pattern of RuBisCOs in Synechococcus β-carboxysome. *Structure (London, England: 1993)*.

Long, B. M., Hee, W. Y., Sharwood, R. E., Rae, B. D., Kaines, S., Lim, Y.-L., Nguyen, N. D., et al. (2018). Carboxysome encapsulation of the CO_2-fixing enzyme Rubisco in tobacco chloroplasts. *Nature Communications, 9*(1), 3570.

McGurn, L. D., Moazami-Goudarzi, M., White, S. A., Suwal, T., Brar, B., Tang, J. Q., Espie, G. S., et al. (2016). The structure, kinetics and interactions of the β-carboxysomal β-carbonic anhydrase, CcaA. *The Biochemical Journal, 473*(24), 4559–4572.

Melnicki, M. R., Sutter, M., & Kerfeld, C. A. (2021). Evolutionary relationships among shell proteins of carboxysomes and metabolosomes. *Current Opinion in Microbiology, 63*, 1–9.

Menon, B. B., Heinhorst, S., Shively, J. M., & Cannon, G. C. (2010). The carboxysome shell is permeable to protons. *Journal of Bacteriology, 192*(22), 5881–5886.

Merlin, C., Masters, M., McAteer, S., & Coulson, A. (2003). Why is carbonic anhydrase essential to Escherichia coli? *Journal of Bacteriology, 185*(21), 6415–6424.

Metskas, L. A., Ortega, D., Oltrogge, L. M., Blikstad, C., Lovejoy, D. R., Laughlin, T. G., Savage, D. F., et al. (2022). Rubisco forms a lattice inside alpha-carboxysomes. *Nature Communications, 13*(1), 4863.

Nguyen, N. D., Pulsford, S. B., Förster, B., Rottet, S., Rourke, L., Long, B. M., & Price, G. D. (2024). A carboxysome-based CO_2 concentrating mechanism for C3 crop chloroplasts: Advances and the road ahead. *The Plant Journal, 118*(4), 940–952.

Nguyen, N. D., Pulsford, S. B., Hee, W. Y., Rae, B. D., Rourke, L. M., Price, G. D., & Long, B. M. (2023). Towards engineering a hybrid carboxysome. *Photosynthesis Research*.

Ni, T., Sun, Y., Burn, W., Al-Hazeem, M. M. J., Zhu, Y., Yu, X., Liu, L.-N., et al. (2022). Structure and assembly of cargo Rubisco in two native α-carboxysomes. *Nature Communications, 13*(1), 4299.

Peña, K. L., Castel, S. E., de Araujo, C., Espie, G. S., & Kimber, M. S. (2010). Structural basis of the oxidative activation of the carboxysomal gamma-carbonic anhydrase, CcmM. *Proceedings of the National Academy of Sciences of the United States of America, 107*(6), 2455–2460.

Price, G. D., & Badger, M. R. (1989). Expression of human carbonic anhydrase in the cyanobacterium Synechococcus PCC7942 creates a high CO(2)-requiring phenotype: Evidence for a central role for carboxysomes in the CO(2) concentrating mechanism. *Plant Physiology, 91*(2), 505–513.

Price, G. D., Coleman, J. R., & Badger, M. R. (1992). Association of carbonic anhydrase activity with carboxysomes isolated from the cyanobacterium Synechococcus PCC7942. *Plant Physiology, 100*(2), 784–793.

Prywes, N., Phillips, N. R., Tuck, O. T., Valentin-Alvarado, L. E., & Savage, D. F. (2022). Rubisco function, evolution, and engineering. *Annual Review of Biochemistry, 92*, 385–410.

Pulsford, S. B., Outram, M. A., Förster, B., Rhodes, T., Williams, S. J., Badger, M. R., Price, G. D., et al. (2024). Cyanobacterial α-carboxysome carbonic anhydrase is allosterically regulated by the Rubisco substrate RuBP. *Science Advances, 10*(19), eadk7283.

Rae, B. D., Long, B. M., Badger, M. R., & Price, G. D. (2013). Functions, compositions, and evolution of the two types of carboxysomes: polyhedral microcompartments that facilitate CO_2 fixation in cyanobacteria and some proteobacteria. *Microbiology and Molecular Biology Reviews, 77*(3), 357–379.

Raven, J. A., Cockell, C. S., & De La Rocha, C. L. (2008). The evolution of inorganic carbon concentrating mechanisms in photosynthesis. *Philosophical Transactions of the Royal Society of London. Series B, Biological Sciences, 363*(1504), 2641–2650.

Roberts, E. W., Cai, F., Kerfeld, C. A., Cannon, G. C., & Heinhorst, S. (2012). Isolation and characterization of the Prochlorococcus carboxysome reveal the presence of the novel shell protein CsoS1D. *Journal of Bacteriology, 194*(4), 787–795.

Rowlett, R. S., Gargiulo, N. J., Santoli, F. A., Jackson, J. M., & Corbett, A. H. (1991). Activation and inhibition of bovine carbonic anhydrase III by dianions. *The Journal of Biological Chemistry, 266*(2), 933–941.

Sawaya, M. R., Cannon, G. C., Heinhorst, S., Tanaka, S., Williams, E. B., Yeates, T. O., & Kerfeld, C. A. (2006). The structure of beta-carbonic anhydrase from the carboxysomal shell reveals a distinct subclass with one active site for the price of two. *The Journal of Biological Chemistry, 281*(11), 7546–7555.

Shively, J. M., Ball, F., Brown, D. H., & Saunders, R. E. (1973). Functional organelles in prokaryotes: Polyhedral inclusions (carboxysomes) of Thiobacillus neapolitanus. *Science (New York, N. Y.), 182*(4112), 584–586.

So, A. K.-C., Espie, G. S., Williams, E. B., Shively, J. M., Heinhorst, S., & Cannon, G. C. (2004). A novel evolutionary lineage of carbonic anhydrase (epsilon class) is a component of the carboxysome shell. *Journal of Bacteriology, 186*(3), 623–630.

Sun, Y., Harman, V. M., Johnson, J. R., Brownridge, P. J., Chen, T., Dykes, G. F., Lin, Y., et al. (2022). Decoding the absolute stoichiometric composition and structural plasticity of α-carboxysomes. *mBio, 13*(2), e0362921.

Supuran, C. T. (2008). Carbonic anhydrases: Novel therapeutic applications for inhibitors and activators. *Nature Reviews. Drug Discovery, 7*(2), 168–181.

Supuran, C. T. (2016). Structure and function of carbonic anhydrases. *The Biochemical Journal, 473*(14), 2023–2032.

Supuran, C. T. (2023). Carbonic anhydrase versatility: From pH regulation to CO_2 sensing and metabolism. *Frontiers in Molecular Biosciences, 10*, 1326633.

Tripp, B. C., & Ferry, J. G. (2000). A structure–function study of a proton transport pathway in the γ-class carbonic anhydrase from *Methanosarcina thermophila*. *Biochemistry, 39*(31), 9232–9240.

Tripp, B. C., Smith, K., & Ferry, J. G. (2001). Carbonic anhydrase: New insights for an ancient enzyme. *The Journal of Biological Chemistry, 276*(52), 48615–48618.

Vishniac, W., & Santer, M. (1957). The thiobacilli. *Bacteriological Reviews, 21*(3), 195–213.

Zang, K., Wang, H., Hartl, F. U., & Hayer-Hartl, M. (2021). Scaffolding protein CcmM directs multiprotein phase separation in β-carboxysome biogenesis. *Nature Structural & Molecular Biology, 28*(11), 909–922.

Zhou, R.-Q., Jiang, Y.-L., Li, H., Hou, P., Kong, W.-W., Deng, J.-X., Chen, Y., et al. (2024). Structure and assembly of the α-carboxysome in the marine cyanobacterium Prochlorococcus. *Nature Plants, 10*(4), 661–672.

CHAPTER FOURTEEN

Radiometric determination of rubisco activation state and quantity in leaves

Catherine J. Ashton, Rhiannon Page, Ana K.M. Lobo, Joana Amaral, Joao A. Siqueira, Douglas J. Orr, and Elizabete Carmo-Silva*

Lancaster Environment Centre, Lancaster University, Lancaster, United Kingdom
*Corresponding author. e-mail address: e.carmosilva@lancaster.ac.uk

Contents

1.	Introduction	324
	1.1 Method principles	326
	1.2 Materials	328
	1.3 Stock solutions to prepare in advance	331
2.	Methods	333
	2.1 Leaf sampling	333
	2.2 Preparations for leaf extractions and rubisco activity assays	335
	2.3 Protein extraction	336
	2.4 Rubisco activity assays	337
	2.5 Calculation of rubisco activation state	340
	2.6 Rubisco quantification	342
3.	Summary and conclusion	347
	Acknowledgments	348
	References	348
	Further reading	351

Abstract

Rubisco is the key enzyme in photosynthesis, catalyzing fixation of carbon dioxide from the atmosphere into energy storage molecules. Several inefficiencies in Rubisco limit the rate of photosynthesis, and, therefore, the growth of the plant. Rubisco is sensitive to light, making deactivation of the enzyme upon sampling likely. Moreover, the indirect methods often used to study its activity make obtaining reliable data difficult. In this Chapter, we describe an approach to generate reliable and repeatable data for Rubisco activities, activation state and abundance in plant leaves. We include methods to sample and extract proteins, minimizing Rubisco degradation and deactivation. We describe radiometric techniques to measure Rubisco activities and calculate its activation state at the time of sampling, and to quantify its abundance.

Methods in Enzymology, Volume 708
ISSN 0076-6879, https://doi.org/10.1016/bs.mie.2024.10.018
Copyright © 2024 Elsevier Inc. All rights are reserved, including those for text and data mining, AI training, and similar technologies.

1. Introduction

Ribulose-1,5-bisphosphate carboxylase/oxygenase (Rubisco) is the most abundant enzyme in nature (Ellis, 1979; Bar-On & Milo, 2019). Its role is of high importance, catalyzing the carboxylation of ribulose-1, 5-bisphosphate (RuBP) to generate two molecules of 3-phosphoglycerate (3-PGA), a triose-phosphate intermediate of the Calvin-Benson-Bassham (CBB) cycle. However, Rubisco's efficiency is limited by several factors including interaction with inhibitors; complex regulation by interaction with ancillary proteins, most notably Rubisco activase; and its frequent catalysis of oxygenation, instead of carboxylation, which initiates photorespiration. Due to Rubisco's abundance and its limited efficiency, it often limits the rate of photosynthesis, restricting crop yields. For this reason, many research projects include analyses of Rubisco in an attempt to identify strategies to improve the efficiency of photosynthesis and make crop production more sustainable.

In higher plants, Rubisco is a hexadecamer, composed of eight large subunits (52 to 55 kDa, depending on species) and eight small subunits (12 to 15 kDa). To become activated, Rubisco catalytic sites must first become carbamylated on lysine-201 of the large subunit, and then subsequently bind Mg^{2+}, Fig. 1A, (Badger and Sharwood, 2022; Mueller-Cajar, 2017). The positively charged active site of Rubisco is key to this process, facilitating loss of a hydrogen atom from the lysine-NH_3, to form lysine-NH_2. Through the reaction of this NH_2 group with an activator carbon dioxide (CO_2) molecule, a carbamate is formed (Cleland et al., 1998; Badger & Sharwood, 2022). The coordination of a magnesium metal ion to one of the carbonyl groups of the carbamate stabilizes the structure and completes activation of the Rubisco catalytic sites. The second negative oxygen atom of the carbamate is then available to participate in catalytic activity through abstraction of a hydrogen atom from RuBP, facilitating the reaction of RuBP with a second CO_2 molecule, and generating two molecules of 3-PGA (reaction mechanism detailed in Cleland et al., 1998; Prywes et al., 2023; Tcherkez, 2013). Neighboring amino acid residues in the Rubisco catalytic site stabilize the carbamate through non-covalent interactions, Fig. 1B.

When determining the activity of Rubisco in a leaf extract, there are several types of measurements to consider:

1. The **initial activity**, V_i, is an approximation of the activity that Rubisco is doing in the moment that the leaf was sampled. In more

Radiometric determination of rubisco activation state and quantity in leaves 325

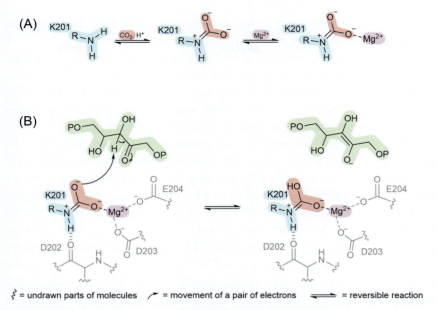

Fig. 1 (A) Formation of the carbamylated active Rubisco. The amine group of Lysine residue 201 (blue) is targeted by a carbon dioxide molecule (red), to generate a carbamate group (blue + red). One of the negatively charged carbamate oxygen atoms is stabilized through coordination to a magnesium ion. The second negatively charged oxygen atom remains available. (B) The carbamate-Mg^{2+} complex is stabilized through non-covalent interactions with surrounding amino acid residues (gray structures). Hydrogen bonding from the main chain carbonyl oxygen of D202 affords stabilization of the carbamate through interaction with the amide hydrogen (blue). As well as coordinating to the carbamate, the magnesium ion (pink) is also coordinated by D203 and E204 carboxyl groups (as described by Cleland et al., 1998). The free O^- on the carbamate initiates enolization of RuBP (green) by abstracting a hydrogen from RuBP's C_3 carbon. *Image produced using ChemDraw.*

specific terms, initial activity measures the activity of Rubisco which is already carbamylated and ready to carboxylate RuBP. Initial activity can be influenced by factors such as the quantity of Rubisco, the carbamylation status at the time of sampling, as well as the catalytic and kinetic properties of the Rubisco being analyzed (Croce et al., 2024).

2. The **total activity**, V_t, is the activity of available Rubisco catalytic sites measured after an incubation with CO_2 and Mg^{2+} to allow full carbamylation of the enzyme. Therefore, unlike V_i, total activity is not influenced by the carbamylation status at the time of sampling, as the incubation period should allow all available catalytic sites to become

carbamylated. However, total activity does not take into account Rubisco catalytic sites which are bound to an inhibitor.[1]

3. The ratio of V_i/V_t gives the **activation state** of the sample and relates to the regulation of the enzyme by carbamylation (Parry et al., 1997). However, the activation state does not consider regulation of Rubisco activity by post-translational modifications or in response to other changes in the chloroplast environment (Amaral et al., 2024).

For around half a century, radiometric assays have been used to determine Rubisco activity. The assays quantify the carboxylation of RuBP with radio-labeled CO_2 ($^{14}CO_2$), yielding a stable labeled product in the form of ^{14}C-3-PGA (Lorimer et al., 1977; Parry et al., 1997). The assays yield accurate and reliable results due to direct tracking of the carbon fixation reaction. Microtiter plate-based assays are also a viable way to determine Rubisco activity; three NADH-linked assays which use alternative coupling enzymes were tested by Sales et al. (2020). Whilst the results of these assays generally correlated with the results of the radiometric method and can be a valuable addition to some studies, some researchers have shown an underestimation of Rubisco activity in spectrophotometric assays. Therefore, the radiometric method is recommended for a thorough characterization of Rubisco properties requiring measurements of the absolute rates of carboxylation (Sales et al., 2020; Sharwood et al., 2016).

When determining Rubisco activity, the sampling method and preparation of the sample for analysis is just as important as the assay itself. Rubisco deactivates very quickly in shade and is reactivated relatively slowly upon returning to the light (Taylor et al., 2022). For this reason, changes in irradiance level upon sampling can have a large effect on Rubisco activity levels.

1.1 Method principles

In general, when extracting proteins from leaf samples, we aim to lyse the cells without degrading the Rubisco protein being released (Carmo-Silva et al., 2024). A buffered extraction solution is used to maintain the pH at 8.2,

---

[1] An additional activity value outside the scope of this chapter is **Maximal activity**, V_{max}. Maximal activity can be used to describe the impact of bound inhibitors. The assay for maximal Rubisco activity requires an additional incubation step with sodium sulfate, which is designed to remove inhibitors from Rubisco, allowing complete carbamylation in the following incubation with carbon dioxide and magnesium ions. Comparing the maximal activity to the initial activity and the total activity can give details on the number of Rubisco enzyme catalytic sites which were inactive at the time of sampling, or the number of active sites which were inhibited at the time of sampling, respectively (Parry et al. 1997). The V_{max} assay requires either spin desalting the sample or protein precipitation followed by a series of washes (Carmo-Silva et al. 2010); it is more time-consuming than V_i or V_t.

which is close to Rubisco's native environment in the chloroplast stroma. Ethylenediaminetetraacetic acid (EDTA) binds to divalent cations which are cofactors for many proteases, hence, reducing protease activity. Protease inhibitors such as benzamidine, phenylmethylsulphonyl fluoride (PMSF), and ε-aminocaproic acid also prevent the hydrolysis of proteins by proteases released in the cell lysis procedure. 2-Mercaptoethanol and DTT are reducing agents used to both prevent formation of disulfide bonds between proteins which may cause aggregation, and to protect existing protein thiol groups. Magnesium chloride ($MgCl_2$) is added as Mg^{2+} is a cofactor in the catalysis mechanism of Rubisco and is found to have a stabilizing effect on Rubisco. Where possible, all stages of the protein extraction process are performed on ice (close to 4 °C), using pre-chilled pestle and mortars or glass homogenizers for grinding the samples and a refrigerated centrifuge pre-chilled to 4 °C. As soon as the clarified extract is obtained, the supernatant is used in the Rubisco activity assays. A more detailed discussion on protein extraction from leaf samples can be found in Carmo-Silva et al. (2024).

Before conducting a larger experiment, it is useful to first determine the incubation time required for the Rubisco total activity assays. To establish this, preliminary assays can be carried out with increasing incubation times, with a suggested range of 3 to 7 min. Depending on species, it is possible that the shorter times tested may not allow full carbamylation of the available Rubisco catalytic sites. If this is the case, as the incubation time increases the Rubisco activity will also increase. However, once the incubation time is sufficient, increasing the time further should not have a significant impact on the Rubisco activity value. An incubation time which sits within this plateau should be chosen to measure Rubisco total activity.

Throughout the assay, the assay buffer, supernatant and RuBP stock solution are kept on ice until pipetted into the assay vials. The assay follows strict timings, to ensure that the length of incubation of Rubisco with CO_2 and Mg^{2+} is consistent (in the case of V_t) and that each assay has a reaction time of 30 s. The assay includes the following components:

- Leaf sample supernatant containing Rubisco
- Assay buffer (pH 8.2, containing $^{14}CO_2$ and Mg^{2+})
- RuBP substrate

The reaction is initiated by the addition of substrate or enzyme, when all of the above components are present, and is quenched by the addition of formic acid (CH_2O_2). The vials are then dried, allowing residual radioactive $^{14}CO_2$ to evaporate (following safe operating procedures, see[2]), and leaving behind organic sugars with radioactive carbon atoms incorporated. Once the radioactivity is quantified, the rate of carbon incorporation and therefore Rubisco activity can be determined.

Radiometric methods are also used to quantify the amount of Rubisco present in a leaf extract. This requires radioactive 2-carboxy-D-arabinitol 1,5-bisphosphate (^{14}CABP), a synthetic inhibitor which binds very strongly to Rubisco catalytic sites (Pierce et al., 1980). By incubating Rubisco with an excess of ^{14}CABP, then separating the Rubisco-^{14}CABP complex from the unbound ^{14}CABP, the amount of Rubisco can be quantified (Parry et al., 1997; Whitney et al., 1999) by measuring the amount of radioactivity using a liquid scintillation counter (Staff, 2004; L'Annunziata et al., 2020; Birks, 2013).

The ^{14}CO$_2$ and other radiolabeled materials must be handled with caution, ensuring that safe systems of work are in place and following the relevant safety and regulatory agencies for your situation and institution. This includes safe and appropriate disposal of any ^{14}C-containing materials. Throughout this protocol, several other hazardous chemicals are used, e.g. formic acid, 2-mercaptoethanol, and DTT, therefore a thorough risk assessment is recommended prior to implementing the procedures described.

The combination of these three techniques: leaf extraction, Rubisco activity assays and Rubisco quantification, coupled with careful sampling, provide a large amount of information on the absolute rates of carboxylation and assessment of carbamylation levels as well as accurate determination of Rubisco content.

1.2 Materials

1.2.1 Materials and equipment
1.2.1.1 Leaf sampling materials
- Light meter, to determine the amount of light irradiating the plant samples.
- Thermometer, to ensure that the sampling temperature is accurately recorded and remains consistent throughout the sampling.
- Cork borer or cutting device of known area and a cork block, Fig. 2A, to allow quick and accurate sampling of leaves.

---

[2] Before undertaking work with radiolabeled materials, the appropriate safe systems of work must be in place and regulated through an appropriate regulatory agency, including any necessary permits for disposal. For example, when evaporating radiolabeled ^{14}CO$_2$, this should be performed in an approved and permitted fume hood to avoid releasing ^{14}CO$_2$ into the laboratory.

Radiometric determination of rubisco activation state and quantity in leaves

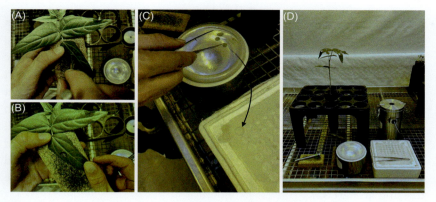

Fig. 2 Leaf sampling directly from the plant for Rubisco assays. (A) Leaf disc samples are taken using a cork borer and a cork block, on leaves which have been fully illuminated for at least 45 min. (B) Samples are taken either side of the main vein where possible. (C) Leaf disc samples are immediately dropped into liquid nitrogen. Once all leaf discs for that leaf are taken and in liquid nitrogen, these are collected using fine tweezers and placed into a microtube in a rack sat in liquid nitrogen. (D) The tube can then be placed in a closed dewar of liquid nitrogen. It helps to have all equipment in close proximity of the plant being sampled to reduce delays in freezing. *Created in BioRender. Ashton, C. (2024) BioRender.com/z77g314.*

- Dewars and liquid nitrogen. It is useful to have both a shallow dewar flask to drop leaf samples into, and a larger (>1 L), sealable dewar flask for storage of the samples in tubes. Use caution – wear appropriate clothing and approved personal protective equipment when working with liquid nitrogen.
- 1.5 mL tubes, labeled, and pierced in the cap with a syringe needle or similar, to store the collected leaf samples. The hole prevents pressure build up.
- Tweezers, including fine blunt forceps (e.g. from Watkins & Doncaster) for handling frozen leaf discs and long tweezers for handling sample tubes in liquid nitrogen.
- Optional: using a controlled irradiance and temperature system for sampling (e.g., light rig) enables stable conditions and maximizes accuracy of results.

1.2.1.2 Protein extraction and rubisco activity assay materials
- Purified water, which has been demineralized. We use RO water (reverse-osmosis water). This is used to prepare all the buffers and to wash equipment.
- Liquid nitrogen and dewar to store samples before extraction.
- Long tweezers, to remove samples from the liquid nitrogen dewar.

- Nitrogen gas source to purge buffer solutions. Since the assay reaction should occur in the presence of a known amount of ^{14}C-$NaHCO_3$, dissolved CO_2 from the atmosphere should be removed from the basic assay buffer. When preparing the buffer solution, all components apart from 2-mercaptoethanol are added and the solution volume is brought up to 90 % of the total desired volume. The solution is then purged with nitrogen gas for 5 min per 100 mL, before adjusting to the desired pH, adding the remaining components and making up to the total required volume. If preparing a large batch, this is then distributed into pre-labeled 50 mL falcon tubes, taking care to avoid aeration of the solution, before freezing at $-20\,°C$.
- A pestle and mortar are used for the leaf extraction and should be chilled before and during use. It is useful to keep several pestle and mortars in the fridge; a chilled pestle and mortar is used for each sample, which is then washed and swapped for the next sample. During the extraction, the pestle and mortar are sat in ice. It is also possible to use pre-chilled glass "Wheaton" style tissue homogenizers.
- Ice boxes, typically foam. It's often convenient to use multiple boxes: one for the pestle and mortar, one for remaining supernatant samples and pellets, and one for the reagents being used in the assay.
- Pipettes and tips are used for addition of buffers and reagents during the protein extractions and assays. Wide bore tips (or cut tips) can be useful for collecting the homogenate if this has a thick consistency after grinding the leaf sample in buffer.
- A refrigerated bench-top centrifuge is used to clarify the leaf sample; the temperature is set to $4\,°C$ to prevent degradation of the sample. A rotor for 1.5 mL tubes is used.
- Plastic microcentrifuge tubes, 1.5 mL, for homogenate centrifugation and for the supernatants.
- Glass screw top vials, 7 mL, these are used for the assays and are sufficiently large to hold both the assay mixture (600 μL) and once the assay mixture is dried off, the water-scintillation cocktail mixture (4 mL).
- A fume hood is required for this assay to ensure that any released $^{14}CO_2$ gases are not released into the laboratory.
- Heat blocks are used throughout the assay procedure, to incubate the sample during the reaction ($30\,°C$) and to dry the assay mixture after the assay is complete ($100\,°C$).
- It's good practice to use a beta-monitor to check workspaces and equipment after assays to ensure no contamination of the work area has occurred.

Radiometric determination of rubisco activation state and quantity in leaves

- Absorbent tissue should be used to clean any contaminated areas after use. In the case of the work areas used for the radioactive assay, the area should be cleaned with a water-based solution and the tissue used be placed in the fume cupboard until dry (by which point, any radioactive $^{14}CO_2$ should have evaporated) and checked with the beta-monitor before disposal.
- A repeater pipette is used as a convenient way to add water to each of the dried assay samples in an efficient manner, but a regular pipette is also suitable.
- Scintillation cocktail is added to the rehydrated assay samples to allow quantification of the radioactivity once placed into the scintillation counter.
- A vortex is required to mix the sample, water and scintillation cocktail mixture.
- A scintillation counter is required to measure the radioactivity levels of each assay vial.
- A timer is necessary to track the extraction and assay times.

1.2.1.3 Rubisco quantification materials
- Ice boxes to store samples before incubation.
- Plastic microcentrifuge tubes, 1.5 mL, for sample incubation with CABP.
- Low pressure columns of internal diameter 0.7 cm, height of 30 cm, 12 mL volume, to fractionate the sample-CABP solution (e.g. Econo-Column Chromatography Columns, Bio-Rad or equivalent).
- A column rack, this allows multiple columns to be run at once.
- Plastic screw-top scintillation vials, 7 mL, for fraction collection.
- Plastic Pasteur pipettes or a syringe and tubing can be useful for loading the Sephadex into the columns.
- A repeater pipette makes adding known volumes of buffer to the column more efficient.

1.3 Stock solutions to prepare in advance

1.3.1 Protein extraction stock solutions
1. Basic protein extraction buffer stock solution [1 ×]: Bicine (50 mM, pH 8.2 adjusted with NaOH), magnesium chloride (20 mM), ethylenedia-minetetraacetic acid (EDTA, 1 mM), benzamidine (2 mM), ε-amino-caproic acid (5 mM). The solution should then be purged with nitrogen gas and the pH re-adjusted to 8.2, before addition of 2-mercaptoethanol (50 mM). This can be prepared in bulk (e.g. 1 L) and then dispensed into

40 mL aliquots and stored at $-20\,°C$. On the day of use, remember to complete the buffer as described in Section 3.2.

2. DL–Dithiothreitol (DTT, 1 M) in water, stored in 1 mL aliquots at $-20\,°C$.

3. Phenylmethylsulphonyl fluoride (PMSF, 100 mM) in ethanol, stored in 1 mL aliquots at $-20\,°C$.

4. Plant protease inhibitor cocktail (e.g. Sigma–Aldrich), decanted into 200 µL aliquots and stored at $-20\,°C$.

1.3.2 Rubisco activity assay stock solutions

1. Basic assay buffer stock solution [2 ×]: Bicine (200 Mm, pH 8.2 adjusted with NaOH), magnesium chloride (40 mM), purged with nitrogen gas and then pH adjusted to 8.2. This can be prepared in bulk (1 L) and then dispensed into 40 mL aliquots and stored at $-20\,°C$. Before use, remember to complete the buffer as described in Section 3.2.

2. RuBP ($\geq$30 mM). This is synthesized in the lab using a protocol adapted from Wong et al. (1982). RuBP may be purchased, however, the reagent's purity must be assessed carefully before use.[3] It is stored at $-80\,°C$ long-term in 1 mL aliquots but is then moved to $-20\,°C$ storage between uses.

3. ^{14}C-NaHCO$_3$ (18.5 kBq/µmol, 0.1 M, 10 mL aliquots).

4. KH$_2$PO$_4$ (0.5 M, 1 mL aliquots in water).

5. Formic acid (5 M, 50-100 mL bottle).

1.3.3 Rubisco quantification stock solutions

1. [2^1-^{14}C]-CABP: 37 kBq/µmol, 12 mM, dispensed and frozen in 125 µL aliquots. ^{14}C-CABP is prepared from RuBP and ^{14}C-KCN as described by Pierce et al. (1980). Before binding to Rubisco, "Neutralized" ^{14}C-CABP must be generated.

2. "Neutralized" ^{14}C-CABP: ^{14}C-CABP (37 kBq/mmol, 12 mM, 125 µL) is combined with a bicine buffer solution (pH 8.2, 0.1 mM, 500 µL) to generate a neutralized solution of ^{14}C-CABP (37 kBq/µmol, 2.4 mM, 625 µL).

3. CABP Basic Binding buffer [2 ×]: Bicine (400 mM, pH 8 adjusted with NaOH), magnesium chloride (80 mM), sodium bicarbonate (40 mM), pH is readjusted to pH 8 before addition of DTT (14 mM). This can be

---

[3] Whether synthesized or purchased, it is important that RuBP used in assays is of high purity. It has been shown that the presence of impurities (which are often also Rubisco inhibitors) can cause a rapid decline in Rubisco activity and therefore an underestimation of the carbamylation status (Kane et al., 1998; Andralojc et al., 2012; Sharwood et al., 2016).

prepared in bulk (e.g. 300 mL) and then dispensed into 12 mL aliquots and stored at −20 °C. On the day of use, remember to complete the buffer as described in Section 3.2.
4. Na$_2$SO$_4$ (1 M, 10 mL aliquots in water). Heating to 30 °C helps ensure full dissolution.
5. Column Buffer: Bicine (20 mM, pH 8.0 adjusted with NaOH), NaCl (75 mM).
6. Gel filtration media: Sephadex G-50 Fine. Column preparation: swell the Sephadex in column buffer at room temperature for at least three hours (or overnight in the fridge). The swelled slurry of Sephadex in buffer can then be poured into the columns, taking care to avoid bubbles and to avoid letting the previous portion settle before adding more, which could create breaks in the Sephadex. The final bed volume should be 10 mL. Once poured, the Sephadex must be packed down by washing with column buffer, 4 × 15 mL.

Note: To determine the amount of dry Sephadex needed, use 1 g of Sephadex per approximately 10 mL of column bed volume required.

2. Methods
2.1 Leaf sampling
2.1.1 In-situ method
It is possible to obtain accurate Rubisco activation states of plants sampled directly in the growth space. However, this should be done with much care:
1. Irradiance and temperature levels must be checked and any differences in certain areas noted. Samples should be taken towards the start of the photoperiod, when leaves have been exposed to a stable irradiance level for at least 45 min, and sampling should be at a consistent time of the day for all samples (we typically collect samples 3-5 h after the start of the photoperiod).
2. Leaf samples are taken from the plant using either a cork borer of known diameter (useful for dicotyledons) or two blades set a specific distance apart (useful for monocotyledons), Fig. 2A and B.
3. For each leaf, around 4 leaf disc samples are taken (4 × 0.55 cm^2) or 1 leaf segment (1 × ~3 cm^2), avoiding any large veins where possible.
   Critical: If not using a light rig, leaves at the top of the canopy should be sampled to ensure that all sampled leaves are fully illuminated and have been for at least the last 45 min.

4. Leaf samples must be collected and immediately frozen in liquid nitrogen (less than 15 s). To achieve this, we usually cut one leaf disc at a time and immediately drop it into an open dewar with liquid nitrogen, Fig. 2C.
5. We then repeat this for the rest of the leaf discs for that leaf sample, i.e. for 4 leaf discs from the same leaf. Then, using light-weight tweezers at liquid nitrogen temperature, we collect the leaf discs from the liquid nitrogen and add them to a 1.5 mL tube in a rack immersed in liquid nitrogen. The tubes are pre-labeled and are pierced to prevent pressure build up, Fig. 2C.
6. Once collected, the tubes can be placed in a sealed dewar of liquid nitrogen before storing in a −80 °C freezer, Fig. 2D.

Pause Point: It is not necessary to use the samples immediately, samples can be stored at −80 °C. In our experience, samples analyzed two years after storage at −80 °C yielded the same results as samples analyzed within two weeks of collection.

2.1.2 Light rig method

The use of a light rig, Fig. 3 and see Taylor et al., 2022, provides a more rigorous method to ensure stable irradiance and temperature levels. Sampling of the plant is performed using the same method as above. However, instead of quickly placing the leaf discs in liquid nitrogen, the leaf samples are placed on wet filter paper. This method ensures the leaf discs of all samples are exposed to specific and constant irradiance and temperature conditions before freezing.

1. Leaf samples are taken from the plant using either a cork borer or two blades set a specific distance apart. In this example, we will assume a cork borer has been used to generate leaf disc samples, but the same principles can be used for strips of leaves if required.
2. For each leaf, 4 leaf discs are usually taken, avoiding any large veins where possible. The leaf discs are quickly placed on a pre-labeled plastic tray containing a wet sheet of filter paper to ensure that the discs remain moist, Fig. 3A.
3. Once all the samples are collected from the plants in their growing environment, the tray of leaf discs is taken to the light rig. The light rig setup contains a water bath set to a specific temperature. Above the water bath is a controllable light source which is used to irradiate the leaf samples. The plastic tray containing the leaf discs on a wet filter paper is suspended on top of the water in the water bath, ensuring that the tray remains in contact with the water, Fig. 3B and C.

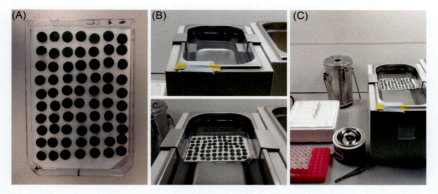

Fig. 3 (A) Leaf samples arranged on a wet filter paper in a plastic tray. The leaf discs are arranged in a known order to keep those from the same leaf sample together. (B) A water bath is used with specific holders to allow the leaf disc tray to sit suspended in the water. (C) The water bath is positioned inside a light rig, with tunable overhead lighting. (D) The space next to the water bath is designed to allow rapid freezing of leaf discs in liquid nitrogen (small open dewar). The frozen leaf discs can then be added to a pre-cooled tube in liquid nitrogen (rack in Styrofoam box), and once sealed, the tube is placed into the closed liquid nitrogen dewar (at the back of the light rig).

4. Throughout the irradiation time in the light rig, care is taken to ensure that the filter paper in the tray does not dry out, and that no leaf discs are momentarily shaded.
5. Once irradiated for at least 45 min, the leaf discs are briefly tapped dry and immediately frozen by dropping individual discs into a small open dewar of liquid nitrogen, Fig. 3D. Once all leaf discs for a particular sample are in liquid nitrogen, these are collected into a labeled tube, which is pre-cooled in a rack of liquid nitrogen (white Styrofoam box), Fig. 3D.

Note: To avoid shading the samples, the leaf discs can be collected from the front to the back of the tray.

Pause Point: Once sampling is complete, samples can be stored at −80 °C until analyzed (and in our experience are stable for at least two years).

2.2 Preparations for leaf extractions and rubisco activity assays

1. Complete buffer solutions and RuBP stocks are prepared the morning of the assays and kept on ice:
    a. Complete extraction buffer: 1 × Basic extraction buffer, 10 mM DTT, 1 mM PMSF, 1% (v/v) plant inhibitor cocktail. 500–700 μL used per extraction (Table 1).

Table 1 Components of the complete extraction buffer.

Reagent (stock solution concentration)	Volume per sample (µL)
Basic extraction buffer [1 ×]	600
Protease Inhibitor cocktail	6
DTT (1 M)	6
PMSF (100 mM)	6

b. Complete assay buffer: $1 \times$ Basic assay buffer, 10 mM ^{14}C-NaHCO$_3$ (18.5 kBq/µmol), 2 mM KH$_2$PO$_4$. 465 µL used per assay vial (Table 2), addition of supernatant (25 µL) and RuBP (10 µL) during the assay increases the total volume to 500 µL.

c. 30 mM RuBP: dilution of $>$ 30 mM RuBP with RO water.

2. Equipment is turned on to give heat blocks time to warm (30 °C, 100 °C) and the centrifuge time to cool to 4 °C.

3. Other equipment is organized: racks for tubes and vials, pipettes and pipette tips positioned in easy reach of extractions and assays, pestle and mortars cooled in the fridge, absorbent tissue near the sink to allow fast washing and drying of used pestle and mortars.

4. The samples should be collected from the −80 °C freezer and immediately placed into a dewar of liquid nitrogen.

2.3 Protein extraction

1. Extraction buffer is added to the ice-cold mortar, sat in an ice box. A few milligrams of acid-washed sand (50−70 mesh particle size) can be added at this point if the leaf samples being used are challenging to homogenize, Fig. 4(1).

2. A sample is then selected at random from the dewar, and the leaf discs immediately added to the mortar before starting to grind thoroughly for 45 s or until a homogenate is formed, Fig. 4(2). The labeled tube which contained the leaf discs should be kept on ice.

3. The homogenate is collected using a pipette with a wide-bore pipette tip if required, and placed back into the ice-cold labeled tube, Fig. 4(3).

4. The tube is then centrifuged at 4 °C for 1 min at 14,000 xg, Fig. 4(4). In the meantime, the sample details can be noted down, the pestle and mortar washed, dried and placed in the fridge, and a cool pestle and mortar retrieved for use with the next sample.

Table 2 Components of the complete assay buffer.

Reagent (stock solution concentration)	Volume per sample (μL)
Basic assay buffer [2 ×]	250
NaHCO$_3$ (^{14}C) (100 mM)	50
KH$_2$PO$_4$ (500 mM)	2
Water	163

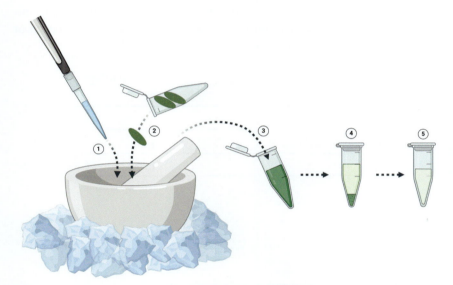

Fig. 4 An overview of the leaf extraction protocol, including (1) addition of buffer solution to the ice-cold mortar, (2) addition of all leaf discs of a given leaf sample and grinding for 30–45 s until a homogenate is formed. (3) Collection of the homogenate into an ice-cold tube, (4) centrifugation at 4 °C and, finally, (5) recovery of the supernatant for use in the subsequent Rubisco activity assays. *Created in BioRender. Ashton, C. (2024) BioRender.com/z14d165.*

5. The supernatant can then be retrieved, placed in a labeled tube and used immediately in a Rubisco activity assay, Fig. 4(5).

2.4 Rubisco activity assays

1. Before starting the assay, four 7 mL glass scintillation vials (two technical replicates for each type of activity) are placed into the 30 °C heat block and assay buffer (465 μL) added to each, providing excess

radioactive carbon dioxide (added in the form of ^{14}C-NaHCO$_3$), and magnesium ions (provided by the MgCl$_2$). Two vials are labeled I$_1$ and I$_2$, representing the two repeats of the initial activity measurement, and the two other vials, T$_1$ and T$_2$, are two repeats for the total activity measurement. It is important to let the vial containing the assay solution reach the desired temperature before starting the assay.

2. For the initial activity assay vials (I$_1$ and I$_2$), RuBP (10 μL of 30 mM stock) should be added to the vials before the start of the assay (Fig. 5).
3. As soon as the leaf extract supernatant is ready, the assay timer is started (T = 0). Delays here could cause inaccurate results due to protein degradation. The supernatant is kept on ice and 25 μL added to each of the four assay vials at 15 s intervals in a specific order:

Time of SN addition	Vial number
00:15	T$_1$
00:30	T$_2$
00:45	I$_1$
01:00	I$_2$

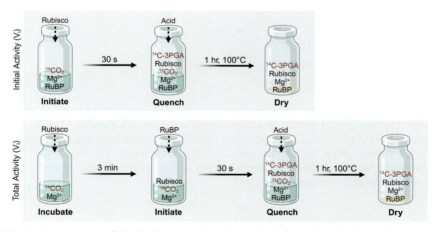

Fig. 5 An overview of the Rubisco initial and total activity assays. Initial activity assay solutions (top) already contain all the cofactors and substrates necessary for Rubisco catalysis, hence, upon addition of the Rubisco-containing supernatant, the assay begins. In the total activity assay (bottom), the Rubisco-containing supernatant is first incubated with ^{14}CO$_2$ and Mg^{2+} to allow carbamylation of all free Rubisco catalytic sites before RuBP is added to initiate the reaction. *Created in BioRender. Ashton, C. (2024) BioRender.com/u20y994.*

Note: Upon addition of reagents to the assay vials, the solution is both pipette-mixed and quickly shaken to ensure complete homogeneity.

4. The initial activity assays are quenched with formic acid ($100\,\mu L$ of a 5 M solution), 30 s after addition of the supernatant:

Time of formic acid addition	Vial number
01:15	I_1
01:30	I_2

At this point, the initial activity assay vials are complete and can be taken out of the 30 °C heat block and placed in a vial rack at room temperature.

5. After 3 min of incubation, the total activity assays are initiated by addition of RuBP ($10\,\mu L$ of 30 mM stock solution):

Time of RuBP addition	Vial number
03:15	T_1
03:30	T_2

6. The total activity assays are allowed 30 s of reaction time, at which point, the reactions are quenched with formic acid ($100\,\mu L$ of 5 M solution) and placed in the vial rack of completed assays.

Time of formic acid addition	Vial number
03:45	T_1
04:00	T_2

7. Blanks should be generated to determine the background radiation for each batch of buffer. These can be produced by adding complete assay buffer ($465\,\mu L$) and formic acid ($100\,\mu L$, 5 M) to a vial, and then processing alongside the sample vials. In theory blanks could also include $25\,\mu L$ of extraction buffer and $10\,\mu L$ of RO water, however this can be omitted as in practice it does not impact the background radiation value of the buffer, and allows blanks to be prepared for each set of assays done in a day rather than for each sample.

8. Once assays are complete, the assay vials and blank vials should be dried thoroughly in a 100 °C heat block inside a fume hood permitted and

approved for ^{14}C gas disposal (this usually takes one hour). Drying in this manner ensures that all radioactive carbon dioxide is evaporated, leaving behind only radioactive ^{14}C atoms which have been incorporated into 3-PGA.

9. Once dried, water (400 μL) is added to each of the vials (the use of a repeat dispenser pipette here can be useful). The vials are left for around 5 min to allow re-solubilization of the dried material, before addition of scintillation cocktail (3.6 mL).
10. The vials can then be capped and vortexed to ensure that the aqueous solution and the scintillation cocktail are fully mixed, before being analyzed in a scintillation counter.

An overview of the timings of the Rubisco initial and total activity assay timings can be seen in Fig. 6.

2.5 Calculation of rubisco activation state

The following measurements and parameters are used or calculated in the determination of initial and total Rubisco activities and the activation state of Rubisco in each sample (Table 3).

1. The radioactivity values are blank-corrected by deducting the average disintegrations per minute (DPM) of the blank samples from all other assay DPM values.

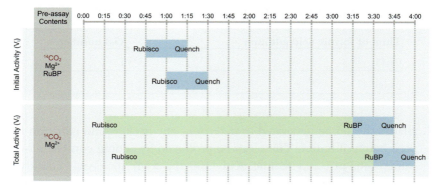

Fig. 6 A timeline of the initial and total activity assays with two repeats of each. Substrates and cofactors (CO_2, Mg^{2+}, and RuBP in the case of the Initial Activity assay) are pre-loaded into the vials before T = 0. The Rubisco-containing supernatant is added at T = 0:15, 0:30, 0:45, 1:00. The Total Activity assays have 3 min of incubation time with the assay buffer (CO_2 and Mg^{2+}) before the addition of RuBP. Samples are quenched 30 s after the start of the reaction with formic acid. *Created in BioRender. Ashton, C. (2024) BioRender.com/i21w538.*

Radiometric determination of rubisco activation state and quantity in leaves 341

Table 3 Abbreviations used in Rubisco activity calculations.

Measurements, parameters (and their units)	Abbreviation
Volume of extraction buffer used (μL)	Vol_{EB}
Leaf Area (cm^2)	A
Volume of supernatant used (μL)	Vol_{SN}
Assay time (s)	t
Disintegration per minute[a]	DPM
Average DPM of blank samples	DPM_{Blk}
Blank-corrected DPM value	DPM_c
Kilo Becquerels	kBq
Initial activity, in the sample ($\mu mol\ min^{-1}\ mL^{-1}$)	$V_{mL,i}$
Total activity, in the sample ($\mu mol\ min^{-1}\ mL^{-1}$)	$V_{mL,t}$
Initial activity, in the leaf ($\mu mol\ m^{-2}s^{-1}$)	$V_{A,i}$
Total activity, in the leaf ($\mu mol\ m^{-2}s^{-1}$)	$V_{A,t}$
Activation State	Act. St.

[a]The measure of radioactivity given by the scintillation counter.

$$DPM_c = DPM - DPM_{Blk}$$

2. The blank-corrected radioactivity, DPM_c, can be converted from DPM to kilo Becquerels (kBq):

$$kBq = \frac{DPM_c}{60,000}$$

3. To determine the number of micromoles of carbon dioxide fixed by Rubisco, we consider the specific activity of the ^{14}C-NaHCO$_3$ used, in this case: 18.7 kBq/μmol. Therefore, for every 18.7 kBq of activity there was one micromole of CO_2 fixed.

$$\mu mol = \frac{kBq}{18.7}$$

4. The initial and total activity can now be calculated in the units: μmol min^{-1} mL^{-1}. The equation starts by determining the μmol of carbon incorporated into sugars by Rubisco in one minute: in this case t = 30 s. To then consider the activity per mL of supernatant, the fraction $\frac{1000}{Vol_{SN}}$ is used, which allows the activity of the 25 μL supernatant to be multiplied up to 1 mL.

$$V_{mL,i} \text{ or } V_{mL,t} \ (\mu mol \ min^{-1} mL^{-1}) = \mu mol \times \frac{60}{t} \times \frac{1000}{Vol_{SN}}$$

5. To calculate initial and total activity per leaf area from the above, i.e. convert to units: μmol m^{-2} s^{-1}, which is required for comparative purposes with other studies, the following equation is used.

$$V_{A,i} \text{ or } V_{A,t} \ (\mu mol \ m^{-2}s^{-1}) = \frac{V_{mL} \times \frac{Vol_{EB}}{1000}}{A} \times \frac{10000}{60}$$

The previously obtained value of activity in terms of $\mu mol \ min^{-1} mL^{-1}$ is scaled for the volume of extraction buffer used, since this is the approximate total volume of homogenized leaf extract (theoretically, the leaf material itself slightly increases this volume). The total activity for the total extraction buffer used is then divided by the leaf area (A) of the sample, giving the activity per cm^2. To convert this value to activity per m^2, we can multiply this value by 10,000 (100 cm $\times$ 100 cm = 1 m^2). Finally, rather than having the activity per minute (this would be a large value per m^2!), we divide by 60 to convert the units to per second.

6. The values from steps 4 and 5 can then be averaged for the two repeats of initial and total activity measurements for each sample.

7. The activation state of each sample can then be calculated by dividing the average initial Rubisco activity by the average total Rubisco activity.

$$Act. \ St. = \frac{V_{mL,i}}{V_{mL,t}} \ \text{ or } Act. \ St. = \frac{V_{A,i}}{V_{A,t}}$$

2.6 Rubisco quantification

This quantification of Rubisco relies on the strong binding of the synthetic Rubisco inhibitor CABP to the catalytic site. Here, a solution containing an excess of radioactive neutralized ^{14}C-CABP is incubated with the supernatant to ensure complete binding of ^{14}C-CABP to every Rubisco catalytic site. The unbound ^{14}C-CABP molecules are then separated from the Rubisco-^{14}C-CABP complexes by passing through a Sephadex column and the amount of radioactivity of the Rubisco-^{14}C-CABP complex used to determine the Rubisco concentration (Fig. 7).

1. Add 100 μL leaf extract to 100 μL ^{14}C-CABP binding solution and incubate at room temperature for 25–30 min.

    Note: If the samples being tested are expected to have higher concentrations of Rubisco (e.g. on the basis of containing high protein concentrations), more ^{14}C-CABP should be added to ensure it is in a large excess. An excess of ^{14}C-CABP can later be confirmed by using the scintillation counter to detect unbound ^{14}C-CABP in Vial 5 (see Rubisco quantification calculations).

2. The solution is then briefly spun to ensure all solution is at the bottom of the microtube.

    Note: The amount of supernatant-^{14}C-CABP solution loaded onto the column will have a large influence on the calculated concentration. Therefore, it is important to load the full contents of the tube to ensure accurate results.

3. Before applying the sample to the columns, open the column outlet valve and allow the column buffer to reach the top of the Sephadex, just enough so that the top of the Sephadex can be seen but without allowing the Sephadex to dry out.

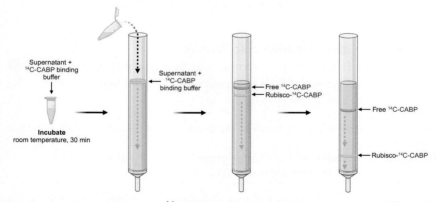

Fig. 7 After being incubated in ^{14}C-CABP binding solution, the supernatant–^{14}C-CABP binding solution mixture is gently pipetted onto the top of the Sephadex (left-hand column). After repeated additions of column buffer, the sample begins to move down the column. Movement through the column allows the Rubisco-^{14}C-CABP complex and the free ^{14}C-CABP to become separated. The Rubisco-^{14}C-CABP complex should be eluted first. The Rubisco-^{14}C-CABP and free ^{14}C-CABP are collected into different vials for scintillation counting. *Image produced using BioRender. Created in BioRender. Ashton, C. (2024) BioRender.com/p05q917.*

4. The supernatant-^{14}C-CABP solution is then added to the top of the Sephadex column, one column is used per sample. The column tap is opened and allowed to run until the sample is loaded onto the Sephadex.

5. The sample is pushed further onto the Sephadex by applying 200 µL of column buffer to the column and allowing this to run onto the Sephadex.

   Note: When applying sample or column buffer to the columns it is important to apply this gently to avoid disturbing the Sephadex resin bed. Disturbing the resin bed could cause the sample to run down the column unevenly or cause the sample to resuspend in the buffer above the resin bed – both of which could cause a decreased separation of the Rubisco-^{14}C-CABP complex from the unbound ^{14}C-CABP.

6. At this point a larger amount of column buffer is added to move the sample through the column: 2250 µL.

7. For the next sets of column buffer applied to the columns, the eluate is collected in five separate vials, used later in analysis (Table 4).

8. Once all the vials have been collected, 3.6 mL of scintillant is added to each vial, the vials are then capped and vortexed before measuring the amount of radioactivity (disintegrations per minute) using a scintillation counter.

   Note: After use, the columns can then be rinsed with several column volumes of buffer, capped and stored for re-use. Once the columns have been used 4–6 times (depending on the type of samples being used), they will become slower and should be replaced with fresh Sephadex.

2.6.1 Rubisco quantification calculations

The amount of Rubisco in a sample can be quantified using the DPM values measured for each of the collected vials. The predicted elution of the

Table 4 Application of column buffer to the quantification columns and the expected eluate collected.

Vial number	Volume of wash buffer added/ eluate collected in vial (µL)	Predicted contents
1	750	background
2	1500	Rubisco-^{14}C-CABP
3	750	Rubisco-^{14}C-CABP tail
4	750	background
5	2250	unbound ^{14}C-CABP

Radiometric determination of rubisco activation state and quantity in leaves   345

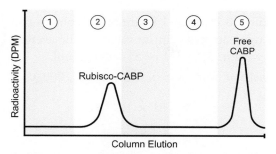

Fig. 8 A theoretical graph showing the change in radioactivity in DPM as the column is eluted. The numbers across the top of the graph denote the vial numbers. *Created in BioRender. Ashton, C. (2024) BioRender.com/h51z544.*

Rubisco-^{14}C-CABP complex and free ^{14}C-CABP molecules are described below (Fig. 8).

- **Vial 1 (V_1)** should only contain the column buffer and therefore any radiation seen in this vial will be background radiation, <100 DPM: the 'blank'.
- **Vial 2 (V_2)** contains most of the Rubisco-^{14}C-CABP complex, and **Vial 3 (V_3)** may contain the tail end of this peak.
- **Vial 4 (V_4)** usually contains only background radiation, and so should anything go wrong with V_1, this can also be used as a blank if required.
- **Vial 5 (V_5)** should contain the excess unbound CABP. This should be a relatively large DPM value (i.e. $V_5 \geq V_2$) to ensure that the quantification is accurate. Should the radiation in vial 5 (V_5) be low, we cannot be sure that all the Rubisco enzymes are saturated with ^{14}C-CABP, and therefore the concentration of Rubisco calculated could be an underestimate.

For each sample, the following procedure can be followed to determine the Rubisco concentration:

1. Subtract the DPM value of the "blank" vial (usually V_1) from the DPM value of each of the other vials to obtain blank corrected values (DPM$_{xc}$). Where x is the vial number.

$$DPM_{xc} = DPM_x - DPM_1$$

2. Sum (blank corrected) vials DPM$_{2c}$, DPM$_{3c}$, and DPM$_{4c}$ to give the total DPMs associated with Rubisco-^{14}C-CABP.

$$DPM_{total} = DPM_{2c} + DPM_{3c} + DPM_{4c}$$

3. In order to determine the mass of Rubisco in a sample, DPM$_{total}$ must be divided by the DPMs associated with 8 ^{14}C-CABP molecules, as a single Rubisco enzyme will bind to 8 CABP molecules.

a. 1 mg of Rubisco contains 1.81 nmol of Rubisco complexes. This is calculated using an approximate value for the average L8S8 Rubisco molecular weight: 552 kDa or 552,000 g/mol.

$$moles\ of\ Rubisco = 0.001\ g \div 552,000\ g/mol$$

$$= 1.81 \times 10^{-9}\ mol$$

$$= 1.81\ nmol$$

b. Every one Rubisco complex will bind to eight CABP molecules.

$$moles\ of\ CABP\ binding\ to\ 1\ Rubisco = 8 \times 1.8\ nmol$$

$$= 14.5\ nmol$$

c. Using the known specific activity of the CABP solution (in this case 37 kBq/μmol), the radioactivity of 14.5 nmol of CABP (i.e. the amount of radioactive CABP bound to one Rubisco enzyme) can be determined:

$$if\ 1\ \mu mol\ CABP = 37\ kBq,$$

$$then,\ 14.5\ nmol\ of\ CABP = 37\ kBq \times \frac{14.5\ nmol}{1000} = 0.5365\ kBq$$

d. The radioactivity can be converted into DPM using the conversion 1 kBq = 60,000 DPM:

$$0.5365\ kBq \times 60,000 = 32190\ DPM$$

e. By dividing the DPM_{total} by 32,190, an estimate mass of Rubisco in the sample applied to the column can be determined.

$$Mass\ of\ Rubisco\ (mg) = DPM_{total} \div 32190\ DPM$$

4. To determine the concentration of the supernatant, the calculated mass of Rubisco can be divided by the volume (μL) of the supernatant mixed with CABP binding solution and applied to the column.

$$Rubisco\ concentration\ (\mu g/\mu L) = (mg\ of\ Rubisco \times 1000) \div \mu LSN$$

3. Summary and conclusion

This chapter describes comprehensive methods to determine the rate of carboxylation by Rubisco, the activation state and amount of the enzyme in leaf extracts. As the most abundant enzyme on the planet and with a critical role for carbon assimilation, Rubisco is the subject of much study (Carmo-Silva and Sharwood, 2023, and references therein). The activity of Rubisco in a leaf is a result of its abundance, catalytic properties and activity regulation (Croce et al., 2024). Adjustment of Rubisco activity involves interaction with regulatory proteins such as Rubisco activase, post-translational modifications and interaction with metabolites and ions in the chloroplast stroma (Amaral et al., 2024; Lobo et al., 2024). Such regulation enables fine tuning Rubisco activity in response to changes in the environment surrounding the leaf, including dynamic changes in irradiance levels, temperature and CO_2 availability. These methods are useful to characterize Rubisco function in plants, one of the key targets for improving photosynthesis and crop yields (Croce et al., 2024).

Methods akin to those described here have been used to understand the impact of environmental conditions on Rubisco carboxylation activity in plant leaves (e.g., Campbell et al., 1988; Jiang et al., 1993; Parry et al., 2002; Carmo-Silva et al., 2010, Galmés et al., 2013; Perdomo et al. 2017; Correia et al. 2021; Taylor et al., 2022) and to establish useful natural variation in Rubisco properties across different plant species and crop varieties for crop breeding (e.g., Driever et al., 2014; Sharwood et al., 2016; Carmo-Silva et al., 2017; De Souza et al., 2020; McAusland et al., 2020; Silva-Pérez et al., 2020; Sales et al., 2022; Sharwood et al., 2022). Some of these studies have highlighted that, in many highly productive crops, Rubisco represents a large fraction of the total soluble protein in leaves (Carmo-Silva et al., 2015; Evans & Clarke, 2019). With plant engineering approaches showing that increased abundance of Rubisco may result in increased crop productivity (Yoon et al., 2020; Salesse-Smith et al., 2024), the methods described will likely aid future photosynthesis improvement research.

When studying Rubisco activities in plant leaves, the sample collection is critical. Many studies that involve measuring photosynthetic traits use excised leaves, which can yield reliable and reproducible data provided that adequate preliminary tests are carried out and robust experimental designs are adopted (Ferguson et al., 2023). Given that Rubisco in leaves deactivates upon only a few seconds of a transition from high to low light, of fully illuminated to shaded conditions (Taylor et al., 2022), it is recommended that leaves are subject to stable irradiance prior to sampling. All procedures for Rubisco activity analysis, from sampling to measuring, should be carried out within a timeframe that is kept as short as feasible and implementing the recommendations suggested in the above methods description, including using appropriate solutions and temperatures to maintain protein integrity.

Throughout this Chapter, we describe detailed methods of taking plant leaf samples and extracting proteins, in addition to a thorough description of Rubisco activity and quantification assays. This contribution provides a basis for many Rubisco studies and can be adopted to study the impact of external factors. The procedure can also be easily adapted where there are equipment limitations, for example, whilst the use of a light rig to incubate leaf samples may be a more rigorous approach to ensure consistent irradiation, a cautious approach to leaf sampling at the start of the photoperiod and choosing un-shaded samples will likely achieve reliable results. Use of these Rubisco protocols will aid further understanding of such an important carboxylase.

Acknowledgments

This work was supported by the project Realizing Increased Photosynthetic Efficiency (RIPE), that is funded by Bill & Melinda Gates Agricultural Innovations grant investment 57248, awarded to Lancaster University by the University of Illinois, USA, and by the European Union's Horizon2020 research and innovation program projects CAPITALISE (grant no. 862201; JA, EC-S) and PhotoBoost (grant no. 862127; AKML, EC-S).

References

Amaral, J., Lobo, A. K. M., & Carmo-Silva, E. (2024). Regulation of Rubisco activity in crops. *New Phytologist, 241*, 35–51.

Andralojc, P. J., Madgwick, P. J., Tao, Y., Keys, A., Ward, J., Beale, M. H., ... Parry, M. A. J. (2012). 2-Carboxy-D-arabinitol 1-phosphate (CA1P) phosphatase: Evidence for a wider role in plant Rubisco regulation. *Biochemical Journal, 442*, 733–742.

Badger, R. B., & Sharwood, R. E. (2022). Rubisco, the imperfect winner: It's all about the base. *Journal of Experimental Botany, 74*, 562–580.

Bar-On, Y. M., & Milo, R. (2019). The global mass and average rate of Rubisco. *Proceedings of the National Academy of Sciences of the United States of America, 116*, 4738–4743.

Birks, J. B. (2013). *The theory and practice of scintillation counting: International series of monographs in electronics and instrumentation, Vol. 27*. Elsevier.

Campbell, W. J., Allen, L. H., Jr, & Bowes, G. (1988). Effects of CO_2 concentration on rubisco activity, amount, and photosynthesis in soybean leaves. *Plant Physiology, 88*, 1310–1316.

Carmo-Silva, A. E., Keys, A. J., Andralojc, P. J., Powers, S. J., Arrabaça, M. C., & Parry, M. A. (2010). Rubisco activities, properties, and regulation in three different C4 grasses under drought. *Journal of Experimental Botany, 61*, 2355–2366.

Carmo-Silva, E., Scales, J. C., Madgwick, P. J., & Parry, M. A. (2015). Optimizing Rubisco and its regulation for greater resource use efficiency. *Plant, Cell & Environment, 38*, 1817–1832.

Carmo-Silva, E., Andralojc, P. J., Scales, J. C., Driever, S. M., Mead, A., Lawson, T., & Parry, M. A. (2017). Phenotyping of field-grown wheat in the UK highlights contribution of light response of photosynthesis and flag leaf longevity to grain yield. *Journal of Experimental Botany, 68*, 3473–3486.

Carmo-Silva, E., & Sharwood, R. E. (2023). Rubisco and its regulation—Major advances to improve carbon assimilation and productivity. *Journal of Experimental Botany, 74*, 507–509.

Carmo-Silva, E., Page, R., Marsden, C. J., Gjindali, A., & Orr, D. J. (2024). Extraction of soluble proteins from leaves. In S. Covshoff, & J. M. Walker (Eds.). *Photosynthesis: Methods and protocols* (pp. 391–404). Springer US.

Cleland, W. W., Andrews, J. T., Gutteridge, S., Hartman, F. C., & Lorimer, G. H. (1998). Mechanism of Rubisco: The carbamate as general base. *Chemical Reviews, 98*, 549–562.

Correia, P. M. P., Da Silva, A. B., Vaz, M., Carmo-Silva, E., & Marques Da Silva, J. (2021). Efficient regulation of CO_2 assimilation enables greater resilience to high temperature and drought in maize. *Frontiers in Plant Science, 12*, 182–196.

Croce, R., Carmo-Silva, E., Cho, Y. B., Ermakova, M., Harbinson, J., Lawson, T., & Zhu, X. G. (2024). Perspectives on improving photosynthesis to increase crop yield. *The Plant Cell*, koae132.

De Souza, A. P., Wang, Y., Orr, D. J., Carmo-Silva, E., & Long, S. P. (2020). Photosynthesis across African cassava germplasm is limited by Rubisco and mesophyll conductance at steady state, but by stomatal conductance in fluctuating light. *New Phytologist, 225*, 2498–2512.

Driever, S. M., Lawson, T., Andralojc, P. J., Raines, C. A., & Parry, M. A. J. (2014). Natural variation in photosynthetic capacity, growth, and yield in 64 field-grown wheat genotypes. *Journal of Experimental Botany, 65*, 4959–4973.

Ellis, J. R. (1979). The most abundant protein in the world. *Trends in Biochemical Sciences, 4*, 241–244.

Evans, J. R., & Clarke, V. C. (2019). The nitrogen cost of photosynthesis. *Journal of Experimental Botany, 70*, 7–15.

Ferguson, J. N., Jithesh, T., Lawson, T., & Kromdijk, J. (2023). Excised leaves show limited and species-specific effects on photosynthetic parameters across crop functional types. *Journal of Experimental Botany, 74*, 6662–6676.

Galmés, J., Aranjuelo, I., Medrano, H., & Flexas, J. (2013). Variation in Rubisco content and activity under variable climatic factors. *Photosynthesis Research, 117*, 73–90.

Jiang, C. Z., Rodermel, S. R., & Shibles, R. M. (1993). Photosynthesis, Rubisco activity and amount, and their regulation by transcription in senescing soybean leaves. *Plant Physiology, 101*, 105–112.

Kane, H. J., Wilkin, J.-M., Portis, A. R., Jr., & Andrews, T. J. (1998). Potent inhibition of ribulose-bisphosphate carboxylase by an oxidized impurity in ribulose-1,5-bisphosphate. *Plant Physiology, 117*, 1059–1069.

L'Annunziata, M. F., Tarancón, A., Bagán, H., & García, J. F. (2020). *Liquid scintillation analysis: Principles and practice. Handbook of radioactivity analysis.* Academic Press, 575–801.

Lobo, A. K. M., Orr, D. J., & Carmo-Silva, E. (2024). Regulation of Rubisco activity by interaction with chloroplast metabolites. *Biochemical Journal, 481*, 1043–1056.

Lorimer, G. H., Badger, M. R., & Andrews, T. J. (1977). D-Ribulose-1,5-bisphosphate carboxylase-oxygenase. *Analytical Biochemistry, 78*, 66–75.

McAusland, L., Vialet-Chabrand, S., Jauregui, I., Burridge, A., Hubbart-Edwards, S., Fryer, M. J., & Murchie, E. H. (2020). Variation in key leaf photosynthetic traits across wheat wild relatives is accession dependent not species dependent. *New Phytologist, 228*, 1767–1780.

Mueller-Cajar, O. (2017). The diverse AAA+ machines that repair inhibited rubisco active sites. *Frontiers Molecular Biosciences, 4*, 31.

Parry, M. A. J., Andralojc, P. J., Parmar, S., Keys, A. J., Habash, D., Paul, M. J., ... Servaites, J. C. (1997). Regulation of Rubisco by inhibitors in the light. *Plant, Cell and Environment, 20*, 528–534.

Parry, M. A., Andralojc, P. J., Khan, S., Lea, P. J., & Keys, A. J. (2002). Rubisco activity: Effects of drought stress. *Annals of Botany, 89*, 833–839.

Perdomo, J. A., Capó-Bauçà, S., Carmo-Silva, E., & Galmés, J. (2017). Rubisco and rubisco activase play an important role in the biochemical limitations of photosynthesis in rice, wheat, and maize under high temperature and water deficit. *Frontiers in Plant Science, 8*, 490.

Pierce, J., Tolbert, N. E., & Barker, R. (1980). Interaction of ribulose bisphosphate carboxylase/oxygenase with transition-state analogs. *Biochemistry, 19*, 934–942.

Prywes, N., Phillips, N. R., Tuck, O. T., Valentin-Alvarado, L. E., & Savage, D. F. (2023). Rubisco function, evolution and engineering. *Annual Review of Biochemistry, 92*, 385–410.

Sales, C. R. G., Bernardes da Silva, A., & Carmo-Silva, E. (2020). Measuring Rubisco activity: Challenges and opportunities of NADH-linked microtiter plate-based and ^{14}C-based assays. *Journal of Experimental Botany, 71*, 5302–5312.

Sales, C. R., Molero, G., Evans, J. R., Taylor, S. H., Joynson, R., Furbank, R. T., & Carmo-Silva, E. (2022). Phenotypic variation in photosynthetic traits in wheat grown under field versus glasshouse conditions. *Journal of Experimental Botany, 73*, 3221–3237.

Salesse-Smith, C. E., Adar, N., Kannan, B., Nguyen, T., Guo, M., Ge, Z., & Long, S. P. (2024). Adapting C4 photosynthesis to atmospheric change and increasing productivity by elevating Rubisco content in Sorghum and Sugarcane. *bioRxiv*, 2024-05.

Sharwood, R. E., Sonawane, B. V., Ghannoum, O., & Whitney, S. M. (2016). Improved analysis of C_4 and C_3 photosynthesis via refined in vitro assays of their carbon fixation biochemistry. *Journal of Experimental Botany, 67*, 3137–3148.

Sharwood, R. E., Quick, W. P., Sargent, D., Estavillo, G. M., Silva-Perez, V., & Furbank, R. T. (2022). Mining for allelic gold: Finding genetic variation in photosynthetic traits in crops and wild relatives. *Journal of Experimental Botany, 73*, 3085–3108.

Silva-Pérez, V., De Faveri, J., Molero, G., Deery, D. M., Condon, A. G., Reynolds, M. P., & Furbank, R. T. (2020). Genetic variation for photosynthetic capacity and efficiency in spring wheat. *Journal of Experimental Botany, 71*, 2299–2311.

Staff, N. D. L. (2004). Principles and applications of liquid scintillation counting. *National Diagnostics*.

Taylor, S. H., Gonzalez-Escobar, E., Page, R., Parry, M. A. J., Long, S. P., & Carmo-Silva, E. (2022). Faster than expected Rubisco deactivation in shade reduces cowpea photosynthetic potential in variable light conditions. *Nature Plants, 8*, 118–124.

Tcherkez, G. (2013). Modelling the reaction mechanism of ribulose-1,5-bisphosphate carboxylase/oxygenase and consequences for kinetic parameters. *Plant, Cell & Environment, 36*, 1586–1596.

Whitney, S. M., Von Caemmerer, S., Hudson, G. S., & Andrews, T. J. (1999). Directed mutation of the Rubisco large subunit of tobacco influences photorespiration and growth. *Plant Physiology, 121*, 579–588.

Wong, C. H., Pollak, A., McCurry, S. D., Sue, J. M., Knowles, J. R., & Whitesides, G. M. (1982). Synthesis of ribulose 1,5-bisphosphate: Routes from glucose 6-phosphate (via 6-phosphogluconate) and from adenosine monophosphate (via ribose 5-phosphate). In W. A. Wood (Ed.). *Methods in enzymology, carbohydrate metabolism—Part D* (pp. 108–121). Academic Press Inc.

Yoon, D. K., Ishiyama, K., Suganami, M., Tazoe, Y., Watanabe, M., Imaruoka, S., & Makino, A. (2020). Transgenic rice overproducing Rubisco exhibits increased yields with improved nitrogen-use efficiency in an experimental paddy field. *Nature Food, 1*, 134–139.

Further reading

Amaral, J., Lobo, A. K., Carmo-Silva, E., & Orr, D. J. (2024). Purification of rubisco from leaves. *Photosynthesis: Methods and Protocols* (pp. 417–426). New York, NY: Springer US.

Sales, C. R. G., Silva, A., & Carmo-Silva, E. (2020). Protocols from Sales et al. (2020) Rubisco activity: challenges and opportunities of NADH-linked microtiter plate-based and ^{14}C-based assays. *Protocols.io*. https://doi.org/10.17504/protocols.io.bf8djrs6.

CHAPTER FIFTEEN

Computational methods for the study of carboxylases: The case of crotonyl-CoA carboxylase/reductase

Rodrigo Recabarren[a,1], Aharon Gómez Llanos[b,1], and Esteban Vöhringer-Martinez[a,*]

[a]Departamento de Físico-Química, Facultad de Ciencias Químicas, Universidad de Concepción, Concepción, Chile
[b]Departamento de Ciencias Biológicas y Químicas, Facultad de Medicina y Ciencia, Universidad San Sebastian, Lientur, Concepción, Chile
*Corresponding author. e-mail address: evohringer@udec.cl

Contents

1. Introduction	354
2. Molecular dynamics simulations identify conformational changes in crotonyl-CoA carboxylase/reductase increasing the local CO_2 concentration in the active site	357
2.1 Methods to quantify local CO_2 concentration and its dependence on protein conformation	359
2.2 Calculation of average CO_2 residence times	363
2.3 Simulation details and validation	366
3. QM/MM molecular dynamics simulations reveal the reaction mechanism of the chemical CO_2 fixation step	369
3.1 QM/MM calculations and free energy profiles	369
3.2 The adaptive string method as an efficient tool for studying complex carboxylation reactions	372
3.3 Corrections to free energy profiles	375
3.4 Dissecting reactivity by comparing free energy profiles of different reaction mechanisms	376
3.5 Model building for QM/MM MD simulations	379
3.6 Simulation details for QM/MM MD simulations	379
4. Summary	382
References	382

[1] Contributed equally.

Methods in Enzymology, Volume 708
ISSN 0076-6879, https://doi.org/10.1016/bs.mie.2024.10.025
Copyright © 2024 Elsevier Inc. All rights reserved, including those for text and data mining, AI training, and similar technologies.

Abstract

The rising levels of atmospheric CO_2 and its impact on climate change call for new methods to transform this greenhouse gas into beneficial compounds. Carboxylases have a significant role in the carbon cycle, converting gigatons of CO_2 into biomass annually. One of the most effective and fastest carboxylases is crotonyl-CoA carboxylase/reductase (Ccr). To understand its underlying mechanism, we have developed computational methods and protocols based on all-atom molecular dynamics simulations. These methods provide the CO_2 binding locations and free energy inside the active site, dependent on different conformations adopted by Ccr and the presence of the crotonyl-CoA substrate. Furthermore, the adaptive string method and quantum mechanics/molecular mechanics (QM/MM) molecular dynamics simulations outline the CO_2 fixation reaction via two different mechanisms. The direct mechanism involves a hydride transfer creating a reactive enolate, which then binds the electrophilic CO_2 molecule, resulting in the carboxylated product. Alternatively, another mechanism involves the formation of a covalent adduct. Our simulations suggest that this adduct serves to store the enolate in a much more stable intermediate avoiding its reduction side reaction, explaining the enzyme's efficiency. Overall, this work presents computational methods for studying carboxylation reactions using Ccr as a model, providing general principles that can be applied to modeling other carboxylases.

1. Introduction

In the grand scheme of life on Earth, nature has ingeniously harnessed the potential of CO_2, an abundant carbon source, to construct biomass. This process has been refined over billions of years, with all three kingdoms of life evolving and acquiring carboxylases, enzymes that facilitate the use of CO_2 through diverse mechanistic pathways. Recently, this natural process has been adapted for human applications, with CO_2 serving as a sustainable alternative to fossil-derived building blocks in biotechnological and synthetic endeavors. In this context, carboxylases from a variety of microbial sources have been utilized to facilitate synthetic CO_2 use (Bierbaumer et al., 2023).

Carboxylation reactions are of interest in this field due to their ability to incorporate C1 units into a molecule of interest, utilizing CO_2 as a carbon source. Despite its abundance and accessibility, the use of CO_2 as a building block is not without challenges. The carbon atom in CO_2 is in its maximum oxidation state, necessitating reductive functionalization. Moreover, the current atmospheric CO_2 levels, around 419 ppm in 2023 (Climate Change, 2024), are not conducive for direct use in carboxylation chemistry. Consideration must also be given to factors such as the transfer of gaseous CO_2 into solution and its solubility, which is up to 1.7 $g \cdot L^{-1}$ at atmospheric pressure and 20 °C. The availability of CO_2 in solution is largely pH dependent, with bicarbonate being the dominant form at neutral to alkaline pH.

To address these challenges, biocatalytic carboxylation has been employed. The efficiency of enzymes enables the acceleration of carboxylation reactions, and their substrate affinity allows for control at low atmospheric CO_2 concentrations. Carbonic anhydrases, enzymes that utilize CO_2, are a prime example of this. They are among the fastest catalysts known, enhancing the equilibrium of bicarbonate and CO_2 (aq) by six orders of magnitude. Over time, evolution has yielded a plethora of carboxylases, each with unique methods of binding and activating CO_2. The electrons required for CO_2 reduction are supplied either by the substrate itself or in the form of reduced coenzymes, specifically, nicotinamides (NAD(P)H) or ferredoxin.

One such carboxylase, Ribulose-1,5-bisphosphate carboxylase/oxygenase (RuBisCO), is globally the most abundant protein and plays a pivotal role in the Calvin-Benson–Bassham cycle and photosynthesis (Bar-On & Milo, 2019). It facilitates the addition of CO_2 to a ribulose bisphosphate sugar (RuBP), generating two 3-phosphoglycerate (3PG) molecules. Unique in its ability to react with O_2, RuBisCO generates 3PG and a 2-phosphoglycolate molecule that is recycled in photorespiration. In-depth analysis of RuBisCO's reaction mechanism reveals a complex process, starting with the coordination of RuBP to Mg^{2+}, followed by a reversible removal of a proton from the C3 carbon. The resulting enediolate then interacts directly with CO_2 to form an intermediate, which undergoes hydrolysis to produce two 3PG molecules (Douglas-Gallardo et al., 2022; Prywes et al., 2023). RuBisCO's efficiency is hampered by the O_2 side reaction and low turnover frequency caused by the complexity of the reaction mechanism. Plants produce RuBisCO abundantly to boost overall carboxylation, offsetting its inefficiency. Efforts to engineer RuBisCO have been met with unique challenges, as any improvement in one biochemical aspect might negatively impact another (Flamholz et al., 2019).

RuBisCO operates at a pace up to 10 times slower than the fastest carboxylase, crotonyl-CoA carboxylase/reductase (Ccr) from *Kitasatospora setae* (Erb et al., 2009; Stoffel et al., 2019). Ccr, part of the enoyl-CoA carboxylases/reductases (Ecr's) class of enzymes, lacks side reactions with oxygen, a characteristic that has enabled Ecr's to be successfully utilized in the development of effective artificial CO_2-fixation pathways in vitro (Luo et al., 2023; Schwander et al., 2016). Ecr's facilitate the reductive carboxylation of α,β-unsaturated acyl–CoA thioesters, using nicotinamide adenine dinucleotide phosphate (NADPH) as a reducing agent. They play a role in the central carbon metabolism of bacteria within acetyl-CoA assimilation pathways and secondary metabolism in producing α-carboxyl-acyl-thioesters to diversify the backbone of polyketide natural products. Specifically, Ccr aids the reductive carboxylation of the crotonyl-CoA substrate (Crot-CoA), resulting in the carboxylated product (2S)-ethylmalonyl-CoA.

Through biochemical characterization and computer simulations, we established that the catalytic activity of Ccr relies on four key amino acids in the active site: Phe170, Asn81, Glu171, and His365 (see Fig. 1B) (DeMirci et al., 2022; Gomez et al., 2023, 2024; Recabarren et al., 2023; Stoffel et al., 2019). Phe170 shields against water and Asn81 stabilizes the CO_2 molecule for the addition reaction. A conserved water molecule connects Glu171 and His365, preserving the hydrogen bond network. We recently demonstrated that His365's pK_a is influenced by the distance

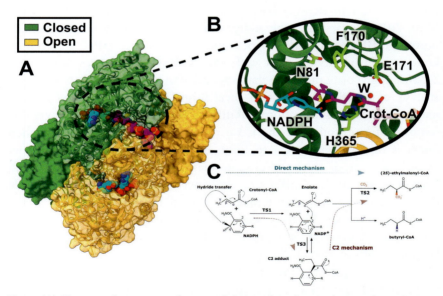

Fig. 1 (A) The crystal structure of crotonyl-CoA carboxylase/reductase from *K. setae* (PDB ID: 6NA4) is shown as a tetramer in a dimer of dimers configuration, featuring closed (green) and open (orange) subunits. Substrate analogue in magenta and NADPH in cyan. (B) Close-up view of the closed active site displaying key residues for the carboxylation reaction in light green, and the coordinated water molecule (W) in red. (C) Reaction scheme of the direct and C2 mechanisms of carboxylation and formation of the reduced side product butyryl-CoA.

to Glu171, causing varying protonation states at pH 8. The distance His365-Glu171 is modulated by the conformational changes of the active site, which fluctuates between open and closed states. The active Ccr oligomer is a tetramer, with X-ray structures showing that its active site can adopt both open and closed conformations in a dimer of dimers arrangement (see Fig. 1A) (DeMirci et al., 2022). Each dimer evidences an active site in a closed conformation, which holds both the substrate analogue and NADPH, and is the only form capable of catalysis. The other active site in the same dimer has an open form and only contains NADPH.

The catalytic cycle stages of Ccr, combined with empirical and biochemical kinetics, offer crucial data for studying the carboxylation reaction in Ccr. Understanding Ccr's carboxylation efficiency requires addressing two main questions: i) How does CO_2 bind to Ccr's active site and how is this binding associated with its structural changes between the open and closed state, ii) Which CO_2 fixation mechanism does Ccr employ that renders these carboxylases superior in efficiency compared to others? We have created computational strategies using molecular dynamics simulations to study CO_2 binding to the active site in different states (Gomez et al., 2023, 2024). These techniques also allowed us to explore the fundamental steps of the reaction mechanism via QM/MM molecular dynamics simulations using the adaptive string method (Recabarren et al., 2023). In the following two sections, we will describe the computational methods employed and the fundamental insights about CO_2 fixation in Ccr that we have acquired.

2. Molecular dynamics simulations identify conformational changes in crotonyl-CoA carboxylase/reductase increasing the local CO_2 concentration in the active site

Molecular dynamics (MD) simulations are a versatile tool used to explore biochemical processes from an atomistic perspective. They use Newton's equations of motion to describe the displacement of every atom in the simulation, with interactions generally based on molecular mechanics (MM). Molecular mechanics aims to reproduce thermodynamic or quantum chemical properties of a molecular system, such as density, equilibrium geometries, or interaction energies, through a combination of simple additive potentials.

The high complexity of biochemical systems, combined with the relatively long timescales of their processes, such as substrate binding, conformational changes, and chemical reactions, has driven the development of myriad simulation techniques to address these challenges. The simplest MD technique, hereafter referred to as traditional MD, involves describing the system using MM, meaning no bond breakage or formation and without the addition of any external biasing forces. In this first section, we will discuss the use of this type of simulation to understand CO_2 binding in the active site of crotonyl-CoA carboxylase/reductase (Ccr). The next section will focus on QM/MM molecular dynamics simulations to understand the reaction mechanism inside the active site.

The use of traditional MD simulations to understand biochemical systems is well documented in the literature, with hemoglobin being one of the most well-known examples (Karplus & McCammon, 2002). Nowadays, the accelerated rise in computational capacities has enabled the handling of larger systems and more challenging research questions, such as ribosome function (Bock et al., 2023), drug development (Khalak et al., 2022), and neurodegenerative diseases (Robustelli et al., 2022). This growth is driven by innovations in hardware, such as multi-core central processing units (CPU) and graphical processing units (GPU), and improvements in software algorithms and parallel processing techniques (Kutzner et al., 2022). With the large availability and rapid generation of data, it is essential to design better analysis tools capable of extracting key information from multidimensional data collections.

To address our first question—how CO_2 binds to the Ccr active site—first, we identified conformational changes of Ccr dependent on the presence of the crotonyl-CoA substrate in the active site. Then, we determined in which conformation it is most probable for the CO_2 molecule to enter the active site.

Crystal structures of Ccr (PDB ID: 6NA4) (DeMirci et al., 2022) evidenced conformational dependence on the presence of a substrate analogue (butyryl-CoA). As mentioned in the introduction, Ccr presents as a tetramer in a dimer-of-dimers configuration (Fig. 1A). Each dimer has one closed subunit with the substrate analogue position well resolved, while the open subunit exhibits no electron density of the thioester moiety of the substrate analogue in the active site. The adenine binding site of the crotonyl-CoA substrate, however, had evidence of electron density only for the adenine base in the open subunit, pointing to a highly flexible crotonyl part of the substrate that is not bound to the active site. To address if the presence of the substrate analogue in the active site induced the conformational change from the open to closed active site, we carried out extensive MD simulations and principal component analysis (PCA). PCA is a dimensional reduction scheme used to identify the largest correlated motion in a multidimensional array (David & Jacobs, 2014). Our results showed that the substrate keeps the active site closed, while its absence leads to the opening of the subunit on a timescale of tens of nanoseconds (DeMirci et al., 2022; Gomez et al., 2024). MD simulations showed that the tetrameric structure of Ccr presents correlated motions associated with its high efficiency of the carboxylation reaction. These correlated motions are initiated by substrate binding to the active site that induces the closing of the subunit, followed by the chemical reaction (DeMirci et al., 2022).

While substrate binding and conformational changes are related to the high efficiency of Ccrs, CO_2 binding to the active site is essential for the carboxylation reaction of the enzyme. Based on its properties as a small apolar molecule capable of fast diffusion in water, we hypothesized that its presence doesn't affect the conformational landscape of the protein. Having established that substrate in the active site induces the closing of the subunit, two new questions arise: Does the amount of CO_2 in the active site depend on the conformation of the subunit, whether open or closed? Does CO_2 enter the active site before or after substrate binding or even at the end of the conformational change? To answer these questions, we performed all-atom MD simulations adding CO_2 molecules to the protein-solvent system. Simulations were performed at higher CO_2 concentrations (55 mM) than in the experiment to ensure sufficient sampling of the CO_2 positions.

To study the role of protein conformation on CO_2 binding, we applied position restraints on the protein backbone atoms to maintain the conformation fixed and evaluated the effect of the substrate's presence: the Open (−) which corresponds to the open conformation without Crot-CoA, Open (+) is an *in-silico* model of the open conformation with a modeled Crot-CoA, and the Closed (+) corresponding to the closed subunit with the Crot-CoA. These systems are representative of the preparation of the active site for the reaction. After setting up each system (see step-by-step methods below), 20 μs long MD simulations were performed.

2.1 Methods to quantify local CO_2 concentration and its dependence on protein conformation

To quantify and compare the CO_2 concentration in the different conformations of the active site, we calculated the excess CO_2 concentration inside a predefined active site volume. This metric compares the CO_2 concentration inside the active site with the solution concentration as a reference state, allowing us to estimate a binding free energy:

$$\Delta G_{bind} = -k_b T \ln\left(\frac{C_{ActiveSite}}{C_{Solution}}\right) \quad (1)$$

We defined an active site volume that contains the reactive parts of the substrate and NADPH molecules and the four key residues for catalysis mentioned above (Fig. 2A). Concentrations in the active site and solution are calculated as an ensemble average from the MD simulations with the *gromaρs* tool (Briones et al., 2019). *Gromaρs* builds synthetic atomic densities by convoluting the atomic positions of the CO_2 molecules with 3-D Gaussians to represent atomic densities. The Gaussian's widths are defined by experimental atomic scattering factors. The atomic density generation was carried out at every step of the trajectory on a grid with a resolution of 0.1 nm and was time-averaged over various 100 ns trajectories.

To estimate the solution concentration ($C_{Solution}$) or the active site concentration ($C_{Active\ Site}$), we calculated the number of carbon dioxide molecules inside the solvent accessible volume (SAV) for the whole system or only in the active site (Gomez et al., 2023, 2024). The SAV represents the regions in space that carbon dioxide molecules can access, excluding

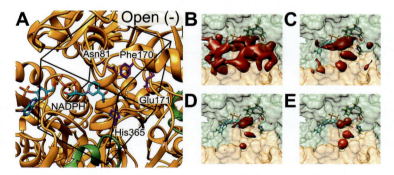

Fig. 2 (A) Close-up view of the Open (−) conformation displaying NADPH (cyan) and the four catalytic residues in purple. The black box outlines the active site volume used to calculate the concentration ratios that contains NADPH (cyan) and the four catalytically relevant residues. (B) The red surface highlights regions in the closed (−) active site where the CO_2 concentration is 2, 5, and 20 times higher than in the bulk, as obtained from time-averaged density maps (B)-(D), respectively. The positions of the binding sites, shown as dark red spheres in (E), match the center of isosurface lobules obtained with 20 times the bulk CO_2 concentration (D).

the volume occupied by the protein, NADPH, and Crot-CoA within the simulation box. To account for changes in the SAV caused by the movement of side chains during the MD simulation, we used the time-averaged ⟨SAV⟩ to estimate the solution concentration:

$$C_{Solution} = \frac{N_{CO_2}}{<SAV>} \tag{2}$$

To calculate ⟨SAV⟩, we selected a threshold isosurface value of 0.1 au, which captured the overall shape of the side chains, cofactors, and substrates in the protein complex. The solvent-accessible volume of each trajectory frame was defined by all grid points with a density below this threshold, corresponding to the volume not occupied by the protein, cofactor, and substrate atoms, thus accessible to the solvent and carbon dioxide molecules. ⟨SAV⟩ was obtained as the time average of 2×10^5 equally spaced frames from a 20 μs trajectory. We defined the local concentrations at the active sites ($C_{Active\ Site}$) by the number of $CO_{2,AS}$ molecules inside the box divided by the active site's accessible volume ⟨SAV⟩$_{AS}$.

We obtained ⟨SAV⟩$_{AS}$ following the same procedure described for the whole protein but considering only the volume inside the active site box:

$$C_{Activesite} = \frac{N_{CO_{2,AS}}}{<SAV>_{AS}} \tag{3}$$

For the number of carbon dioxide molecules inside the active site ($CO_{2,AS}$) we summed up all the atomic densities inside the active site box and divided them by the density of one CO_2 molecule. This concentration was derived as the time average of the various configurations of CO_2 molecules inside the active site.

CO_2 binding sites were identified from the CO_2 density maps near the active sites. From the entire trajectory, we selected 100 ns time windows where the average CO_2 concentration in the active site box was at least twice the bulk concentration (see Method Validation). The configurations within these time windows presented sufficient binding events to define binding sites and unbinding kinetics. Binding sites were determined using isosurfaces of the CO_2 concentration with isovalues 20 times higher than the bulk concentration (see Fig. 2B–E for a comparison of different isovalues). The centers of these isosurface volumes define the binding sites, as shown in Fig. 3B–D.

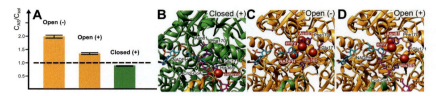

Fig. 3 (A) CO_2 excess inside Ccr's active site in either the open or closed conformation and in the presence (+) or absence (−) of substrate. Error bars are estimated using the bootstrap method. (B) In the closed (+) conformation, binding sites are shown as red spheres outside the active site, far from the four key catalytic residues (purple): Phe170, Asn81, His365, and Glu171. (C) In the open conformation, binding sites are located inside the protein near the Asn81 and His365 residues, below NADPH, and at the entry. (D) Substrate binding to the open configuration (open (+)) removes the binding site below the NADPH molecule while maintaining the other two near Asn81 and His365 in the protein interior and the one at the entry.

The main motivation for developing these computational methods was to understand how the CO_2 concentration depends on the subunit conformation and crot-CoA's presence. By quantifying CO_2 concentration differences in the active sites, we aim to elucidate the biological function of conformational changes in relation to CO_2 binding. Fig. 3A shows that the presence of a substrate and the closed compared to the open conformation of the active site (closed (+) and open (−)) reduces CO_2 binding, while the open conformation presents the largest CO_2 affinity. The derived concentration ratios are similar to those reported for H_2 and O_2 molecules in the *Desulfovibrio fructosovorans* [NiFe]-hydrogenase, ranging from 1.7 to 2.8, respectively (Wang et al., 2011). Then, we addressed the effect of substrate binding in the open active site by the *in-silico* model open (+). Substrate binding to the open active site in the ternary complex slightly decreases the CO_2 affinity in open (+) compared to open (−).

Our results demonstrate that the CO_2 concentration in the active site is modulated by the conformation of the enzyme. As the enzyme is a dimer of dimers with an open and a closed subunit in each dimer, the closed subunit prepares the opposing open active site with CO_2 while the substrate reacts in the closed active site, according to the enzyme's most favorable binding affinity to open (−). The binding affinities demonstrate that CO_2 binds prior to the closing of the active site, which triggers the chemical reaction in a subsequent step.

Next, we attempted to identify the specific CO_2 binding sites in the closed and open subunits as shown in Fig. 3B–D. Binding sites in closed (+) are outside the active site (external), far from the reaction center located

Computational methods for the study of carboxylases 363

between the NADPH coenzyme (cyan) and substrate (magenta) (see Fig. 3B). Both open ($-$) and open ($+$) active sites share three out of four binding sites: substrate binding in open ($+$) eliminates the fourth binding site below NADPH in open ($-$). Interestingly, the two binding sites in the protein interior are near residues Asn81 and His365 (violet), which are crucial for carboxylation activity. One of the binding sites is solvent-exposed at the entrance to the active site. In conclusion, the binding site analysis shows that CO_2 prefers the open active site near important residues His365 and Asn81, which have been experimentally linked to carboxylation efficiency (Stoffel et al., 2019).

Our analysis of the binding thermodynamics reveals that CO_2 binds preferentially to the open active site and that substrate addition maintains binding sites close to those of His365 and Asn81. Next, we examined the binding kinetics to determine if substrate binding modifies the residence time of CO_2. Longer residence times would allow the enzyme to close its active site before CO_2 leaves and start the reaction. With this motivation, we developed a new methodology to calculate the average CO_2 residence time in the previously identified binding sites.

2.2 Calculation of average CO_2 residence times

CO_2 residence times were obtained based on the average duration that a CO_2 molecule remained bound to a specific binding site (Gomez et al., 2023). To estimate how long a binding event is, we counted the time span that the center of mass of each CO_2 molecule stayed within a sphere with a radius twice the $C{=}O$ bond distance (2.4 Å). The closest binding sites are separated by twice that distance. To get the time length, we built a binding state matrix $S_{N,c}$ for each of the active site conformations (open ($+$), open ($-$), and closed ($+$)). This is a two-dimensional matrix for the total number of CO_2 molecules and the total trajectory frames. The matrix elements $s_{N,c}$ took values of $i = 1, 2, 3, \ldots, n$ when a specific CO_2 molecule n was bound to the binding site i or 0 otherwise. For each CO_2 molecule, we recorded the length of the binding events in which the molecule was bound to a specific binding site. The length of the binding events for each site were averaged to obtain the residence time of each binding site. In some cases, CO_2 molecules registered an unbinding event followed by new binding to the same binding site. These recrossing events result from the geometric criteria used to define the bound state of a CO_2 molecule and lead to an underestimation of the residence time. To account for this effect, we performed an iterative protocol in which we used the previous residence time as a threshold for recrossing events: if two binding events were separated by less than the residence time, they were appended and counted as only one event. The iterative cycles ended when there was no difference between residence times in two consecutive iterations.

Table 1 shows that the most solvent-exposed binding site, at the entrance of the active site, has the shortest residence time due to its fast exchange with the solvent. The binding site near NADPH in open (−) has the longest residence time. Interestingly, substrate binding increases the residence time near the His365 residue, where CO_2 is best positioned to attack the enolate species once it is formed by hydride transfer. The binding site near Asn81 presents residence times of several nanoseconds and is slightly reduced upon substrate binding. From the residence time analysis, we conclude that substrate binding increases to several tens of nanoseconds the time CO_2 is positioned in the reactive His 365 binding site. This longer period would give the enzyme the time to close its active site and carry out the reaction.

To prove whether a CO_2 molecule would remain in the binding site throughout the conformational change, we estimated the time required to close the active site considering the conformational dynamics of Ccr. We took advantage of the principal component analysis that describes the conformational change from the closed to the open state of the subunits (DeMirci et al., 2022). The opening rate constant, assuming first-order kinetics, is 3.2×10^{-2} ns^{-1}. We calculated a half-life between 1 and 21.7 ns for closing the active site, depending on the equilibrium constants between the closed and open states (Gomez et al., 2024). Both values are considerably smaller than the residence time of CO_2 in binding site His365 from open (+), indicating that the conformational change occurs faster than CO_2 unbinding.

Table 1 Residence times τ in nanoseconds obtained for the binding sites shown in Fig. 3 (error estimation obtained with bootstrap analysis).

Binding site	Open (−)	Open (+)
Asn 81	8 ± 2	5 ± 2
His 365	11 ± 2	70 ± 20
NADPH	60 ± 20	—
entry	0.2 ± 0.1	0.2 ± 0.1

Through these computational methods, we studied the thermodynamics and kinetics of the CO_2 binding process. Based on the affinity of the active sites, the CO_2 molecule binds before the closing of the active site of Ccr, with the highest affinity in the absence of the substrate. Considering the kinetic information for CO_2 binding and the conformational change of the active site, our results show that CO_2 remains bound to the active site throughout the conformational change (Fig. 4). Based on these conclusions we can respond to the questions that motivated the study: the conformation of the subunit modulates the amount of CO_2 in the active site, with the open (−) subunit having the highest CO_2 affinity. Thus, the CO_2 molecule binds to the open subunit before the substrate. The substrate triggers the conformational change to the closed conformation initiating the carboxylation reaction.

The computational methods developed for studying Ccr are fully applicable for investigating the binding affinity of small molecules to a protein's catalytic site. This is particularly relevant for carboxylases, an active research field It remains unclear why RuBisCO, the most abundant enzyme on earth, can react with both molecular oxygen and CO_2. Additionally, the functional role of addition of the small subunits in RuBisCO as an evolutionary step is still unknown. It is uncertain whether this addition is related solely to the binding affinity, serving as a CO_2 reservoir as previously described (Van Lun et al., 2014), or if this subunit affects the protein's conformational landscape, potentially modulating the active site's affinity for CO_2 and O_2 and contributing to its current selectivity (Schulz et al., 2022).

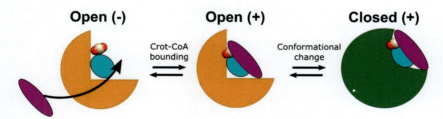

Fig. 4 Scheme of CO_2 and Crot-CoA binding in Ccr followed by conformational change: Initially, CO_2 is bound to the active site in the open (−) conformation containing NADPH represented by the cyan circle. In the second stage, the substrate (magenta ovoid) binds to the open subunit forming the open (+) state, which triggers the conformational change to the closed (+) state and subsequently the chemical reaction to form the enolate and the carboxylated product.

2.3 Simulation details and validation

2.3.1 Step-by-step method details

The computational models were derived from the crystal structure of the ternary complex of *K. setae* Ccr (KsCcr, PDB ID: 6NA4) (DeMirci et al., 2022), which forms a tetrameric structure forming a dimer of dimers configuration. Each dimer includes one closed subunit and one open subunit (subunits A/C and B/D), with the closed state representing the catalytically active conformation. Each subunit contains an NADPH cofactor, and the closed subunits additionally hold the substrate analogue butyryl–CoA (subunits A and B). This analogue was modified to restore the original substrate crotonyl-CoA by removing one hydrogen atom from both the β-carbon and the α-carbon.

To construct the binary complex that represents the open (−) and closed (−) subunits, substrate molecules were removed from the closed subunits, resulting in coordinates similar to the X-ray structure of the binary complex (PDB-ID: 6NA6) (DeMirci et al., 2022; Gomez et al., 2023). To look deeper into the effect of the substrate in the open subunit, we built the ternary in-silico complex representing the open (+) and closed (+) subunits. For the in-silico complexes, substrate and complementary NADPH molecules were modeled into the open subunit to match the oligomerization domains for the closed subunits.

Hydrogen addition and topology generation were performed using the GROMACS 2019.3 software package (Abraham et al., 2015). Protonation states of titratable residues, except His365, were assigned using the Propka 3.1 program (Olsson et al., 2011). The CHARMM36m (Huang et al., 2017) force field and the CHARMM TIP3P water model (Boonstra et al., 2016), combined with the CHARMM general Force Fields (CGenFF) (Vanommeslaeghe & MacKerell, 2012; Vanommeslaeghe et al., 2010, 2012) for describing CO_2, substrate, and cofactor interactions, were utilized. Force field parameters for CO_2 were validated using heptane/water partition coefficients and free energy calculations. Specific parameters for NADPH and CoA fragments were sourced from Pavelites *et al.* (Pavelites et al., 1997) and Aleksandrov and Field (Aleksandrov & Field, 2011).

The protein complex was solvated with a 1 nm water region between the protein and the box limits, KCl concentration of 125 mM was added to neutralize the system. Energy minimization was conducted using the steepest descent algorithm, followed by a 1 ns NVT equilibration with position restraints on the protein backbone and ligand, followed for a 5 ns NPT

Computational methods for the study of carboxylases   367

equilibration at 1 atm pressure and 298 K, with a 2 fs integration time step. The velocity rescaling thermostat (Bussi et al., 2007) and Berendsen pressure coupling (Berendsen et al., 1984) were applied with coupling coefficients of $\tau = 0.1$ and 2 ps, respectively. All bond lengths of the protein and ligands were constrained using the LINCS (Hess et al., 1997) algorithm with an expansion order of 4. Electrostatic interactions were calculated using particle mesh Ewald (Darden et al., 1993) with a 1.2 nm real space cutoff and 0.16 nm Fourier spacing. van der Waals interactions used a 1.2 nm cutoff and a switching function starting at 1 nm.

100 CO_2 molecules were added, replacing water molecules to achieve a concentration approximately 60 times higher than the experimental concentration of CO_2 (55 mM) (Stoffel et al., 2019). This elevated concentration was necessary to sample all potential CO_2 positions during the simulation. The minimization and equilibration protocol were repeated. After pressure equilibration, production dynamics were performed in the NVT ensemble with position restraints on protein alpha carbons and NADPH heavy atoms, maintaining a force constant of 1000 kJ nm^{-2}·mol^{-1}.

In the binary complex (open (−), closed (−)), the distance between Arg276 and NADPH's adenine moiety was restrained to their initial values. For the ternary complex (open (−), closed (+)), distances between Arg99 and crotonyl-CoA's phosphate groups, Arg303 and crotonyl-CoA's adenine moiety, the reactive crotonyl-CoA's C$_\beta$ with NADPH's C4 atoms, and His409 with NADPH's phosphate group in the closed subunits were restrained to their initial values. For the ternary complex (open (+), closed (+)), additional position restraints were added to the open (+) crotonyl-CoA molecules to explore the CO_2 distribution in the enzyme's defined catalytic state. Multiple replicas were conducted, yielding a total of 20 µs of simulation time for the systems. Coordinates were saved every 100 ps, resulting in up to 2×10^5 configurations available for analysis.

2.3.2 Method validation

We analyzed the simulation trajectories in the presence of explicit CO_2 molecules to estimate its concentration in different active sites. We selected 100 ns (1000 frames) time windows to generate the density maps for the entire system as described above. The 100 ns time windows were chosen to reduce computational costs compared to analyzing the whole trajectory. We then averaged the density inside each of the active sites over 1000 frames and calculated the CO_2 concentration in the active site. The CO_2 excess or depletion at the active site was calculated as the ratio between the

CO_2 concentration in the active site and in the bulk solution. Fig. 5 shows the CO_2 excess or depletion for each of the 100 ns time windows, considering twelve 1600 ns trajectories for each active site of the ternary system. Each point corresponds to a time and space averaged 100 ns window for the different colored trajectories connected with dotted lines. The open site shows multiple instances of excess CO_2 (concentration ratio > 1), with several instances much higher than the mean concentration of the site (blue horizontal line). In the closed sites, most points show a depletion of CO_2 concentration (concentration ratio < 1) with very few samples showing excess CO_2. In the tetramer, both open and closed active sites present similar trends and were therefore averaged for further analysis. The horizontal blue line in Fig. 5 represents the mean value of CO_2 excess, and the width represents the error estimation obtained with bootstrap analysis (Nishikawa et al., 2021). Some of the 100 ns windows showed CO_2 excesses more than twice the concentration in the bulk. A concentration ratio of two corresponds to a free energy difference between the active site and the solution of 0.7 k_BT. Furthermore, this free energy difference and CO_2 excess also match the average value observed for the open active sites without substrate, with maximal CO_2 affinity (see horizontal light blue line and green line for the open site 1 in Fig. 5). Based on these observations, we decided to use this threshold to select 100 ns time windows for the identification of binding sites. In Fig. 5, the horizontal

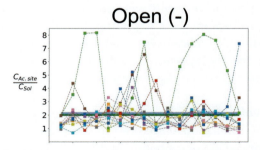

Fig. 5 CO_2 excess in the open(+) active site of Ccr of the ternary complex. A total of 12 trajectories, each in a different color, are displayed for every active site box. The colored dots correspond to a 100 ns time average and are connected with dashed lines representing the time evolution of every simulation. The horizontal blue line represents the mean value of CO_2 excess, and the width corresponds to the error estimation obtained by bootstrap analysis. The horizontal green line displays the threshold used to select the 100 ns time windows for binding site analysis.

green line represents the excess CO_2 threshold value of 2. The 100 ns trajectories with CO_2 excess above this threshold were then used to identify CO_2 binding sites displayed in Fig. 3.

3. QM/MM molecular dynamics simulations reveal the reaction mechanism of the chemical CO_2 fixation step

In the previous section, we described computational methods to quantify local CO_2 concentration at the active site of Ccr and to identify CO_2 binding sites. With these methods we derived CO_2 residence times at specific binding sites proposing a catalytic mechanism. Our results show that CO_2 binds most favorably in the open conformation without the substrate Crot-CoA, with the binding of the latter triggering the conformational change to the closed state of the active site. This closed state precedes the CO_2 fixation step. Overall, these results highlight the importance of computational methods in describing CO_2 binding events in carboxylases, using Ccr as a model. In this section, we will delve into computational methods to characterize enzymatic reaction mechanisms and their application to the reductive carboxylation reaction in Ccr.

3.1 QM/MM calculations and free energy profiles

We studied the chemical step using the QM/MM approach. QM/MM techniques are at the forefront of simulating enzymatic reaction mechanisms (Groenhof, 2013; Senn & Thiel, 2009). They balance accuracy and computational cost by treating part of the system at the quantum mechanical level, where bond breaking and bond forming processes take place, and the rest of the system at the molecular mechanical (MM) level. QM/MM studies of carboxylation reactions catalyzed by carboxylases have been reported in the literature (Douglas-Gallardo et al., 2022; Sheng et al., 2019; Zhang & Bruice, 2007).

One approach to model enzymatic reactions is the static method (Lonsdale et al., 2012). Here, sequential energy minimization calculations are performed along a distinguished reaction coordinate or the intrinsic reaction coordinate, leading to the creation of a potential energy surface (PES). While widely used in the literature, these methods do not account for entropy, which is essential for estimating free energies. To address this, molecular dynamics at the QM/MM level are generally conducted to encompass all relevant atomic configurations, generating statistical distributions for each significant state of the reaction. The free energy change for a specific transformation between two thermodynamic states of a system can be determined by comparing the relative probabilities of the system being in each state during the simulation. In the static approach, in contrast, the statistical distribution of the reactant state is replaced by a single representative structure at the PES minimum. In this case, estimates for Gibbs free energies are accessible by applying simple models like the rigid-rotor and harmonic-oscillator approximations for the rotational and vibrational degrees of freedom. These models offer analytical expressions for partition functions from which Gibbs energies can be approximated. Nevertheless, for molecular systems in condensed phases, such as solutions, biomolecules, or solids, the free energy cannot be determined with these simple models because of the inherent anharmonicity and the coupling of vibrational and rotational degrees of freedom due to intermolecular interactions (Tuñón & Williams, 2019).

In the case of carboxylation reactions, entropic effects are very important since the fixation of a CO_2 molecule creates an entropic penalty that must be considered in the computation of the free energy profile. The translational entropy of a free CO_2 molecule has been estimated to be 11.1 kcal/mol at room temperature (Sheng et al., 2015). Ideally, QM/MM molecular dynamics simulations using density functional theory (DFT), or *ab-initio* methods, are the best options, as they allow for the inclusion of entropic effects with an accurate representation of the interactions between atoms. However, they are still computationally very costly because calculation of the forces on each atom is required for each step of the MD simulation. The high computational cost often results in very limited sampling (few picoseconds) of the configurational space, hindering convergence of the free energy profiles. As an alternative to DFT, semi-empirical methods are widely used. These are up to three orders of magnitude faster than DFT methods, reducing computational costs considerably. This translates into sampling more conformations, usually accessing sampling times in the order of hundreds of picoseconds. Thus, well-converged free energy profiles can be obtained where entropic contributions are properly considered.

In our previous studies on Ccr, the QM region has consisted of the CO_2 molecule, reactive parts of the crotonyl-CoA substrate and NADPH cofactor, residues His365, Glu171, and a conserved water molecule (see Fig. 6C). The selection of the QM region was based on the

Computational methods for the study of carboxylases

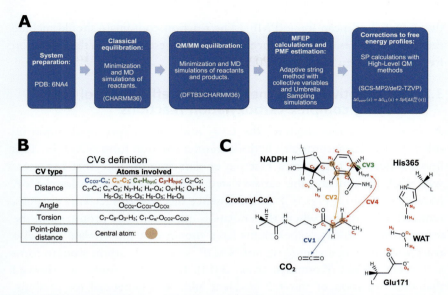

Fig. 6 (A) General protocol used in the study of the enzymatic reaction mechanism in Ccr. (B) Definition of collective variables (CVs) used in string calculations. (C) QM region used in QM/MM calculations depicting the four most important CVs for describing the carboxylation reaction. L denotes the link atom dividing QM and MM regions (Recabarren et al., 2023).

initial hypothesis of a potential water-mediated proton transfer between His365 and Glu171, though it was later found that this was not spontaneous based on the associated free energies (Recabarren et al., 2023). We chose the DFTB3 semi-empirical method (Gaus et al., 2011) as the QM method to study the reductive carboxylation reaction in Ccr due to its wide applicability in (bio)organic reactions. DFTB3 generally provides good geometries but tends to underestimate activation barriers, introducing a potential error in QM/MM calculations (Gruden et al., 2017). Subsequent energy corrections are thus necessary as a posteriori adjustment to the free energy profile (Ramos-Guzmán et al., 2022; Ruiz-Pernía et al., 2004).

To explore possible reaction mechanisms by QM/MM molecular dynamics simulations, enhanced-sampling techniques are the preferred methods (Hénin et al., 2022). These methods introduce biasing potentials to circumvent high free energy barriers. The primary decision at this stage is the selection of the reaction coordinate, as its quality will dictate the accuracy of the final free energy profile along that reaction coordinate, what constitutes the potential of mean force (PMF). The reaction coordinate should include the slower degrees of freedom needed for proper estimation of the activation free energy (Chen & Chipot, 2022).

3.2 The adaptive string method as an efficient tool for studying complex carboxylation reactions

As stated in the introduction, the CO_2 molecule is very inert requiring substrate activation through a hydride transfer reaction from NADPH. This initial reduction generates a reactive enolate intermediate that can react with CO_2 to generate the carboxylated product (2S)-ethylmalonyl-CoA (direct mechanism, Fig. 1C). In the absence of CO_2, this enzyme creates a reduced side product where the enolate is quenched by a proton from the solvent. Alternatively, experiments have shown that this enolate may also react with the nicotinamide ring of $NADP^+$ at the C2 carbon atom to form a covalent adduct (Rosenthal et al., 2014). In a recent study, we showed that the formation of this C2 covalent adduct is beneficial for catalysis, opening a way to "store" the reactive enolate in a more stable intermediate when no CO_2 is present in the active site (Recabarren et al., 2023). The enolate can be regenerated by dissociation of the C2 covalent adduct to form the carboxylated product once a CO_2 molecule is available in the active site (C2 mechanism, Fig. 1C).

The CO_2 molecule in Ccr's active site interacts with different residues, while its interaction with Asn81 is characterized as crucial. Though this interaction is mostly conserved through weak hydrogen bonding, molecular dynamics simulations have shown that it fluctuates considerably in the reactant state becoming stronger in the course of the carboxylation reaction (Recabarren et al., 2023).

To describe all possible reaction paths in the Ccr reductive carboxylation mechanism, we have chosen the adaptive string method (Zinovjev & Tuñón, 2017). In this method, there is no need to define a priori the reaction coordinate to be used; instead, different collective variables (CVs) representing reactive distances, angles, dihedrals, or hybridization changes can be selected, and the minimum free energy path (MFEP), the path with lowest free energy that connects reactant and product states, is obtained in CV space. CVs representing reactive bond distances should be included, and other CVs can be included based on previous chemical knowledge of the reaction, such as an angle describing the rotation of a bond needed for the reaction to happen. The MFEP is then used to define a mathematical function that describes the progress of the reaction and serves as a one dimensional reaction coordinate where the PMF can be obtained (the so-called path CV or *s* reaction coordinate). Focusing on the MFEP and sampling only regions of phase space around it is an efficient strategy that greatly reduces the computational cost compared to traditional 2D Umbrella Sampling simulations, where two reaction coordinates are explored simultaneously and the complete free energy landscape needs to be reconstructed (Zinovjev & Tuñón, 2018).

The most important CVs for the study of the reductive carboxylation reaction in Ccr included reactive distances describing bond breaking and forming for the hydride transfer step, the bond between CO_2 and the α-carbon of crotonyl-CoA, and the bond forming between α-carbon of crotonyl-CoA and the C2 carbon of $NADP^+$ (Recabarren et al., 2023). One of the main advantages of the string methodology is that inclusion of many CVs does not affect computational efficiency (Maragliano et al., 2006), and therefore a total of 24 CVs were used in this case (see Fig. 6B and C for detailed description of all CVs). Though many CVs were used, it was confirmed that those describing reactive bond distances contribute most to the free energy profile(Recabarren et al., 2023).

Using the adaptive string method, free energy profiles for the different competing mechanisms were compared. Fig. 7 shows a simplified scheme with the free energies obtained for the direct and C2 mechanisms (Recabarren et al., 2023). In the former, hydride transfer is followed by subsequent carboxylation, and in the latter, formation of the C2 adduct proceeds after enolate formation. The C2 covalent adduct can dissociate, regenerating the enolate for subsequent carboxylation. The string method allowed exploring these complex reaction mechanisms in Ccr, providing important chemical insights into the reaction mechanisms.

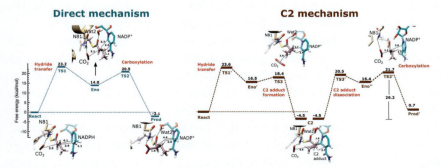

Fig. 7 Schematic representation of the free energy profiles for both direct and C2 mechanisms including representative structures of stationary points displaying characteristic bond distances (Recabarren et al., 2023).

One important aspect of this reductive carboxylation reaction is the synchronicity of the different elementary steps involved, i. e., hydride transfer and carboxylation in the direct mechanism, and hydride transfer and C2 adduct formation in the C2 mechanism. To address the synchronicity in the direct mechanism, one string calculation was performed using the reactant and product states as the initial and final points of the string. This implies that exploring the reaction mechanism requires obtaining equilibrated structures of both states. This is performed by initial equilibration of the reactant state at the MM level, and subsequent equilibration of the reactant and product states at the QM/MM level (see Fig. 6A). Construction of the product state was performed by running QM/MM steered molecular dynamics simulations starting from the reactant state. Alternatively, a QM/MM potential energy scan could be employed to construct an approximate product state following the C–C bond distance between CO_2 and the α-carbon of crotonyl-CoA.

To initiate the reaction path calculation with the adaptive string method, a linear interpolation of CVs values between reactant and product states was carried out. The string is divided into several nodes where an independent QM/MM MD simulation is performed applying a harmonic biasing potential that keeps the system close to the node's position in CV space. The nodes are maintained equidistant through continuous reparametrization of their positions. They drift towards regions of lower free energy, ultimately converging at the minimum-free energy path (MFEP). In addition to the harmonic biasing potential along the string, which preserves the node positions, an orthogonal biasing potential is applied to keep the system close to the path. This method has the advantage of automatically obtaining the force constants for both biasing potentials, minimizing the need for prior user knowledge. More details of the adaptive string method can be found elsewhere (Zinovjev & Tuñón, 2017).

Avoiding the definition of a reaction coordinate a priori allows for the natural characterization of the synchronicity of the reaction. In the case of the direct mechanism, the CO_2 molecule starts approaching the α-carbon of crotonyl-CoA after the hydride transfer step has been completed, showing a high degree of asynchronicity (Fig. 7). In the C2 mechanism, one string calculation was performed for the reactant–C2 segment and another one for the C2–product segment. Both segments also show a high degree of asynchronicity, hydride transfer is well decoupled of C2 adduct formation and C2 adduct dissociation from carboxylation.

In principle, one could study the complete C2 reaction mechanism in one string calculation starting from the reactant state to the product state going through the C2 adduct intermediate. Yet, it should be noted that this would require a large number of nodes to provide the string with enough flexibility to describe the PMF profile. As shown in Fig. 7, the C2 adduct is a clear intermediate, and therefore the partition of the complete C2 mechanism into two string calculations is well justified. In terms of computational resources, it is advisable to use at least 20 nodes per elementary step, which would correspond to 40 nodes for each segment of the reaction mechanism. Our calculations utilized one CPU core per node, meaning that 40 string nodes required 40 CPU cores running in parallel. More efficient calculations could be achieved if multiple cores were used per string node.

The adaptive string method consists of two stages: string optimization for finding the MFEP and therefore the one dimensional reaction coordinate, and a second stage using this path CV to sample all relevant regions of conformational space close to the MFEP for accurate prediction of the free energy profile. This second stage is carried out by Umbrella Sampling (US) simulations (Kästner, 2011) using converged node positions and force constants from the string optimization stage. Each node represents an US window. It is important to set the orthogonal biasing potential to a value other than zero, since in the case of the direct mechanism, the string initially defined for this path can easily "fall" in the C2 mechanism, since C2 adduct formation has a very low free energy barrier from the enolate. In our experience a soft force constant of $5\,kcal{\cdot}mol^{-1}{\cdot}amu^{-1}{\cdot}\text{Å}^{-2}$ for the orthogonal potential is enough to keep the system close to the direct mechanism.

3.3 Corrections to free energy profiles

By using a semi-empirical method such as DFTB3, sampling times of 100 ps per US window are easily accessible, resulting in well-converged free energy profiles. Free energy profiles can be generated by integration techniques such as Umbrella Integration (default method) (Kästner & Thiel, 2005) or WHAM (Kumar et al., 1992), the latter was chosen in our case. As stated above, corrections to the converged free energy profiles are generally required when using more approximate semi-empirical methods. Free energy profiles were adjusted using a dual-level approach (Ramos-Guzmán et al., 2022; Ruiz-Pernía et al., 2004):

$$\Delta G_{corr}(s) = \Delta G_{LL}(s) + Spl\left[\Delta E_{LL}^{HL}(s)\right] \qquad (4)$$

In this method, $\Delta G_{corr}(s)$ is the corrected free energy profile as a function of the path CV (reaction coordinate s). $\Delta G_{LL}(s)$ is the original free energy profile derived from the low-level (LL) QM method (DFTB3 in this case). The Spl function denotes a cubic spline, and $\Delta E_{LL}^{HL}(s)$ represents the energy difference of the QM region between the LL method and a high-level (HL) method (SCS-MP2/def2-TZVP in this case) at various reaction coordinate values (Recabarren et al., 2023). $\Delta E_{LL}^{HL}(s)$ is calculated by running restrained QM/MM geometry optimizations starting from a representative structure of the first transition state in the studied reaction mechanisms. These optimizations were performed following the converged CVs values from the string nodes and conducted iteratively in both forward and backward directions until converged potential energy profiles were obtained. At each scan point (node), single point calculations were performed at both LL and HL methods. Fig. 6A includes a schematic representation of the computational protocol used to study the reductive carboxylation mechanism in Ccr summarizing the main steps.

3.4 Dissecting reactivity by comparing free energy profiles of different reaction mechanisms

Corrected free energy profiles can be used to evaluate differences between the reaction mechanisms (see Fig. 7). In the direct mechanism, the hydride transfer is the rate-limiting step, with the enolate characterized as a reactive intermediate. Carboxylation from this species occurs with a relatively low activation free energy (6.5 kcal/mol). The carboxylated product is predicted to be thermodynamically stable, due to a stabilizing hydrogen bond interaction with Asn81 (Fig. 7, Prod). In the case of the C2 mechanism, the formation of the C2 adduct occurs with an even lower activation free energy (1.9 kcal/mol), explaining its experimental detection even in the presence of CO_2 (Rosenthal et al., 2014). The covalent C2 adduct is predicted to be a very stable intermediate, corroborating experimental observations that suggest its formation is part of the catalytic cycle. Comparing both reaction mechanisms, the activation free energy for the rate-limiting hydride transfer step in the direct mechanism is comparable to the total activation free energy for carboxylation from the C2 adduct, indicating that both processes exhibit similar kinetics, as confirmed experimentally (Rosenthal et al., 2014). Kinetic isotope effect (KIE) experiments showed that both catalytic steps contribute to the measured catalytic rate constant, suggesting that the C2 covalent adduct functions as a "storage" form of the more reactive enolate intermediate (Recabarren et al., 2023). This finding would help to explain the greater efficiency of Ccr as a carboxylase enzyme.

As described above, carboxylase enzymes utilize substrate activation to generate reactive intermediates called enolates, which are strong carbon nucleophiles capable of attacking CO_2 to form stable C–C bonds. In aqueous solutions, enolates are easily protonated, but carboxylases stabilize them through interactions with the protein environment (Hamed et al., 2008). The formation of enolates is found in substrates like acyl-thioesters (Pandey et al., 2011; Tong, 2005), ketones (Boyd & Ensign, 2005), and α,β–unsaturated enoyl-thioesters (Erb et al., 2009). Different carboxylases use various strategies to generate enolates: in RuBisCO, the sidechain of an active site lysine, carboxylated to form a carbamate, abstracts a proton from C3 of the substrate ribulose 1,5-bisphosphate, generating a dienolate stabilized by a magnesium ion in the active site (Cleland et al., 1998). In acetone carboxylase, enolate formation relies on ATP and Mg^{2+}, producing phosphoenol acetone as the activated nucleophile that attacks CO_2 (Boyd & Ensign, 2005). As described above, in the case of Ccr, the reactive enolate is generated by a reduction reaction from NADPH. We found that the energy of this reactive intermediate can be greatly modulated by local interactions in the active site. The active site of Ccr does not contain metal cofactors to stabilize this reactive intermediate but local hydrogen bonding interactions can partially fulfill this role. An OH group of the ribose fragment of NADPH (O_3-H_3, Fig. 6C) is well positioned to interact with the carbonyl oxygen (O_1, Fig. 6C) that becomes negatively charged when the enolate is formed.

Our QM/MM MD simulations predicted that this hydrogen bond interaction can modulate the enolate's stability (Recabarren et al., 2023). When this interaction is not present and the ribose O–H group is rotated away from O_1 oxygen, a water molecule is positioned to interact with the carbonyl oxygen (see Fig. 7, Wat2). A dihedral angle describing the rotation of this O–H group (C_7-C_8-O_3-H_3, Fig. 6C) was incorporated as a CV in the string calculations to account for any contribution to the definition of the path CV. During the equilibration simulations at the DFTB3/CHARMM36 level for the reactant state, this interaction was not stable, and therefore, to study its direct effect on the reaction mechanism, a distance restraint in the QM/MM MD simulations was incorporated. This restraint was applied to the O_1-O_3 distance using a force constant of $200\ \text{kcal·mol}^{-1}\text{·Å}^{-2}$ for distances longer than 3 Å. The same protocol for the adaptive string calculations was used and the PMF for the direct mechanism was estimated.

Fig. 8 shows the comparison between the free energy profiles obtained without (original simulations) and with the distance restraint. Incorporating the distance restraint to preserve the hydrogen bond between O1 and O3 significantly stabilizes the enolate intermediate with respect to the reactants (~8 kcal/mol). This modification does not greatly affect the rate-limiting hydride transfer step nor the reactivity towards carboxylation, though it increases the activation free energy for carboxylation from the enolate by 2.2 kcal/mol. At the studied QM/MM level, this interaction did not appear stable during QM/MM MD simulations of the reactant state and was therefore not considered in the simulations of the C2 mechanism. However, the substantial reactivity modulation of the enolate intermediate underscores the importance of local interactions in the Ccr active site. This principle is expected to apply to other carboxylases that produce nucleophilic enolates to react with CO_2. Our simulations highlight the importance of using computational methods that correctly account for the dynamic nature of enzymes and the utilization of different initial conformations of the reactant state, where crucial intermolecular interactions may be important for reactivity modulation. Distance restraints may be useful for preserving a molecular interaction in the active site, but the interpretation of the results should be taken with care because of the artificial bias that may hinder the exploration of other relevant states.

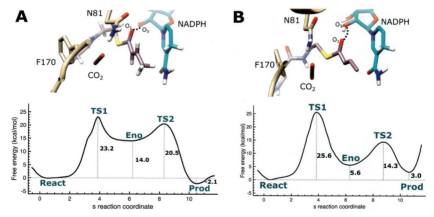

Fig. 8 (A) Original free energy profile at the SCS-MP2/def2-TZVP//DFTB3/CHARMM36 level of theory for the direct mechanism without imposing a distance restraint at the O_1-O_3 distance (dashed line). (B) Free energy profile at the SCS-MP2/def2-TZVP//DFTB3/CHARMM36 level of theory for the direct mechanism imposing a distance restraint at the O_1-O_3 distance (dashed line) for preserving the stabilizing hydrogen bond interaction (Recabarren et al., 2023).

3.5 Model building for QM/MM MD simulations

A CO_2 molecule was positioned 4.0 Å from the α-carbon of crotonyl-CoA within the closed subunits A and B, replicating the reactant state. The reaction mechanism and analysis were focused on the closed subunit B. Missing hydrogen atoms were added, topology was generated, and the system was constructed using the *psfgen* tool and VMD program (Humphrey et al., 1996). Protonation states of titratable residues were assigned using the Propka 3.1 program (Olsson et al., 2011). The CHARMM36 force field (Best et al., 2012) was employed for protein interactions, while substrate parameters, as described in previous sections, were derived from CGenFF (Vanommeslaeghe & MacKerell, 2012; Vanommeslaeghe et al., 2010, 2012). The protein system was immersed in a rectangular box of TIP3P water molecules (Jorgensen, 1981), ensuring a 14 Å buffer between the protein and the box edges. Neutralization was achieved by adding the appropriate amount of Na^+ counterions.

The system underwent a two-stage minimization process: first, allowing water molecules to relax around the protein (1500 steps), followed by a complete minimization of the solvated protein complex without restraints (3000 steps). The minimized structure was then used for equilibration MD simulations. Initial thermalization was performed for 50 ps (2 fs time step) at 100 K under NVT conditions, using a Langevin thermostat ($\gamma = 1$ ps^{-1}) and restraining the heavy atoms of the protein, cofactors, and substrates (k = 3 kcal·mol^{-1}·Å^{-2}). The system's density was equilibrated with a 2 ns MD simulation under NPT conditions at 300 K and 1 bar, using a Monte Carlo barostat while maintaining the same restraints. A final 100 ns NVT simulation was conducted, keeping the same simulation conditions but lightly restraining the heavy atoms of substrates and cofactors (k = 2 kcal·mol^{-1}·Å^{-2}). Simulations were executed under periodic boundary conditions, with long-range electrostatic interactions handled by the Particle Mesh Ewald method (Darden et al., 1993). Nonbonded interactions were calculated within an 8 Å cutoff. All simulations were performed using Amber18 software (Case et al., 2018). The last frame from the classical MD simulations was used as the starting point for further QM/MM MD simulations.

3.6 Simulation details for QM/MM MD simulations

To complete the valences at the QM/MM interface, we used the link atom approach, resulting in a total of 84 atoms in the QM region. This region was treated using the DFTB3 semiempirical method (3ob-3–1 parameter set) (Elstner et al., 1998; Gaus et al., 2011), while the MM region utilized the CHARMM36 force field and parameters previously mentioned. The system underwent initial minimization (1500 steps) and a brief QM/MM MD equilibration (100 ps, NVT ensemble), maintaining the conditions from prior classical MD simulations, without position restraints, and using a 1 fs time step. A distance restraint was applied between the carbon atom of CO_2 and the α-carbon of crotonyl-CoA for distances over 4 Å ($k = 200$ kcal·mol^{-1}·Å^{-2}) to keep the CO_2 molecule in the active site. The carboxylated product, (2S)-ethylmalonyl-CoA, was generated from the reactant state via QM/MM steered MD simulations, followed by further equilibration for 100 ps. The C_β-H_{hyd} and C_{CO2}-C_α distances were driven from 3.1 to 1.1 Å and from 3.3 to 1.5 Å ($k = 150$ kcal·mol^{-1}·Å^{-2}, spline mode), respectively, over a total simulation time of 25 ps. All QM/MM MD simulations were conducted using the sander module from AmberTools18 (Case et al., 2018; Walker et al., 2008).

The first and last nodes, representing the reactant and product states respectively, were left unrestrained during MFEP optimization calculations. We used a time step of 0.5 fs for all string calculations. To enhance sampling during string optimization and PMF stages, we employed replica exchange as implemented by default in the adaptive string method in Amber (https://github.com/kzinovjev/string-amber). Exchanges were attempted every 50 fs. After constructing the initial string, we relaxed the nodes for 1 ps, during which the string bias potential was gradually increased. The convergence of the string optimization stage was monitored by measuring the average difference between the current nodes' positions and all previous nodes' positions until no variation was detected and a near-zero plateau was observed. Once convergence was achieved, we used the last 5 ps of the string optimization stage to define the reaction coordinate s.

US windows and force constants were defined using the average positions and force constants of nodes from the final 5 ps of the string optimization stage. Importantly, the adaptive string method dynamically optimizes node positions and force constants during simulations, ensuring uniform sampling along the reaction coordinate and allowing for the correction of any initial parameter issues. All other parameters remained at their default settings. The final free energy profiles were generated using the WHAM method (Kumar et al., 1992). These profiles were considered converged when the superimposed profiles from the last 30 ps of US simulations exhibited no variations exceeding 0.5 kcal/mol, and the statistical error calculated via bootstrapping (200 bootstrap samples) was also below 0.5 kcal/mol for any point along the reaction coordinate s. Sampling durations for obtaining the final free energy profiles were 70 ps for the direct mechanism, 106 ps for C2 adduct formation from reactants, and 84 ps for carboxylation from the C2 adduct.

3.6.1 Details of corrections to free energy profiles

We selected a representative structure of the first transition state for each reaction mechanism (i.e., direct mechanism, first and second steps of the C2 mechanism) from the converged MFEP. Subsequent minimization calculations were performed using the LBFGS algorithm (ntmin = 3, convergence criterion 0.005 kcal·mol^{-1}·Å^{-1}) implemented in AmberTools18. To constrain CV values at each node, we utilized averaged CVs from the converged MFEPs with a sufficiently large force constant (k = 5000 kcal·mol^{-1}·Å^{-2}). Residues within a 20 Å radius from the αC of crotonyl-CoA were allowed to move freely during the scan calculations, while other residues remained fixed. We chose the spin-component scaled MP2 (SCS-MP2) method for single-point energy corrections due to its balance of accuracy and computational cost in QM/MM calculations of enzyme-catalyzed reactions (Kaiyawet et al., 2015; Sirirak et al., 2020; Sousa et al., 2020; Van Der Kamp et al., 2010). These calculations were accelerated using the Resolution of Identity approximation (RI) and RI-C auxiliary basis sets (def2-TZVP/C) in Orca 4.2.1 (Neese et al., 2020). All QM/MM electrostatic interactions were considered up to a 40 Å cutoff. All calculations involving the HL SCS-MP2 were carried out with the Amber/Orca interface (Götz et al., 2014).

3.6.2 Validation of semiempirical method DFTB3

To validate the use of DFTB3 geometries, particularly for the hydride transfer step, we conducted a QM/MM scan calculation. We defined the reaction coordinate (Rc) as the antisymmetric combination of distances involved in the hydride transfer (Rc = d(C$_4$-H$_{hyd}$) − d(C$_\beta$-H$_{hyd}$)). Starting with a reactant state structure obtained from previous scans used for free energy profile corrections, we drove the system from the reactant to the enolate species to approximate the potential energy surface. All parameters and options from the previous QM/MM minimization calculations, including the QM region size, were maintained.

We compared the DFTB3 method to the DFT method M06–2X/def2-SVP, which is anticipated to provide more accurate geometries. Geometries obtained from both methods were then used for single-point energy calculations at the SCS-MP2/def2-TZVP level. The corrected potential energy profiles showed good agreement between the DFTB3 and DFT geometries, with discrepancies around the transition state (TS1) region being approximately 1.5 kcal/mol (Recabarren et al., 2023).

4. Summary

The outlined computational methods allow for detailed descriptions of various stages employed by carboxylases for effective CO_2 fixation. Molecular dynamics simulations identify conformational changes in Ccr that increase the local concentration of CO_2 at the unoccupied active site. Subsequent substrate binding initiates the chemical reaction by closing the active site, positioning the coenzyme and the Crot-CoA substrate for hydride transfer. Furthermore, by establishing CO_2 binding sites, we identified four distinct residues that ensure the proper positioning of CO_2 for carboxylation.

QM/MM molecular dynamics simulations helped to investigate the chemical carboxylation step, identifying two competing reaction mechanisms. One mechanism forms a stable C2 covalent intermediate, retaining the reactivity of the enolate for future CO_2 fixation. This alternative mechanism may also explain the superior efficiency of Ccr in comparison to other carboxylases. Overall, we presented methods for characterizing the CO_2 binding and reactivity of carboxylases using Ccr as a model system.

References

Abraham, M. J., Murtola, T., Schulz, R., Páll, S., Smith, J. C., Hess, B., & Lindahl, E. (2015). GROMACS: High performance molecular simulations through multi-level parallelism from laptops to supercomputers. *SoftwareX, 1-2*, 19–25. https://doi.org/10.1016/j.softx.2015.06.001.

Aleksandrov, A., & Field, M. (2011). Efficient solvent boundary potential for hybrid potential simulations. *Physical Chemistry Chemical Physics, 13*(22), 10503. https://doi.org/10.1039/c0cp02828b.

Bar-On, Y. M., & Milo, R. (2019). The global mass and average rate of rubisco. *Proceedings of the National Academy of Sciences, 116*(10), 4738–4743. https://doi.org/10.1073/pnas.1816654116.

Berendsen, H. J. C., Postma, J. P. M., Van Gunsteren, W. F., DiNola, A., & Haak, J. R. (1984). Molecular dynamics with coupling to an external bath. *The Journal of Chemical Physics, 81*(8), 3684–3690. https://doi.org/10.1063/1.448118.

Best, R. B., Zhu, X., Shim, J., Lopes, P. E. M., Mittal, J., Feig, M., & MacKerell, A. D. (2012). Optimization of the additive CHARMM all-atom protein force field targeting improved sampling of the backbone ϕ, ψ and side-chain $\chi 1$ and $\chi 2$ dihedral angles. *Journal of Chemical Theory and Computation, 8*(9), 3257–3273. https://doi.org/10.1021/ct300400x.

Bierbaumer, S., Nattermann, M., Schulz, L., Zschoche, R., Erb, T. J., Winkler, C. K., ... Glueck, S. M. (2023). Enzymatic conversion of CO_2: From natural to artificial utilization. *Chemical Reviews, 123*(9), 5702–5754. https://doi.org/10.1021/acs.chemrev.2c00581.

Bock, L. V., Gabrielli, S., Kolář, M. H., & Grubmüller, H. (2023). Simulation of complex biomolecular systems: The ribosome challenge. *Annual Review of Biophysics, 52*(1), 361–390. https://doi.org/10.1146/annurev-biophys-111622-091147.

Boonstra, S., Onck, P. R., & Van Der Giessen, E. (2016). CHARMM TIP3P water model suppresses peptide folding by solvating the unfolded state. *The Journal of Physical Chemistry. B, 120*(15), 3692–3698. https://doi.org/10.1021/acs.jpcb.6b01316.

Boyd, J. M., & Ensign, S. A. (2005). ATP-dependent enolization of acetone by acetone carboxylase from rhodobacter capsulatus. *Biochemistry, 44*(23), 8543–8553. https://doi.org/10.1021/bi050393k.

Briones, R., Blau, C., Kutzner, C., de Groot, B. L., & Aponte-Santamaría, C. (2019). GROmaps: A GROMACS-based toolset to analyze density maps derived from molecular dynamics simulations. *Biophysical Journal, 116*(1), 4–11. https://doi.org/10.1016/j.bpj.2018.11.3126.

Bussi, G., Donadio, D., & Parrinello, M. (2007). Canonical sampling through velocity rescaling. *The Journal of Chemical Physics, 126*(1), 014101. https://doi.org/10.1063/1.2408420.

Case, D. A., Ben-Shalom, I. Y., Brozell, S. R., Cerutti, D. S., III, Cheatham, T. E., Cruzeiro, V. W. D., ... Kollman, P. A. (2018). *AMBER, 2018* ([En]).

Chen, H., & Chipot, C. (2022). Enhancing sampling with free-energy calculations. *Current Opinion in Structural Biology, 77*, 102497. https://doi.org/10.1016/j.sbi.2022.102497.

Cleland, W. W., Andrews, T. J., Gutteridge, S., Hartman, F. C., & Lorimer, G. H. (1998). Mechanism of rubisco: The carbamate as general base. *Chemical Reviews, 98*(2), 549–562. https://doi.org/10.1021/cr970010r.

Climate Change: *Atmospheric Carbon Dioxide | NOAA Climate.gov*. (2024, april 9). http://www.climate.gov/news-features/understanding-climate/climate-change-atmospheric-carbon-dioxide.

Darden, T., York, D., & Pedersen, L. (1993). Particle mesh Ewald: An N·log(N) method for Ewald sums in large systems. *The Journal of Chemical Physics, 98*(12), 10089–10092. https://doi.org/10.1063/1.464397.

David, C. C., & Jacobs, D. J. (2014). Principal component analysis: A method for determining the essential dynamics of proteins. *Methods in molecular biology (Clifton, N. J.), 1084*, 193–226. https://doi.org/10.1007/978-1-62703-658-0_11.

DeMirci, H., Rao, Y., Stoffel, G. M., Vögeli, B., Schell, K., Gomez, A., ... Wakatsuki, S. (2022). Intersubunit coupling enables fast CO_2-fixation by reductive carboxylases. *ACS Central Science*. https://doi.org/10.1021/acscentsci.2c00057.

Douglas-Gallardo, O. A., Murillo-López, J. A., Oller, J., Mulholland, A. J., & Vöhringer-Martinez, E. (2022). Carbon dioxide fixation in RuBisCO is protonation-state-dependent and irreversible. *ACS Catalysis, 12*(15), 9418–9429. https://doi.org/10.1021/acscatal.2c01677.

Elstner, M., Porezag, D., Jungnickel, G., Elsner, J., Haugk, M., Frauenheim, Th, ... Seifert, G. (1998). Self-consistent-charge density-functional tight-binding method for simulations of complex materials properties. *Physical Review B, 58*(11), 7260–7268. https://doi.org/10.1103/PhysRevB.58.7260.

Erb, T. J., Brecht, V., Fuchs, G., Müller, M., & Alber, B. E. (2009). Carboxylation mechanism and stereochemistry of crotonyl-CoA carboxylase/reductase, a carboxylating enoyl-thioester reductase. *Proceedings of the National Academy of Sciences of the United States of America, 106*(22), 8871–8876. https://doi.org/10.1073/pnas.0903939106.

Flamholz, A. I., Prywes, N., Moran, U., Davidi, D., Bar-On, Y. M., Oltrogge, L. M., ... Milo, R. (2019). Revisiting trade-offs between rubisco kinetic parameters. *Biochemistry, 58*(31), 3365–3376. https://doi.org/10.1021/acs.biochem.9b00237.

Gaus, M., Cui, Q., & Elstner, M. (2011). DFTB3: Extension of the self-consistent-charge density-functional tight-binding method (SCC-DFTB). *Journal of Chemical Theory and Computation, 7*(4), 931–948. https://doi.org/10.1021/ct100684s.

Gomez, A., Erb, T. J., Grubmüller, H., & Vöhringer-Martinez, E. (2023). Conformational dynamics of the most efficient carboxylase contributes to efficient CO_2 fixation. *Journal of Chemical Information and Modeling, 63*(24), 7807–7815. https://doi.org/10.1021/acs.jcim.3c01447.

Gomez, A., Tinzl, M., Stoffel, G., Westedt, H., Grubmüller, H., Erb, T. J., ... Stripp, S. T. (2024). Infrared spectroscopy reveals metal-independent carbonic anhydrase activity in crotonyl-CoA carboxylase/reductase. *Chemical Science, 15*(13), 4960–4968. https://doi.org/10.1039/D3SC04208A.

Götz, A. W., Clark, M. A., & Walker, R. C. (2014). An extensible interface for QM/MM molecular dynamics simulations with AMBER. *Journal of Computational Chemistry, 35*(2), 95–108. https://doi.org/10.1002/jcc.23444.

Groenhof, G. (2013). Introduction to QM/MM simulations. *Methods in Molecular Biology (Clifton, N. J.), 924*, 43–66. https://doi.org/10.1007/978-1-62703-017-5_3.

Gruden, M., Andjeklović, L., Jissy, A. K., Stepanović, S., Zlatar, M., Cui, Q., & Elstner, M. (2017). Benchmarking density functional tight binding models for barrier heights and reaction energetics of organic molecules. *Journal of Computational Chemistry, 38*(25), 2171–2185. https://doi.org/10.1002/jcc.24866.

Hamed, R. B., Batchelar, E. T., Clifton, I. J., & Schofield, C. J. (2008). Mechanisms and structures of crotonase superfamily enzymes—How nature controls enolate and oxyanion reactivity. *Cellular and Molecular Life Sciences: CMLS, 65*(16), 2507–2527. https://doi.org/10.1007/s00018-008-8082-6.

Hénin, J., Lelièvre, T., Shirts, M. R., Valsson, O., & Delemotte, L. (2022). Enhanced sampling methods for molecular dynamics simulations [Article v1.0]. Article 1. *Living Journal of Computational Molecular Science, 4*(1), https://doi.org/10.33011/livecoms.4.1.1583.

Hess, B., Bekker, H., Berendsen, H. J. C., & Fraaije, J. G. E. M. (1997). LINCS: A linear constraint solver for molecular simulations. *Journal of Computational Chemistry, 18*(12), 1463–1472. https://doi.org/10.1002/(SICI)1096-987X(199709)18:12 < 1463::AID-JCC4 > 3.0.CO;2-H.

Huang, J., Rauscher, S., Nawrocki, G., Ran, T., Feig, M., De Groot, B. L., ... MacKerell, A. D. (2017). CHARMM36m: An improved force field for folded and intrinsically disordered proteins. *Nature Methods, 14*(1), 71–73. https://doi.org/10.1038/nmeth.4067.

Humphrey, W., Dalke, A., & Schulten, K. (1996). VMD: Visual molecular dynamics. *Journal of Molecular Graphics, 14*(1), 33–38. https://doi.org/10.1016/0263-7855(96)00018-5.

Jorgensen, W. L. (1981). Quantum and statistical mechanical studies of liquids. 10. Transferable intermolecular potential functions for water, alcohols, and ethers. Application to liquid water. *Journal of the American Chemical Society, 103*(2), 335–340. https://doi.org/10.1021/ja00392a016.

Kaiyawet, N., Lonsdale, R., Rungrotmongkol, T., Mulholland, A. J., & Hannongbua, S. (2015). High-level QM/MM calculations support the concerted mechanism for michael addition and covalent complex formation in thymidylate synthase. *Journal of Chemical Theory and Computation, 11*(2), 713–722. https://doi.org/10.1021/ct5005033.

Karplus, M., & McCammon, J. A. (2002). Molecular dynamics simulations of biomolecules. *Nature Structural Biology, 9*(9), 646–652. https://doi.org/10.1038/nsb0902-646.

Kästner, J. (2011). Umbrella sampling: Umbrella sampling. *Wiley Interdisciplinary Reviews: Computational Molecular Science, 1*(6), 932–942. https://doi.org/10.1002/wcms.66.

Kästner, J., & Thiel, W. (2005). Bridging the gap between thermodynamic integration and umbrella sampling provides a novel analysis method: "Umbrella integration". *The Journal of Chemical Physics, 123*(14), 144104. https://doi.org/10.1063/1.2052648.

Khalak, Y., Tresadern, G., Hahn, D. F., De Groot, B. L., & Gapsys, V. (2022). Chemical space exploration with active learning and alchemical free energies. *Journal of Chemical Theory and Computation, 18*(10), 6259–6270. https://doi.org/10.1021/acs.jctc.2c00752.

Kumar, S., Rosenberg, J. M., Bouzida, D., Swendsen, R. H., & Kollman, P. A. (1992). THE weighted histogram analysis method for free-energy calculations on biomolecules. I. The method. *Journal of Computational Chemistry, 13*(8), 1011–1021. https://doi.org/10.1002/jcc.540130812.

Kutzner, C., Kniep, C., Cherian, A., Nordstrom, L., Grubmüller, H., De Groot, B. L., & Gapsys, V. (2022). GROMACS in the cloud: A global supercomputer to speed up alchemical drug design. *Journal of Chemical Information and Modeling, 62*(7), 1691–1711. https://doi.org/10.1021/acs.jcim.2c00044.

Lonsdale, R. N., Harvey, J. J., & Mulholland, A. (2012). A practical guide to modelling enzyme-catalysed reactions. *Chemical Society Reviews, 41*(8), 3025–3038. https://doi.org/10.1039/C2CS15297E.

Luo, S., Diehl, C., He, H., Bae, Y., Klose, M., Claus, P., ... Erb, T. J. (2023). Construction and modular implementation of the THETA cycle for synthetic CO_2 fixation. *Nature Catalysis, 6*(12), 1228–1240. https://doi.org/10.1038/s41929-023-01079-z.

Maragliano, L., Fischer, A., Vanden-Eijnden, E., & Ciccotti, G. (2006). String method in collective variables: Minimum free energy paths and isocommittor surfaces. *The Journal of Chemical Physics, 125*(2), 024106. https://doi.org/10.1063/1.2212942.

Neese, F., Wennmohs, F., Becker, U., & Riplinger, C. (2020). The ORCA quantum chemistry program package. *The Journal of Chemical Physics, 152*(22), 224108. https://doi.org/10.1063/5.0004608.

Nishikawa, Y., Takahashi, J., & Takahashi, T. (2021). Stationary bootstrap: A refined error estimation for equilibrium time series. *CoRR*. https://doi.org/10.48550/ARXIV.2112.11837.

Olsson, M. H. M., Søndergaard, C. R., Rostkowski, M., & Jensen, J. H. (2011). PROPKA3: Consistent treatment of internal and surface residues in empirical pKa predictions. *Journal of Chemical Theory and Computation, 7*(2), 525–537. https://doi.org/10.1021/ct100578z.

Pandey, A. S., Mulder, D. W., Ensign, S. A., & Peters, J. W. (2011). Structural basis for carbon dioxide binding by 2-ketopropyl coenzyme M oxidoreductase/carboxylase. *FEBS Letters, 585*(3), 459–464. https://doi.org/10.1016/j.febslet.2010.12.035.

Pavelites, J. J., Gao, J., Bash, P. A., & Mackerell, A. D., Jr. (1997). A molecular mechanics force field for NAD+ NADH, and the pyrophosphate groups of nucleotides. *Journal of Computational Chemistry, 18*(2), 221–239. https://doi.org/10.1002/(SICI)1096-987X(19970130)18:2 < 221::AID-JCC7 > 3.0.CO;2-X.

Prywes, N., Phillips, N. R., Tuck, O. T., Valentin-Alvarado, L. E., & Savage, D. F. (2023). Rubisco function, evolution, and engineering. *Annual Review of Biochemistry, 92*(1), 385–410. https://doi.org/10.1146/annurev-biochem-040320-101244.

Ramos-Guzmán, C. A., Velázquez-Libera, J. L., Ruiz-Pernía, J. J., & Tuñón, I. (2022). Testing affordable strategies for the computational study of reactivity in cysteine proteases: The case of SARS-CoV-2 3CL protease inhibition. *Journal of Chemical Theory and Computation, 18*(6), 4005–4013. https://doi.org/10.1021/acs.jctc.2c00294.

Recabarren, R., Tinzl, M., Saez, D. A., Gomez, A., Erb, T. J., & Vöhringer-Martinez, E. (2023). Covalent adduct formation as a strategy for efficient CO_2 fixation in crotonyl-CoA carboxylases/reductases. *ACS Catalysis, 13*(9), 6230–6241. https://doi.org/10.1021/acscatal.2c05250.

Robustelli, P., Ibanez-de-Opakua, A., Campbell-Bezat, C., Giordanetto, F., Becker, S., Zweckstetter, M., ... Shaw, D. E. (2022). Molecular basis of small-molecule binding to α-synuclein. *Journal of the American Chemical Society, 144*(6), 2501–2510. https://doi.org/10.1021/jacs.1c07591.

Rosenthal, R. G., Ebert, M.-O., Kiefer, P., Peter, D. M., Vorholt, J. A., & Erb, T. J. (2014). Direct evidence for a covalent ene adduct intermediate in NAD(P)H-dependent enzymes. *Nature Chemical Biology, 10*(1), 50–55. https://doi.org/10.1038/nchembio.1385.

Ruiz-Pernía, J. J., Silla, E., Tuñón, I., Martí, S., & Moliner, V. (2004). Hybrid QM/MM potentials of mean force with interpolated corrections. *The Journal of Physical Chemistry. B, 108*(24), 8427–8433. https://doi.org/10.1021/jp049633g.

Schulz, L., Guo, Z., Zarzycki, J., Steinchen, W., Schuller, J. M., Heimerl, T., ... Hochberg, G. K. A. (2022). Evolution of increased complexity and specificity at the dawn of form I Rubiscos. *Science (New York, N. Y.), 378*(6616), 155–160. https://doi.org/10.1126/science.abq1416.

Schwander, T., Schada von Borzyskowski, L., Burgener, S., Cortina, N. S., & Erb, T. J. (2016). A synthetic pathway for the fixation of carbon dioxide in vitro. *Science (New York, N. Y.), 354*(6314), 900–904. https://doi.org/10.1126/science.aah5237.

Senn, H. M., & Thiel, W. (2009). QM/MM methods for biomolecular systems. *Angewandte Chemie International Edition, 48*(7), 1198–1229. https://doi.org/10.1002/anie.200802019.

Sheng, X., Hou, Q., & Liu, W. (2019). Computational evidence for the importance of lysine carboxylation in the reaction catalyzed by carboxyl transferase domain of pyruvate carboxylase: A QM/MM study. *Theoretical Chemistry Accounts, 138*(1), 17. https://doi.org/10.1007/s00214-018-2408-8.

Sheng, X., Lind, M. E. S., & Himo, F. (2015). Theoretical study of the reaction mechanism of phenolic acid decarboxylase. *The FEBS Journal, 282*(24), 4703–4713. https://doi.org/10.1111/febs.13525.

Sirirak, J., Lawan, N., Kamp, M. W. V., der, Harvey, J. N., & Mulholland, A. J. (2020). Benchmarking quantum mechanical methods for calculating reaction energies of reactions catalyzed by enzymes. *PeerJ Physical Chemistry, 2*, e8. https://doi.org/10.7717/peerj-pchem.8.

Sousa, J. P. M., Neves, R. P. P., Sousa, S. F., Ramos, M. J., & Fernandes, P. A. (2020). Reaction mechanism and determinants for efficient catalysis by DszB, a key enzyme for crude oil bio-desulfurization. *ACS Catalysis, 10*(16), 9545–9554. https://doi.org/10.1021/acscatal.0c03122.

Stoffel, G. M. M., Saez, D. A., DeMirci, H., Vögeli, B., Rao, Y., Zarzycki, J., ... Erb, T. J. (2019). Four amino acids define the CO_2 binding pocket of enoyl-CoA carboxylases/reductases. *Proceedings of the National Academy of Sciences, 116*(28), 13964–13969. https://doi.org/10.1073/pnas.1901471116.

Tong, L. (2005). Acetyl-coenzyme A carboxylase: Crucial metabolic enzyme and attractive target for drug discovery. *Cellular and Molecular Life Sciences CMLS, 62*(16), 1784–1803. https://doi.org/10.1007/s00018-005-5121-4.

Tuñón, I., & Williams, I. H. (2019). Chapter Two—The transition state and cognate concepts. In E. I. H. Williams, & N. H. Williams (Vol. Eds.), *Advances in physical organic chemistry: 53*, (pp. 29–68). (pp. 29)Academic Press. https://doi.org/10.1016/bs.apoc.2019.09.001.

Van Der Kamp, M. W., Žurek, J., Manby, F. R., Harvey, J. N., & Mulholland, A. J. (2010). Testing high-level QM/MM methods for modeling enzyme reactions: Acetyl-CoA deprotonation in citrate synthase. *The Journal of Physical Chemistry. B, 114*(34), 11303–11314. https://doi.org/10.1021/jp104069t.

Van Lun, M., Hub, J. S., van der Spoel, D., & Andersson, I. (2014). CO_2 and O2 distribution in rubisco suggests the small subunit functions as a CO_2 reservoir. *Journal of the American Chemical Society, 136*(8), 3165–3171. https://doi.org/10.1021/ja411579b.

Vanommeslaeghe, K., Hatcher, E., Acharya, C., Kundu, S., Zhong, S., Shim, J., ... Mackerell, A. D., Jr. (2010). CHARMM general force field: A force field for drug-like molecules compatible with the CHARMM all-atom additive biological force fields. *Journal of Computational Chemistry, 31*(4), 671–690. https://doi.org/10.1002/jcc.21367.

Vanommeslaeghe, K., & MacKerell, A. D. (2012). Automation of the CHARMM general force field (CGenFF) I: Bond perception and atom typing. *Journal of Chemical Information and Modeling, 52*(12), 3144–3154. https://doi.org/10.1021/ci300363c.

Vanommeslaeghe, K., Raman, E. P., & MacKerell, A. D. (2012). Automation of the CHARMM general force field (CGenFF) II: Assignment of bonded parameters and partial atomic charges. *Journal of Chemical Information and Modeling, 52*(12), 3155–3168. https://doi.org/10.1021/ci3003649.

Walker, R. C., Crowley, M. F., & Case, D. A. (2008). The implementation of a fast and accurate QM/MM potential method in Amber. *Journal of Computational Chemistry, 29*(7), 1019–1031. https://doi.org/10.1002/jcc.20857.

Wang, P., Best, R. B., & Blumberger, J. (2011). Multiscale simulation reveals multiple pathways for H_2 and O_2 transport in a [NiFe]-hydrogenase. *Journal of the American Chemical Society, 133*(10), 3548–3556. https://doi.org/10.1021/ja109712q.

Zhang, X., & Bruice, T. C. (2007). Catalytic mechanism and product specificity of rubisco large subunit methyltransferase: QM/MM and MD investigations. *Biochemistry, 46*(18), 5505–5514. https://doi.org/10.1021/bi700119p.

Zinovjev, K., & Tuñón, I. (2017). Adaptive finite temperature string method in collective variables. *The Journal of Physical Chemistry. A, 121*(51), 9764–9772. https://doi.org/10.1021/acs.jpca.7b10842.

Zinovjev, K., & Tuñón, I. (2018). Reaction coordinates and transition states in enzymatic catalysis. *WIREs Computational Molecular Science, 8*(1), e1329. https://doi.org/10.1002/wcms.1329.

CHAPTER SIXTEEN

Carboxylation in *de novo* purine biosynthesis

Marcella F. Sharma and Steven M. Firestine*

Department of Pharmaceutical Sciences, Eugene Applebaum College of Pharmacy and Health Sciences,
Wayne State University, Detroit, MI, United States
*Corresponding author. e-mail address: sfirestine@wayne.edu

Contents

1. Introduction	390
2. Carboxylation in *de novo* purine biosynthesis	392
2.1 AIR carboxylase	394
2.2 N^5-CAIR synthetase and N^5-CAIR mutase	395
3. CAIR synthesis as a drug target	398
4. Expression and purification of *E. coli* $N^{5\text{-CAIR mutase}}$ (Thoden, Holden, & Firestine, 2008; Zhao, 2014)	400
4.1 Equipment	400
4.2 Reagents	401
4.3 Procedure	401
5. Expression and purification of *A. clavatus* $N^{5\text{-CAIR synthetase}}$ (Dewal & Firestine, 2013)	403
5.1 Equipment	403
5.2 Reagents	403
5.3 Procedure	404
6. Synthesis of CAIR and AIR	405
6.1 Notes	406
7. Activity assays	406
7.1 CAIR decarboxylation assay (Note 1,2)	406
7.2 AIR carboxylation assay	409
7.3 SAICAR synthetase coupled assay for AIR carboxylase (Note 1)	410
7.4 SAICAR synthetase coupled assay for N^5-CAIR mutase (Note 1)	412
7.5 N^5-CAIR synthetase coupled assay	413
7.6 Discontinuous phosphate assay for N^5-CAIR synthetase (Firestine, Paritala, McDonnell, Thoden, Holden, 2009)	414
7.7 Isatin-fluorescein enzyme activity assay (Sharma, Sharma, Streeter, Firestine, 2023) (Note 1)	416
7.8 Intrinsic Tryptophan Fluorescence Assay for N^5-CAIR Mutase (Constantine et al., 2006; Lei et al., 2016)	418
7.9 Thermal melting assay for N^5-CAIR mutase	419
8. Summary and conclusions	420
References	421

Methods in Enzymology, Volume 708
ISSN 0076-6879, https://doi.org/10.1016/bs.mie.2024.10.021
Copyright © 2024 Elsevier Inc. All rights are reserved, including those for text and data
mining, AI training, and similar technologies.

Abstract

De novo purine biosynthesis is one of two pathways for the synthesis of purine nucleotides that are critical for numerous biological processes, most notably nucleic acid replication. Within the pathway, there is only one carbon-carbon bond formation which is the carboxylation of 5-aminoimidazole ribonucleotide (AIR) to 4-carboxy-5-aminoimidazole ribonucleotide (CAIR). Interestingly, there are two unique pathways within purine biosynthesis to accomplish this transformation and this divergence is species specific. In humans and higher eukaryotes, CAIR is synthesized directly from AIR and carbon dioxide by the enzyme AIR carboxylase. In bacteria, yeast, fungi and plants, CAIR synthesis requires two steps. In the first, AIR is converted into the unstable carbamate, N^5-CAIR by the enzyme N^5-CAIR synthetase. N^5-CAIR is then converted into CAIR by transfer of the CO_2 group from N5 to C4. This is catalyzed by the enzyme N^5-CAIR mutase. This divergence has provided a biochemical rationale for targeting CAIR synthesis in the development of antimicrobial agents, but recent studies have provided strong evidence that AIR carboxylase plays a critical role in several cancers. Given the significance of these enzymes as drug targets, methods to prepare and evaluate these enzymes is of interest. In this chapter, we have accumulated the most relevant assays and provided methods to synthesize the substrates and purify the enzymes.

1. Introduction

Purines are 5,6-ring fused aromatic heterocycles that are critical biomolecules required for cellular function, growth, and survival (Huang et al., 2021). Purines are incorporated into DNA and RNA, play key roles as the energy molecules of the cell (ATP and GTP), are utilized in signal transduction (ATP, cAMP, etc) and are components of several coenzymes (NADH, etc) (Huang et al., 2021; Pedley & Benkovic, 2017). Purines are biosynthesized as nucleotides and given their essentiality, some form of purine biosynthesis is found in all three domains of life (Chua & Fraser, 2020). Biologically, purines are synthesized by one of two pathways, salvage or de novo. The salvage pathway recycles purine heterocycles obtained from the degradation of nucleic acids back into new purine nucleotides using phosphoribosyl pyrophosphate (PRPP) and a series of heterocycle-specific phosphoribosyltransferases (Moffatt & Ashihara, 2002; Murray, 1971). The salvage pathway is energetically efficient, requiring only one molecule of ATP per synthesized nucleotide (Murray, 1971). Under normal conditions, or in tissues lacking the de novo pathway, cellular purine levels are maintained by the salvage pathway (Murray, 1971).

De novo purine biosynthesis sequentially synthesizes a new purine heterocycle directly onto PRPP. This pathway is energy intensive, requiring between

4–7 ATP molecules depending on the organism (Kappock, Ealick, & Stubbe, 2000; Zhang, Morar, & Ealick, 2008). In de novo synthesis, PRPP is converted into inosine 5′-monophosphate which is then transformed into either AMP or GMP depending on the needs of the cell (Kappock, Ealick, & Stubbe, 2000; Zhang, Morar, & Ealick, 2008). The conversion of PRPP to IMP is different across different domains of life (Firestine & Davisson, 1994; Firestine, Poon, Mueller, Stubbe, & Davisson, 1994; Firestine, Wu, Youn, & Davisson, 2009; Kappock, Ealick, & Stubbe, 2000). In humans and vertebrates, de novo purine biosynthesis consists of 10 steps catalyzed by 6 enzymes, many of them multi-functional (Firestine, Wu, Youn, & Davisson, 2009; Kappock, Ealick, & Stubbe, 2000; Zhang, Morar, & Ealick, 2008). In bacteria, yeast, fungi, and plants, purine biosynthesis occurs in 11 steps (Firestine, Wu, Youn, & Davisson, 2009; Zhang, Morar, & Ealick, 2008). Purine biosynthesis in archaea has not been studied in much biological detail; however, genetic analysis indicates that they appear to have members which follow one or the other pathway (Brown, Hoopes, White, & Sarisky, 2011; Chua & Fraser, 2020).

The de novo pathway is outlined in Fig. 1. (Kappock, Ealick, & Stubbe, 2000; Zhang, Morar, & Ealick, 2008) The pathway starts with the addition of ammonia onto PRPP to generate phophoribosylamine (PRA) catalyzed by the enzyme PRPP amidotransferase (PPAT). This is the first committed step in the pathway (Kappock, Ealick, & Stubbe, 2000; Zhang, Morar, & Ealick, 2008). PRA is reacted with glycine activated by ATP to generate glycinamide ribonucleotide (GAR) by the enzyme GAR synthetase (GARS). Formylation of GAR by N^{10}-formyl-THF generates formylglycinamide

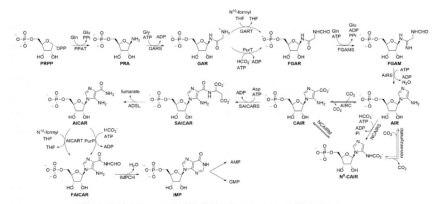

Fig. 1 De novo purine biosynthetic pathway. The figure shows the step-by-step conversion of PRPP to IMP that constitutes the beginning of the de novo purine biosynthetic pathway. This pathway shows variations of the pathway that is found in the three domains of life. Compound names and enzyme abbreviations are defined in the text.

ribonucleotide (FGAR). This step is catalyzed by GAR transformylase (GART). In some organisms, this step can be replaced by a different transformylase, formate-GAR transformylase (PurT), which utilizes ATP and formate as a one carbon source (Marolewski et al., 1994). Reaction of FGAR with ammonia generated from glutamine leads to formylglycinamidine ribonucleotide (FGAM) and this step is catalyzed by FGAR synthetase (FGARS). The first heterocyclic nucleotide is generated in the next step when FGAM is cyclized to generate aminoimidazole ribonucleotide (AIR) by the enzyme AIR synthase (AIRS).

The 6th step of de novo purine biosynthesis is different depending on the class of organism. In bacteria, yeast, fungi, and plants, 4-carboxy-5-aminoimidazole ribonucleotide (CAIR) is synthesized from AIR via the unstable intermediate N^5-carboxyaminoimidazole ribonucleotide (N^5-CAIR) (Mueller et al., 1994). N^5-CAIR is synthesized from AIR by the enzyme N^5-CAIR synthetase (NCAIRS) which requires ATP and bicarbonate (Meyer et al., 1992). N^5-CAIR is converted to CAIR by the transfer of the CO_2 group from N5 to C4 (Meyer et al., 1999). This reaction is catalyzed by the enzyme N^5-CAIR mutase (NCAIRM). In vertebrates, CAIR is synthesized directly from AIR and CO_2 by the action of the enzyme AIR carboxylase (AIRC) (Firestine and Davisson, 1994).

The remainder of the de novo purine biosynthesis pathway is fairly conserved across organisms (Kappock, Ealick, & Stubbe, 2000; Zhang, Morar, & Ealick, 2008). CAIR is converted into succinoaminoimidazolecarboxyamide ribonucleotide (SAICAR) using ATP and aspartic acid by SAICAR synthetase (SAICARS). The elimination of formate from SAICAR by the enzyme adenylosuccinate lyase (ADCL) yields 5-aminoimidazole-4-carboxyamide ribonucleotide (AICAR). AICAR is formylated at the N5-position by the enzyme AICAR transformylase (AICART) to give 5-formylamidoimidazole-4-carboxylamide ribonucleotide (FAICAR). This step can be catalyzed by an enzyme which uses N^{10}-formyl-THF as a cofactor or formate and ATP as the one carbon source (PurP) (Ownby, Xu, & White, 2005). Cyclization of FAICAR generates inosine-5′-monophosphate (IMP). This step is catalyzed by the enzyme IMP cyclohydrolase (IMPCH).

2. Carboxylation in *de novo* purine biosynthesis

Radiolabeled incorporation studies have shown that C6 of purines is derived from CO_2/HCO_3^- indicating that there is a carboxylation in purine biosynthesis (Lukens & Buchanan, 1959). Within the pathway, there is only one carbon-carbon bond formation which is the addition of a CO_2/HCO_3^- to C4 of AIR to give CAIR (Firestine & Davisson, 1994; Zhang, Morar, & Ealick, 2008). The conversion of AIR to CAIR is species specific and reflects their cellular CO_2/HCO_3^- environment. In vertebrates and some microorganisms that live in environments with elevated CO_2, CAIR is directly prepared from AIR and CO_2 by the enzyme AIR carboxylase (Firestine, Poon, Mueller, Stubbe, & Davisson, 1994). In contrast, bacteria, yeast, fungi and plants employ a two-enzyme system which uses HCO_3^- and ATP for CAIR synthesis (Firestine, Poon, Mueller, Stubbe, & Davisson, 1994; Meyer, Leonard, Bhat, Stubbe, & Smith, 1992; Mueller, Meyer, Rudolph, Davisson, & Stubbe, 1994). These differences likely reflect unique strategies in overcoming the inherent challenge with carboxylation, namely using the more reactive, but less abundant CO_2 or the poorly reactive but abundant bicarbonate.

The non-enzymatic chemistry of this system is also interesting. Incubation of AIR with high concentrations of bicarbonate rapidly produces the N5-carbamate, termed N^5-CAIR, followed by CAIR on a slower timescale ($t_{1/2}$ = 69 hr) (Mueller, Meyer, Rudolph, Davisson, & Stubbe, 1994). The carbamate is unstable and readily decarboxylates to generate AIR with a half-life of 0.9 min at pH 7.8 and at 30 °C (Dewal & Firestine, 2013; Firestine & Davisson, 1994). The half-life for N^5-CAIR is both temperature and pH sensitive with the rate of decarboxylation increasing as pH decreases and as temperature increases. N^5-CAIR is stable under very basic conditions (Mueller, Meyer, Rudolph, Davisson, & Stubbe, 1994). The non-enzymatic reaction with AIR indicates that a second nucleotide, N^5-CAIR, is present anytime AIR is added to a solution of HCO_3^-/CO_2. This conversion lowers the concentration of AIR and introduces the presence of a potential inhibitor in the reaction. Lastly, it is important to note that CAIR will also non-enzymatically decarboxylate with a maximal rate at pH 4.8 and a half-life of 84 min at pH 7 (Bhat, Groziak, & Leonard, 1990; Constantine, 2009; Litchfield & Shaw, 1971).

The enzymatic conversion of AIR into CAIR is mediated by three enzymes: AIR carboxylase, N^5-CAIR synthetase and N^5-CAIR mutase (Firestine & Davisson, 1994; Kappock, Ealick, & Stubbe, 2000; Thoden, Holden, & Firestine, 2008; Zhang, Morar, & Ealick, 2008). Technically only AIR carboxylase and N^5-CAIR synthetase are carboxylase enzymes as they utilize CO_2 and HCO_3^-; however, N^5-CAIR mutase is believed to use CO_2 in its reaction mechanism (Firestine & Davisson, 1994). AIR carboxylase and N^5-CAIR mutase are structurally and evolutionarily conserved, produce the same product (CAIR), yet are specific for their own reactions (Firestine and Davisson, 1994). Below, we will discuss each of these enzymes in detail.

2.1 AIR carboxylase

AIR carboxylase catalyzes the reversible carboxylation of AIR to give CAIR (Firestine & Davisson, 1994; Firestine, Poon, Mueller, Stubbe, & Davisson, 1994). A variety of studies, including X-ray crystallography, have shown that no cofactor or metal ion is required for this reaction, classifying this enzyme as belonging to a small class of carboxylases that require no cofactor for catalysis (Firestine & Davisson, 1994; Firestine, Poon, Mueller, Stubbe, & Davisson, 1994). The equilibrium for the reaction lies towards AIR and kinetically, the enzyme is normally measured in the physiologically reversed direction (Meyer, Leonard, Bhat, Stubbe, & Smith, 1992). AIR carboxylase uses CO_2, and no energy requirement is needed for the reaction (Firestine, Poon, Mueller, Stubbe, & Davisson, 1994). The enzyme is specific in that it produces AIR, not N^5-CAIR from the decarboxylation of CAIR and it cannot convert N^5-CAIR into CAIR (Firestine, Poon, Mueller, Stubbe, & Davisson, 1994).

Mechanistic studies on AIR carboxylase have been somewhat limited. A recent computational study of the carboxylase has been conducted (Fig. 2) (Prejano, Skerlova, Stenmark, & Himo, 2022). In a concerted reaction, electrons from the exocyclic amine are donated into the imidazole ring resulting in an attack at C4 on CO_2 bound in the active site (Prejano, Skerlova, Stenmark, & Himo, 2022). The resulting tetrahedral intermediate (iso–CAIR) is deprotonated by a conserved histidine resulting in the re-aromatization of the imidazole ring and the production of CAIR (Prejano, Skerlova, Stenmark, & Himo, 2022). This mechanism is supported by the fact that the enzyme catalyzes deuterium exchange at C4 indicating a facilitation of a reaction at C4. In addition, the tetrahedral intermediate, iso-CAIR, has been observed in the crystal structure of a catalytically inactive mutant of N^5-CAIR mutase (Constantine et al., 2006). Finally, 4–nitro-5–aminoimidazole ribonucleotide (NAIR) is a slow, tight-binding inhibitor of AIR carboxylase and is believed to mimic the transition state of the reaction (Fig. 2) (Firestine & Davisson, 1993; Firestine, Wu, Youn, & Davisson, 2009).

One challenge with AIR carboxylase is the binding of CO_2. The concentration of CO_2 in solution is low ($\sim 10\,\mu M$), yet the estimated K_m for CO_2 is $800\,\mu M$ (Firestine, Poon, Mueller, Stubbe, & Davisson, 1994). The

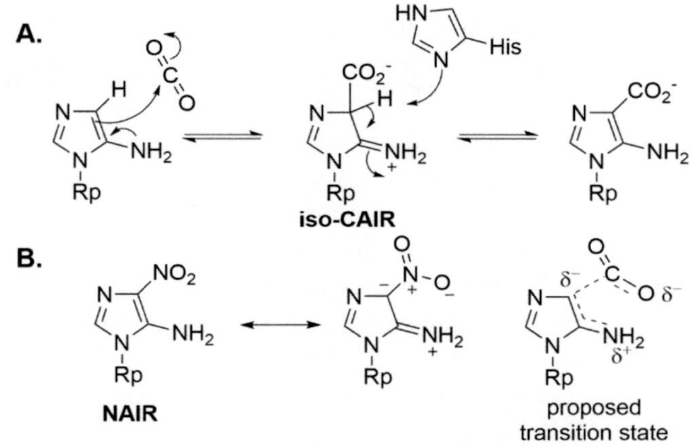

Fig. 2 (A) Proposed mechanism for AIR carboxylase. Rp=ribosyl phosphate. (B) The structure of NAIR and its similarity to the proposed transition state for AIR carboxylase.

identification of Michaelis complexes with CO_2 is limited in carboxylase enzymes. Studies with *Gallus gallus* AIR carboxylase have suggested that a complex may be formed, but definitive data supporting such a complex has not been obtained (Firestine & Davisson, 1994; Firestine, Poon, Mueller, Stubbe, & Davisson, 1994). The crystal structure of human bifunctional AIR carboxylase (also called PAICS) suggested a location for CO_2 binding (Li et al., 2007; Skerlova et al., 2020); however, further processing of the electron density casted doubt on the original assignment of CO_2 at that specific location (Tranchimand et al., 2011). Computational investigation on AIR carboxylase identified two potential binding sites for CO_2 (Prejano, Skerlova, Stenmark, & Himo, 2022). One of these sites was energetically more favorable then the other; however, it was not productive. Calculations indicate that the binding of CO_2 requires 14.9 kcal/mol of energy and that the CO_2 interacts with the amide group of Ala and Gly in the active site. This binding site aligns well with the structure of *A. aceti* N^5-CAIR mutase complexed with AIR and CO_2 (2clj). This site is also occupied by the carboxylate group of CAIR and the NO_2 group of NAIR (Skerlova et al., 2020, Tranchimand et al., 2011). The pathway for binding of CO_2 and the driving force for the binding of CO_2 has not been elucidated.

2.2 N^5-CAIR synthetase and N^5-CAIR mutase

The two-enzyme system employed by most microorganisms and plants synthesizes CAIR from AIR using ATP and bicarbonate. This approach is comparable to the biotin-dependent carboxylases in which ATP is used to activate bicarbonate for the generation of carboxybiotin which is then transferred to a second active site for carboxylation of an acceptor (Tong, 2013). Unlike the biotin-dependent carboxylases, purine biosynthesis utilizes a different carrier to transport CO_2 to a second enzyme, namely N^5-CAIR. This system is unique to purine biosynthesis.

N^5-CAIR synthetase catalyzes the synthesis of N^5-CAIR from AIR. The reaction can be broken into two half reactions (Fig. 3). In the first, ATP and HCO_3^- react to generate the extremely unstable intermediate carboxyphosphate ($t_{1/2} \sim 70$ ms) (Sauers, Jencks, & Groh, 1975). In the second reaction, carboxyphosphate reacts with AIR, either directly or after decomposition to CO_2, to generate the carbamate, N^5-CAIR, ADP, and inorganic phosphate (Dewal & Firestine, 2013). Positional isotope exchange studies show that an oxygen from bicarbonate is transferred to inorganic phosphate demonstrating an intermediate between ATP and bicarbonate (Mueller, Meyer, Rudolph, Davisson, & Stubbe, 1994). One interesting aspect of this system is that N^5-CAIR will non-enzymatically decarboxylate back to AIR. Thus, if N^5-CAIR is not protected or efficiently transported to N^5-CAIR mutase, non-stoichiometric consumption of ATP can occur due to the recycling of AIR. To prevent this, channeling the unstable N^5-CAIR directly to N^5-CAIR mutase would be beneficial. However, no evidence for channeling has been found in in vitro studies. Experiments on the bifunctional N^5-CAIR synthetase-N^5-CAIR mutase enzyme present in yeast and fungi found that a large amount of N^5-CAIR is released from this protein indicating the lack of direct transport to the mutase (Firestine et al., 1998). Examination of these enzymes from *Thermotoga maritima*, where, due to elevated temperatures, N^5-CAIR would be even more unstable, also failed to detect channeling (Zhao, 2014). The available evidence indicates that if channeling exists between these enzymes, other proteins must be involved. A recent study has provided

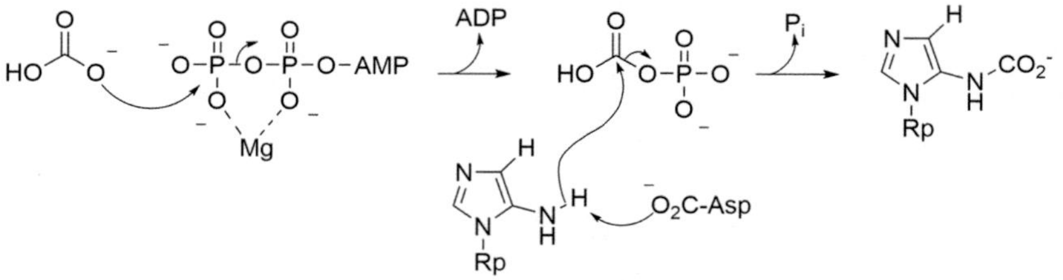

Fig. 3 Proposed mechanism for N^5-CAIR synthetase.

evidence for an extensive interaction network among de novo purine enzymes in vivo and interestingly, N^5-CAIR synthetase was shown to play a central role in the network (Gedeon et al., 2023).

Previous studies have shown that N^5-CAIR synthetase is not required for growth under certain conditions (Firestine, Poon, Mueller, Stubbe, & Davisson, 1994). Studies on N^5-CAIR synthetase-deficient strains show that elevated levels of CO_2 can complement this mutation (Firestine, Poon, Mueller, Stubbe, & Davisson, 1994). This paradox is explained by the fact that N^5-CAIR is readily formed non-enzymatically from AIR and CO_2/ HCO_3^-. The N^5-CAIR thus generated bypasses the need for N^5-CAIR synthetase. This observation suggests that N^5-CAIR synthetase may have evolved because organisms could no longer non-enzymatically produce a sufficient concentration of N^5-CAIR to support purine biosynthesis. Such a scenario could occur if the concentration of environmental CO_2 decreased substantially.

N^5-CAIR synthetase is a member of the ATP-grasp superfamily of enzymes (Fawaz, Topper, & Firestine, 2011). These enzymes are named for their atypical ATP-binding domain. Members of this superfamily catalyze similar reactions which react ATP with a carboxylic acid to generate an electrophilic acylphosphate intermediate which is attacked by a nucleophile to generate the final product (Fawaz, Topper, & Firestine, 2011). The exact nature of the product depends on the carboxylic acid and nucleophile. Interestingly, several members of the family are also carboxylases which utilize bicarbonate as a substrate including acetyl-CoA carboxylase, pyruvate carboxylase, and carbamoyl phosphate synthetase (Fawaz, Topper, & Firestine, 2011).

N^5-CAIR mutase catalyzes the reversible conversion of N^5-CAIR into CAIR. This enzyme is structurally and evolutionarily related to AIR carboxylase and both enzymes produce the same product, CAIR (Firestine, Wu, Youn, & Davisson, 2009; Zhang, Morar, & Ealick, 2008). Despite their similarity, N^5-CAIR mutase cannot convert AIR and CO_2 into CAIR and in the reverse direction, the CO_2 group from C4 of CAIR is retained and transferred to N5 to give N^5-CAIR, not AIR, as the product (Firestine, Poon, Mueller, Stubbe, & Davisson, 1994; Mueller, Meyer, Rudolph, Davisson, & Stubbe, 1994). In the proposed mechanism (Constantine et al., 2006; Ravi, Biswal, Kanagarajan, & Jeyakanthan, 2019), the carbamate at N5 is decarboxylated to generate CO_2 in the active site while protonating N5 by His_{45} to give AIR (Fig. 4). Donation of electrons from N5 with attack at C4 onto the trapped CO_2 yields the

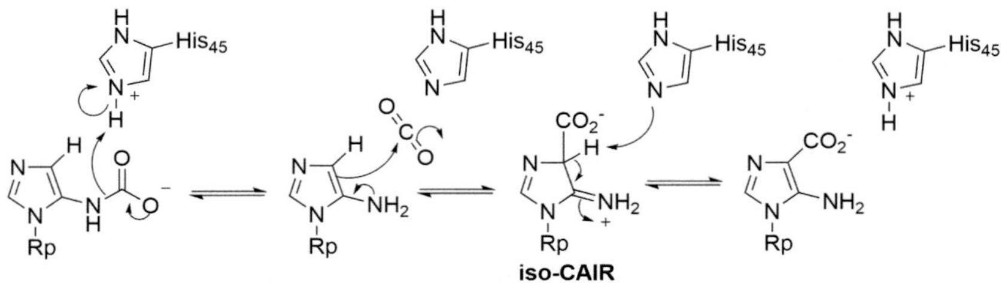

Fig. 4 Proposed mechanism for N^5-CAIR mutase.

tetrahedral intermediate iso-CAIR which is the same intermediate seen in AIR carboxylase. Deprotonation of C4 by His45 gives CAIR. This mechanism is supported by the fact that CO_2 from N^5-CAIR is directly transferred to C4, iso-CAIR has been detected in a catalytically inactive N^5-CAIR mutase, and by computational studies (Constantine et al., 2006; Meyer, Kappock, Osuji, & Stubbe, 1999; Li, Zheng, Zhang, & Zhang, 2011; Ravi, Biswal, Kanagarajan, & Jeyakanthan, 2019).

Why is N^5-CAIR utilized as a CO_2 carrier? A likely explanation comes from the fact that N^5-CAIR can be generated non-enzymatically under conditions of elevated CO_2 or bicarbonate levels (Alenin, Kostikova, & Domkin, 1987; Mueller, Meyer, Rudolph, Davisson, & Stubbe, 1994). Thus, N^5-CAIR exists anytime AIR is produced. N^5-CAIR mutase could have evolved to take advantage of the natural CO_2 capture ability of AIR in generating N^5-CAIR. N^5-CAIR makes an "ideal" substrate since it not only transports CO_2 to the active site, thus bypassing the difficulty of CO_2 acquisition by the enzyme, but also the substrate AIR (or its equivalent). Such a strategy is seen in *Methanoarchaea* which uses methanofuran to capture CO_2 to generate the carbamate N-carboxymethanofuran (Bartoschek, Vorholt, Thauer, Geierstanger, & Griesinger, 2000). This is the first step of the reduction of CO_2 to CH_4 in *Methanoarchaea* (Bartoschek, Vorholt, Thauer, Geierstanger, & Griesinger, 2000).

3. CAIR synthesis as a drug target

De novo purine biosynthesis has been a target of drug discovery with most of the efforts focused on anticancer agents targeting the folate utilizing enzymes and IMP dehydrogenase which is involved in the conversion of IMP to GMP (Christopherson, Lyons, & Wilson, 2002). Recently, however, interest in human AIR carboxylase as an anticancer target has increased.

AIR carboxylase is fused to the next enzyme in the pathway, SAICAR synthetase to give the bifunctional enzyme phosphoribosylaminoimidazole carboxylase-phosphoribosylaminoimidazolesuccinocarboxamide synthetase (PAICS). Recent studies have shown that PAICS is elevated in various cancers including breast, bladder, lung, prostate, and colorectal and that elevation of PAICS expression is correlated with poor survival (Huang et al., 2020). Studies which silence PAICS show cancer cell death and down-regulation of PAICS results in enhanced survivability (Meng, Chen, Jia, Li, & Yang, 2018). Taken together, the evidence supports PAICS as a promising, but unexplored, anticancer target.

The differences between microbes and humans for CAIR synthesis provides a biochemical rationale for the identification of inhibitors targeting N^5-CAIR synthetase and N^5-CAIR mutase as potential antibacterial or antifungal agents. Gene deletion studies have provided support for N^5-CAIR mutase as a viable target in several organisms including *E. coli* and *B. anthracis* (Samant et al., 2008). While N^5-CAIR synthetase is required for microbial growth, evidence suggests that it is likely not essential for growth in a host.

Several groups have published studies on developing inhibitors of these enzymes. The Firestine lab has conducted high-throughput screens to identify inhibitors of *E. coli* N^5-CAIR synthetase; however, the compounds identified reacted with AIR thereby reducing the concentration of the substrate available for the enzyme (Firestine, Paritala, McDonnell, Thoden, & Holden, 2009; Streeter, Lin, & Firestine, 2019). The same lab has developed inhibitors targeting the ATP-binding site of *E. coli* and *Aspergillus clavatus* N^5-CAIR synthetase (Lin, 2018). Compounds with K_i values in the low micromolar were identified, but the selectivity of these compounds for N^5-CAIR synthetase over other ATP-utilizing enzymes was not assessed. Multiple approaches have been taken to identify inhibitors of N^5-CAIR mutase. These involve examining nucleotide analogs, fragment-based screens, high-throughput screens, and in silico screens against N^5-CAIR mutase from *E. coli, B. anthracis, Burkholderia cenocepacia, Legionella pneumophila,* and *Pyrococcus horikoshii* (Belfon et al., 2023; Firestine, Paritala, McDonnell, Thoden, & Holden, 2009; Firestine, Wu, Youn, & Davisson, 2009; Kim et al., 2015; Lei et al., 2016). Compounds with K_d, K_i or IC_{50} values ranging from low micromolar to low millimolar have been identified; however, only a few have shown inhibition of bacterial growth. Further, none have addressed the critical question of selectivity for N^5-CAIR mutase versus AIR carboxylase.

There have been limited studies on the identification of inhibitors of AIR carboxylase (PAICS) and most of these have utilized the enzyme from *G. gallus* instead of *H. sapiens* (Firestine, 1995; Firestine & Davisson, 1993). The nucleotide analog NAIR is a slow, tight-binding inhibitor of *G. gallus* AIR carboxylase with a K_i of 0.34 nM (Firestine & Davisson, 1993; Firestine, Wu, Youn, & Davisson, 2009). In contrast, NAIR is a simple competitive inhibitor of N^5-CAIR mutase with a K_i of 0.5 μM (Firestine, Poon, Mueller, Stubbe, & Davisson, 1994; Firestine, Wu, Youn, & Davisson, 2009). The 1000-fold selectivity of NAIR for AIR carboxylase over N^5-CAIR mutase indicates that it is possible to identify a selective inhibitor between these enzymes. Unfortunately, NAIR has troubling physicochemical properties which result in it being a poor candidate for drug development (Firestine, Wu, Youn, & Davisson, 2009). No other drug discovery efforts targeting AIR carboxylase have been reported in the literature, although a recent publication on a non-targeted in vivo fragment screen identified 3 fragments which bound to PAICS (Offensperger et al., 2024). Whether these agents bind to AIR carboxylase is currently unknown.

In this chapter, we will describe experimental procedures for the examining the activity of N^5-CAIR synthetase, AIR carboxylase, and N^5-CAIR mutase. We will outline both activity-based and biophysical based assays for these enzymes and discuss the challenges and limitations with existing assays. We will also discuss the difficulty in acquiring the substrates for these enzymes.

4. Expression and purification of *E. coli* $N^{5\text{-CAIR}}$ mutase (Thoden, Holden, & Firestine, 2008; Zhao, 2014)

4.1 Equipment

2 L Erlenmeyer flask (use 750 mL media) or 2.8 L Fernbach baffled culture flask (use 1000 mL media).
Incubator (Blue M).
Shaking incubator (Thermo Scientific Max Q or New Brunswick Scientific Excella E24).
Spectrophotometer (Varian Cary 1).
Disposable cuvettes (Fisherbrand Cat. # 14-955-127).
12 mL plastic columns (Gold Biotechnology, Cat. # P-301).
Electrophoresis system (BioRad Mini-Protean Tetra).

4.2 Reagents

pJK057 plasmid. Plasmid pJK057 expressing both N-terminal His-tagged *E. coli* N^5-CAIR mutase and untagged *E. coli* N^5-CAIR synthetase was acquired from Addgene. This changes residues 2–7 from SSRNNP to HHHHHH. The plasmid was provided in bacteria as an agar stab in XL1 Blue competent cells. Plasmid was isolated from a bacterial culture using the Promega A7100 miniprep system according to the manufacturers protocols.

LB media (10 g tryptone, 10 g NaCl, and 5 g yeast extract per liter).

LB agar plates with 100 µg/mL ampicillin (10 g tryptone, 10 g NaCl, 5 g yeast extract, and 15 g agar per liter). Ampicillin was added after autoclaving and cooling to approximately 55 °C.

IPTG (Gold Biotechnology Cat. # I2481C25).

Streptomycin sulfate (Gold Biotechnology Cat. # S-150–100).

B-PER Bacterial Protein Extraction Reagent (Thermo Scientific Cat. # 78248).

BL21 (DE3) chemical competent cells (Thermo Scientific Cat. # EC0114).

Ampicillin stock at 100 mg/mL (sterile filtered, Gold Biotechnology Cat. # A-301).

Cobalt high-density agarose beads (Gold Biotechnology Cat. # H-310).

Buffer A: 50 mM sodium phosphate, 300 mM NaCl, pH 7.4.

Imidazole (Thermo Fisher Cat. # AC122025000).

Coomassie stain (0.1 % Brilliant Blue R-250 (Sigma), 10 % acetic acid, 50 % methanol, 40 % water).

Coomassie destain (125 mL methanol, 25 mL glacial acetic acid, 100 mL distilled water).

Prestained protein ladder (Thermo Scientific cat # 26616).

Dialysis cassette (Slide-A-Lyzer, 15 mL, 10 kDa MWCO, Thermo Scientific Cat. # A52972).

Centrifugal concentrator (Amicon Ultra, MilliporeSigma Cat. # UFC901008, UFC201024).

Bradford Plus Protein Assay Reagent (Thermo Scientific Pierce Cat. # PI23238).

4.3 Procedure

1. Plasmid pJK057 is transformed into BL21(DE3) competent cells according to a manufacturer provided heat shock protocol, plated onto LB/Amp plates and grown overnight at 37°C.

2. A single colony is used to inoculate 10 mL of LB media containing 100 µg/mL ampicillin and grown overnight with shaking at 250 RPM at 37°C. The overnight culture is used to inoculate 750-1000 mL of LB media containing 100 µg/mL ampicillin and the culture is grown with shaking at 250 RPM at 37°C until the OD_{600} is between 0.6-0.8.

3. To initiate protein expression, IPTG is added to a final concentration of 1 mM and the temperature of the shaking incubator is decreased to 18°C. The culture is shaken at 250 RPM overnight and the cells are collected by centrifuging at 3,800 × g at 4°C. The liquid is drained, and the resulting cell pellet is stored in −80°C until use.

4. A high-density cobalt column is prepared by adding 5 mL of cobalt resin to a disposable column and the resin is washed with several column volumes of distilled water followed by equilibrating with 10 column volumes of buffer A containing 10 mM imidazole. All washing and equilibrations are done at 4°C.

5. The cell pellet is thawed on ice and treated with B-PER Bacterial Protein Extraction Reagent according to the manufacturer's instructions. The lysis is incubated on ice for 10-20 min and the lysed bacteria are centrifuged at 4°C at 31,000 × g for 30 min to collect the cell debris. The lysate is removed from the debris and streptomycin sulfate (a final concentration of 1 % w/v) is added to remove nucleic acids. The lysate is incubated on ice for 10 min and gently mixed by rocking by hand to aid precipitation of the nucleic acids. The lysate is centrifuged at 4°C at 31,000 × g for 30 min to pellet the white nucleic acid/streptomycin complex.

6. The lysate is loaded onto the freshly prepared cobalt agarose column at 4°C by gravity flow. The flow through is collected and can be reloaded onto the column to achieve higher yields of the protein.

7. Unbound material is removed by washing the column with 10 column volumes of buffer A containing 25 mM imidazole, followed by 10 column volumes of buffer A containing 50 mM imidazole. N^5-CAIR mutase is eluted using 10-15 mL of buffer A containing 150 mM imidazole (collected as 1.5 mL fractions) and 10-15 mL of buffer A containing 400 mM of imidazole (collected as 1.5 mL fractions). Protein usually eluted in all fractions but with differing purity.

8. Individual fractions are analyzed by SDS-PAGE with Coomassie staining to identify pure protein. These fractions are combined and dialyzed (10 K MWCO, Slide-A-Lyzer) at 4°C against 10 mM Tris, 200 mM NaCl, pH 8 storage buffer. The dialysis buffer is replaced twice over four hours (once every two hours) then dialyzed overnight. The

protein is concentrated by centrifugal concentration using a 10 K MWCO Amicon Ultra centrifugal filter. Concentration of protein is determined using a Bradford assay and protein stocks ranged from 5 mg/mL to 15 mg/mL. The protein is stored as 10 μL aliquots at −80°C until use. Flash freezing is not required.

5. Expression and purification of *A. clavatus* N[5]-CAIR synthetase (Dewal & Firestine, 2013)

5.1 Equipment

2 L Erlenmeyer flask (use 750 mL media) or 2.8 L Fernbach baffled culture flask (use 1000 mL media).
Incubator (Blue M).
Shaking incubator (Thermo Scientific Max Q or New Brunswick Scientific Excella E24).
Spectrophotometer (Varian Cary 1).
Disposable cuvettes (Fisherbrand Cat. # 14-955-127).
12 mL plastic columns (Gold Biotechnology, Cat. # P-301).
Electrophoresis system (BioRad Mini-Protean Tetra).

5.2 Reagents

pET-N[5]-CAIR synthetase plasmid. This plasmid generates an N-terminal His tag of *Aspergilllus clavatus* with the sequence MGSSHHHHHHSSENLYFQGH. This sequence contains a TEV cleavage site.
LB media (10 g tryptone, 10 g NaCl, and 5 g yeast extract per liter).
LB agar plates with 100 μg/mL ampicillin (10 g tryptone, 10 g NaCl, 5 g yeast extract, and 15 g agar per liter). Ampicillin was added after autoclaving and cooling to approximately 55 °C.
IPTG (Gold Biotechnology Cat. # I2481C25).
B-PER Bacterial Protein Extraction Reagent (Thermo Scientific Cat. # 78248).
BL21 (DE3) pLysS chemical competent cells (Invitrogen Cat. # C606010).
Ampicillin stock at 100 mg/mL (sterile filtered, Gold Biotechnology Cat. # A-301).
Streptomycin sulfate (Gold Biotechnology Cat. # S-150–100).
Cobalt high-density agarose beads (Gold Biotechnology Cat. # H-310).
Buffer A: 50 mM sodium phosphate, 300 mM NaCl, pH 7.4.

Imidazole (Thermo Fisher Cat. # AC122025000).

Coomassie stain (0.1 % Brilliant Blue R-250 (Sigma), 10 % acetic acid, 50 % methanol, 40 % water).

Coomassie destain (125 mL methanol, 25 mL glacial acetic acid, 100 mL distilled water).

Prestained protein ladder (Thermo Scientific cat # 26616).

Dialysis cassette (Slide-A-Lyzer, 15 mL, 10KDa MWCO, Thermo Scientific Cat. # A52972).

Centrifugal concentrator (Amicon Ultra, MilliporeSigma Cat. # UFC901008, UFC201024).

Bradford Plus Protein Assay Reagent (Thermo Scientific Pierce Cat. # PI23238).

5.3 Procedure

1. Plasmid is transformed into BL21(DE3) pLysS competent cells according to a manufacturer provided heat shock protocol, plated onto LB/Amp plates and grown overnight at 37°C.

2. A single colony is used to inoculate 10 mL of LB media containing 100 μg/mL ampicillin and grown overnight with shaking at 250 RPM at 37°C. The overnight culture is used to inoculate 1 L of LB media containing 100 μg/mL ampicillin and the culture is grown with shaking at 250 RPM at 37°C until the OD_{600} was 0.2.

3. The culture is cooled to 25°C, and protein expression is induced by the addition of IPTG to a final concentration of 1 mM. The induction is continued with shaking at 250 RPM at 25°C for 5 hrs. The cells are collected by centrifuging at 3,800 × g at 4°C for 15 min. The liquid is drained, and the resulting cell pellet is stored at −20°C until use.

4. A 2 cm × 4 cm high-density cobalt column is prepared by adding the cobalt resin to a disposable column and the resin is washed with several column volumes of distilled water followed by equilibrating with 10 column volumes of buffer A containing 10 mM imidazole. All washing and equilibrations are done at 4°C.

5. The cell pellet is thawed on ice and treated with B-PER Bacterial Protein Extraction Reagent according to the manufacturer's instructions. The lysis is incubated on ice for 10-20 min and the lysed bacteria are centrifuged at 4°C at 31,000 × g for 30 min to collect the cell debris. The lysate is removed from the debris and streptomycin sulfate (a final concentration of 1 % w/v) is added to remove nucleic acids. The lysate is incubated on ice for 30 min with occasional gentle rocking every

Carboxylation in *de novo* purine biosynthesis

405

10 min to aid precipitation. The lysate is centrifuged at 4°C at 31,000 × g for 45 min to pellet the white nucleic acid/streptomycin complex.

6. The lysate is loaded onto the freshly cobalt agarose column at 4°C by gravity flow. The flow through is collected and can be reloaded onto the column to achieve higher yields of the protein.

7. Unbound material is removed by washing the column with 3 column volumes of buffer A containing 25 mM imidazole. The protein is eluted using buffer A containing 100 mM imidazole while collecting 1 mL fractions.

8. Individual fractions are analyzed by SDS-PAGE with Coomassie staining to identify pure protein. These fractions are combined and dialyzed (10 K MWCO, Slide-A-Lyzer dialysis cassette) at 4°C against 3 L of 10 mM Tris-HCl, 200 mM NaCl, pH 8 buffer. The protein is concentrated by centrifugal concentration using a 10 K MWCO Amicon Ultra centrifugal filter. Concentration of protein is determined using a Bradford assay and protein stocks ranged from 5 mg/mL to 15 mg/mL. The protein is stored as 10 μL aliquots at −80°C until use. Flash freezing is not necessary.

6. Synthesis of CAIR and AIR

One of the most significant challenges with the study of these enzymes is the acquisition of the substrates for the enzymes. None of the substrates are commercially available and they can be challenging to synthesize since they are polar and chemically unstable. Two substrates, CAIR and AIR are utilized in these assays. The third substrate, N^5-CAIR can be produced, but is very unstable and is not recommended for routine assays. If N^5-CAIR is needed for a reaction, it is recommended that this substrate be generated in situ from AIR and high concentrations of bicarbonate or from AIR, bicarbonate, ATP and N^5-CAIR synthetase.

The synthesis of CAIR and AIR starts from AICAR monophosphate (Bhat, Groziak, & Leonard, 1990; Firestine & Davisson, 1994; Sullivan, Huma, Mullins, Johnson, & Kappock, 2014). AICAR monophosphate can be acquired commercially or prepared by phosphorylation of the nucleoside (Firestine & Davisson, 1994; Bhat, Groziak, & Leonard, 1990). Several protocols have been published for the conversion of AICAR-mp to CAIR. AICAR monophosphate is saponified with base (NaOH or LiOH) to generate CAIR. A recently published method generates pure CAIR after a series of ethanol washes to remove salts (Sullivan, Huma, Mullins, Johnson, & Kappock, 2014).

Alternatively, CAIR can be purified using an anion exchange column with an ammonium bicarbonate gradient (Firestine & Davisson, 1994) (Note 1). AIR is prepared by decarboxylating CAIR using an ammonium acetate pH 4.8 buffer while bubbling with nitrogen or argon to remove CO_2 (Bhat, Groziak, & Leonard, 1990; Firestine & Davisson, 1994) (Note 2). After multiple lyophilizations, the AIR generated is suitable for assays but usually contains acetate (Note 1,3). AIR can be purified by anion exchange chromatography to remove acetate.

The concentration of CAIR and AIR can be determined by measuring the absorbance at either 250 or 260 nm in 50 mM Tris-HCl, pH 8 buffer (Meyer, Leonard, Bhat, Stubbe, & Smith, 1992). CAIR has a λ_{max} at 260 nM while AIR has a maximum absorbance at 230 nm. The extinction coefficients for CAIR are $\varepsilon_{250} = 10,980$ M^{-1} cm^{-1} and $\varepsilon_{260} = 10,500$ M^{-1} cm^{-1}. The extinction coefficients for AIR are $\varepsilon_{250} = 3,270$ M^{-1} cm^{-1} and $\varepsilon_{260} = 1,570$ M^{-1} cm^{-1}.

6.1 Notes

1. Ammonium bicarbonate and ammonium acetate are used since these salts can be removed by lyophilization. The pH of solutions should be monitored to make sure that they are not acidic which can result in degradation of the nucleotides.
2. Either N^5-CAIR mutase or AIR carboxylase can be used to generate AIR from CAIR, but these enzymes will only produce an equilibrium mixture of both AIR and CAIR. Thus, a challenging purification step is needed to separate the two compounds. The chemical decarboxylation will drive the reaction to completion
3. AIR is photosensitive and flasks should be covered with aluminum foil during lyophilization to prevent degradation.

7. Activity assays

7.1 CAIR decarboxylation assay (Note 1,2)

$$CAIR \underset{NCAIRM}{\rightleftharpoons} N^5\text{-}CAIR \underset{\substack{non-\\enzymatic}}{\rightleftharpoons} AIR + CO_2 \quad \Delta_{260} \quad \varepsilon = 9,365 \text{ M}^{-1} \text{ cm}^{-1}$$

$$CAIR \underset{AIRC}{\rightleftharpoons} AIR + CO_2 \quad \Delta_{260} \quad \varepsilon = 9,000 \text{ M}^{-1} \text{ cm}^{-1}$$

7.1.1 Equipment
UV spectrophotometer or microplate reader capable of UV detection at 260 nm (Varian Cary 1 or Biotek Synergy 2).
1 mL, 1 cm pathlength quartz cuvettes (Fisherbrand Cat. # 14–958-126). Other pathlengths can be used.
UV-transparent microplate (Corning Cat. # 3635).

7.1.2 Reagents
Buffer: 25 mM Tris, pH 8 buffer (Note 3).
CAIR (see above, Note 4).
N^5-CAIR mutase or AIR carboxylase (Firestine & Davisson, 1994; Skerlova et al., 2020) (Note 5).

7.1.3 Procedure
7.1.3.1 Cuvette method
1. In a quartz cuvette with a final volume of 1 mL, buffer, CAIR (typically ranged from 20-100 μM) are incubated at room temperature in the spectrophotometer. The instrument is started and the background rate of non-enzymatic decarboxylation of CAIR is monitored at 260 nm (Note 6).
2. The reaction is initiated by the addition of N^5-CAIR mutase or AIR carboxylase (typically 20-60 ng), the reaction is carefully mixed and the decrease in absorption at 260 nm is monitored over time. The slope of the change in absorption over the first minute after addition of enzyme is determined using the instrument software.
3. The concentration of CAIR consumed over time is calculated using Beer's Law with a $\Delta\varepsilon = 9{,}000\text{-}9{,}365\ M^{-1}\ cm^{-1}$ (Note 7).

7.1.3.2 Microplate method
1. In a non-sterile 96-well flat bottom UV-transparent microplate, buffer and CAIR (ranging from 10-400 μM) are added to each well such that the final volume was 100 μL (Note 6).
2. The reaction is initiated by the addition of N^5-CAIR mutase or AIR carboxylase (20-60 ng) and the progress of the enzymatic reaction is monitored at 260 nm over 40 min using a plate reader.
3. Progress curves are imported into Excel and normalized to zero absorbance at time zero. The concentration of CAIR is calculated using Beer's Law with a $\Delta\varepsilon = 9{,}000\text{-}9{,}365\ M^{-1}\ cm^{-1}$ and the calculated pathlength for the microplate reader (Note 7). To determine the initial

velocity, the normalized data is fit to equation 1 in KaleidaGraph. For Equation 1,

$$P = \frac{V_0}{\eta}(1 - e^{-\eta t}) \tag{1}$$

P represents product formation, V_0 represents the initial velocity, and η is the rate of change for the non-linear portion of the plot (Cao & De La Cruz, 2013) (Note 8).

7.1.4 Notes

1. The decarboxylation assay measures the enzyme activity in the physiologically reverse direction. This is done because the equilibrium lies towards AIR ($K_{eq} = 1.8$) and CAIR is more chemically stable then AIR or N^5-CAIR (Meyer, Leonard, Bhat, Stubbe, & Smith, 1992). In this assay, CO_2 will be produced. The concentration of CO_2 dissolved in the buffer will affect the equilibrium as well as whether the cuvette is open or closed. In general, these factors play a minimal role in routine assays.
2. This assay is highly sensitive to the divalent metal ions Mg^{+2}, Zn^{+2}, and Ni^{+2} although other divalent metal ions may also affect the assay (Firestine & Davisson, 1994; Sullivan, Huma, Mullins, Johnson, & Kappock, 2014). These metals form a complex with CAIR which prevents decarboxylation by the enzyme. Care should be taken, particularly with inhibitors, that these ions are not present as they could lead to false inhibition.
3. Tris buffer at pH 8 is used because CAIR is more stable at this pH. Buffers with reduced pH can be used; however, the non-enzymatic rate of CAIR decarboxylation will increase as the pH decreases (Litchfield & Shaw, 1971; Meyer, Leonard, Bhat, Stubbe, & Smith, 1992; Mueller, Meyer, Rudolph, Davisson, & Stubbe, 1994). Other buffers (HEPES) have been used but this issue has not been extensively studied.
4. CAIR should be stored at $-80°C$ until use and it is better to store this substrate in individual aliquots. Repeated freeze-thawing of this substrate will result in decarboxylation to AIR and the development of a darkly colored solution over time.
5. This assay can also be used for either N^5-CAIR mutase or AIR carboxylase since both use CAIR as a substrate. However, it should be noted that when using AIR carboxylase, AIR is produced. When using N^5-CAIR mutase, N^5-CAIR is produced (Mueller, Meyer, Rudolph, Davisson, & Stubbe, 1994). The N^5-CAIR produced will non-enzymatically decarboxylate to generate AIR.

Carboxylation in *de novo* purine biosynthesis

6. This assay can be used to examine inhibition. The enzymes can tolerate DMSO to approximately 4 % total volume, but lower DMSO levels are better. Triton can also be used in this assay up to a level of 0.01 %. The wavelength of the assay can be altered slightly with a corresponding change in the extinction coefficient. However, many organic compounds will absorb at the wavelengths used in this assay.

7. As mentioned in Note 5, the product of the decarboxylation of CAIR is different between AIR carboxylase versus N^5-CAIR mutase (Mueller, Meyer, Rudolph, Davisson, & Stubbe, 1994). If AIR carboxylase is used, $\Delta\varepsilon_{260}$ is $9,000\ M^{-1}\ cm^{-1}$. If N^5-CAIR mutase is used, $\Delta\varepsilon_{260}$ is $9,365\ M^{-1}\ cm^{-1}$. The difference is due to the lower extinction coefficient of N^5-CAIR relative to AIR.

8. This method is utilized to determine the initial velocity since it is challenging to synchronize the time of enzyme additional across multiple wells in the plate. Thus, determining the initial slope of the reaction at a fixed time introduces error. Fitting of the progress curves using equation 1 eliminates this problem but does add additional steps in the analysis of the data. The time taken for the progress curve can be altered by the amount of enzyme used in the assay.

7.2 AIR carboxylation assay

$$HCO_3^-/CO_2 + AIR \underset{AIRC}{\rightleftharpoons} CAIR \quad \Delta_{260}\ \varepsilon=9{,}000\ M^{-1}\ cm^{-1}$$

$$HCO_3^-/CO_2 + AIR \underset{\substack{\text{non-}\\ \text{enzymatic}}}{\rightleftharpoons} N^5\text{-}CAIR \underset{NCAIRM}{\rightleftharpoons} CAIR \quad \Delta_{260}\ \varepsilon=9{,}000\ M^{-1}\ cm^{-1}$$

7.2.1 Reagents
Buffer (50 mM Tris, pH 8).
AIR (prepared as above, Note 1).
AIR carboxylase (Note 2).
1 M $KHCO_3$.

7.2.2 Equipment
UV spectrophotometer (Varian Cary 1).
1 mL, 1 cm pathlength quartz cuvettes (Fisherbrand Cat. # 14–958-126). Other pathlengths can be used.

7.2.3 Procedure

1. In a 1 mL total volume, buffer, 300 μM AIR, and 200 mM KHCO₃ (Note 3) are added to a cuvette and incubated at room temperature. The instrument is started and the absorption at 260 nm is monitored to determine any background changes in absorption.
4. The reaction is initiated by the addition of AIR carboxylase (20–60 ng), carefully mixed and the increase in absorbance at 260 nm is monitored as AIR is converted into CAIR. The slope of the change in absorption over the first minute after addition of enzyme is determined using the instrument software.
5. The concentration of CAIR produced is calculated using $\Delta\varepsilon = 9{,}000\ \mathrm{M}^{-1}\ \mathrm{cm}^{-1}$.

7.2.4 Notes

1. Solutions of AIR are not stable, and it is recommended that fresh stocks be used. If this cannot be done, stocks of AIR should be stored as aliquots at −80°C to prevent multiple freeze-thaw cycles. AIR solutions remaining after completion of the assays should be discarded. AIR will slowly polymerize over time resulting in a dark purple/black color. This typically does not affect the utility of the stock if the color is minimal and transparent; however, stocks with intense opaque color should be discarded.
2. This assay can be used for N^5-CAIR mutase since N^5-CAIR is produced non-enzymatically from AIR and KHCO₃ (Mueller, Meyer, Rudolph, Davisson, & Stubbe, 1994). However, the concentration of N^5-CAIR in solution is challenging to determine accurately and therefore this assay is not recommended for detailed kinetic analysis which requires known concentrations of N^5-CAIR. If this is needed, a coupled assay system (see below) should be used.
3. The substrate for AIR carboxylase is CO_2; however, the addition of bicarbonate to the buffer produces CO_2 with the concentration dependent on the equilibrium value established by the pH and temperature. The concentration of bicarbonate used in these assays is high enough that the concentration of CO_2 is not rate limiting

7.3 SAICAR synthetase coupled assay for AIR carboxylase (Note 1)

$$HCO_3^-/CO_2 + AIR \xrightleftharpoons[AIRC]{} CAIR \xrightarrow{\overset{Asp}{(\ SAICARS}} SAICAR \quad \Delta_{340}\ \varepsilon = 6{,}200\ \mathrm{M}^{-1}\ \mathrm{cm}^{-1}$$

lactate ← NAD⁺ NADH → pyruvate ← ATP ADP → PEP
(LDH) (PK)

Carboxylation in *de novo* purine biosynthesis

7.3.1 Equipment
UV spectrophotometer (Varian Cary 1).
1 mL, 1 cm pathlength quartz cuvettes (Fisherbrand Cat. #14–958-126).
Other pathlengths can be used.

7.3.2 Reagents
Buffer (50 mM HEPES, 20 mM KCl, pH 7.5).
1 M $MgCl_2$.
100 mM ATP disodium salt (Sigma Cat. # A6419).
20 mM NADH disodium salt (Sigma Cat. # 481913).
200 mM PEP trisodium salt (Sigma Cat. # P97002).
1 M $NaHCO_3$.
AIR.
Pyruvate kinase/lactate dehydrogenase (Sigma Cat. # P0294).
1 M Aspartic acid.
AIR carboxylase (Firestine & Davisson, 1994; Skerlova et al., 2020).
E. coli SAICAR synthetase (Meyer, Leonard, Bhat, Stubbe, & Smith, 1992).

7.3.3 Procedure
1. To a final volume of 1 mL, buffer, 6.0 mM $MgCl_2$, 1.1 mM ATP, 0.2 mM NADH, 2.0 mM PEP, 1.0 mM $NaHCO_3$, 5 units of pyruvate kinase/lactate dehydrogenase, 10 mM aspartic acid, and 7 μg *E. coli* SAICAR synthetase are added to a cuvette and incubated at room temperature (Note 2).
2. The reaction is initiated by the addition of 50 μM AIR and 1 μg of AIR carboxylase and the decrease in absorbance at 340 nm is observed. The change in absorbance over time is determined for the first minute and the ATP consumed is determined using the $\Delta\varepsilon$ of 6,200 M^{-1} cm^{-1}.

7.3.4 Notes
1. The SAICAR coupled assay is needed because AIR carboxylase is reversible with the equilibrium lying towards AIR (Firestine & Davisson, 1994). To drive complete use of AIR, it must be converted into the chemically stable SAICAR using this coupled assay system.
2. This assay can also be done by directly monitoring SAICAR produced. To do this, the concentration of ATP in the assay is lowered to 200 μM. The amount of SAICAR produced is determined by measuring the change in absorbance at 282 nm using the extinction coefficient of 8,480 M^{-1} cm^{-1} (Firestine & Davisson, 1994).

7.4 SAICAR synthetase coupled assay for N^5-CAIR mutase (Note 1)

7.4.1 Equipment

UV spectrophotometer (Varian Cary 1).
1 mL, 1 cm pathlength quartz cuvettes (Fisherbrand Cat. # 14–958-126). Other pathlengths can be used.

7.4.2 Reagents

Buffer (50 mM HEPES, 20 mM KCl, pH 7.5).
1 M $MgCl_2$.
100 mM ATP disodium salt (Sigma Cat. # A6419).
20 mM NADH disodium salt (Sigma Cat. # 481913).
200 mM PEP trisodium salt (Sigma Cat. # P97002).
1 M $NaHCO_3$.
AIR.
Pyruvate kinase/lactate dehydrogenase (Sigma Cat. # P0294).
1 M Aspartic acid.
N^5-CAIR synthetase.
N^5-CAIR mutase.
E. coli SAICAR synthetase (Meyer, Leonard, Bhat, Stubbe, & Smith, 1992).

7.4.3 Procedure

1. In a quartz cuvette, buffer, 6.0 mM $MgCl_2$, 1.1 mM ATP, 0.2 mM NADH, 2.0 mM PEP, 1.0 mM $NaHCO_3$, 5 units of pyruvate kinase/lactate dehydrogenase, 10 mM aspartic acid, 7 µg N^5-CAIR mutase, and 7 µg E. coli SAICAR synthetase is added to a final volume of 1 mL and incubated at room temperature. (Note 2).
2. The reaction is initiated by the addition of 50 µM AIR followed immediately by 1.1 µg N^5-CAIR synthetase. The decrease in absorbance at 340 nm is observed. The change in absorbance over time is

Carboxylation in *de novo* purine biosynthesis

determined for the first minute and the ATP consumed is determined using the $\Delta\varepsilon$ of $6{,}200\,\text{M}^{-1}\,\text{cm}^{-1}$. Note that this assay will consume 2 molecules of ATP and thus 2 molecules of NADH for every molecule of AIR used.

7.4.4 Notes

1. This assay is different for that for AIR carboxylase in that N^5-CAIR is generated in situ from bicarbonate, ATP, and AIR by the action of N^5-CAIR synthetase.
2. This assay can also be done by directly monitoring SAICAR produced. To do this, the concentration of ATP in the assay is lowered to $200\,\mu M$ and NADH is eliminated. The amount of SAICAR produced is determined by measuring the change in absorbance at 282 nm using the extinction coefficient of $8{,}480\,\text{M}^{-1}\,\text{cm}^{-1}$ (Firestine & Davisson, 1994).

7.5 N^5-CAIR synthetase coupled assay

7.5.1 Equipment

UV spectrophotometer (Varian Cary-1).
1 mL, 1 cm pathlength quartz cuvettes (Fisherbrand Cat. # 14–958-126). Other pathlengths can be used.

7.5.2 Reagents

Buffer (50 mM HEPES, 20 mM KCl, pH 7.5).
1 M $MgCl_2$.
100 mM ATP disodium salt (Sigma Cat. # A6419).
20 mM NADH disodium salt (Sigma Cat. # 481913).
200 mM PEP trisodium salt (Sigma Cat. # P97002).
1 M $NaHCO_3$.
AIR.
Pyruvate kinase/lactate dehydrogenase (Sigma Cat. # P0294).
N^5-CAIR synthetase.

7.5.3 Procedure

1. To a final volume of 1 mL, buffer, 6.0 mM $MgCl_2$, 1.1 mM ATP, 0.2 mM NADH, 2.0 mM PEP, 1.0 mM $NaHCO_3$, 5 units of pyruvate kinase/lactate dehydrogenase, and 70-100 ng N^5-CAIR synthetase is added to a cuvette and incubated at room temperature for at least 3 min.
2. The reaction is initiated by the addition of 25 μM AIR (Note 1) and the oxidation of NADH is monitored at 340 nm. The initial velocity is determined for the first minute of the reaction and the amount of ATP consumed is calculated using by calculating using the extinction coefficient of $6,200\,M^{-1}\,cm^{-1}$ (Note 2).

7.5.4 Notes

1. The reaction is initiated by the addition of AIR to minimize the non-enzymatic conversion of AIR into N^5-CAIR. However, this conversion is rapid and so the actual concentration of free AIR in solution is lower than the AIR added.
2. This reaction will produce non-stoichiometric consumption of ATP relative to AIR. This is because the N^5-CAIR produced will non-enzymatically degrade to AIR causing a recycling of the substrate.

7.6 Discontinuous phosphate assay for N^5-CAIR synthetase (Firestine, Paritala, McDonnell, Thoden, & Holden, 2009)

$$HCO_3^-/CO_2 + AIR \xrightarrow[NCAIRS]{ATP\quad ADP} N^5\text{-CAIR} + P_i \xrightarrow[12\,H_2MoO_4]{H^+} H_3PO_4(MoO_3)_{12} \xrightarrow{\text{malachite green (MG)}} MG(H_3PO_4(MoO_3)_{12})$$
$$Abs=620\,nm$$

7.6.1 Equipment

Microplate Reader (Biotek Synergy 2).
96-well plate (USA Scientific 655101).

7.6.2 Reagents

Buffer (50 mM HEPES, 20 mM KCl, pH 7.5).
1 M $MgCl_2$.
100 mM ATP.
1 M $NaHCO_3$.
AIR.
Malachite Green Phosphate Assay Kit (BioAssay Systems Cat. # POMG-25H).
N^5-CAIR synthetase.

Carboxylation in *de novo* purine biosynthesis

7.6.3 *Procedure*

1. In a non-sterile 96-well plate (Note 1), buffer, 6.0 mM $MgCl_2$, 100 µM ATP and 1 mM $NaHCO_3$ is added to a final volume of 100 µL (Note 2,3). Each plate contained multiple wells (3–8) of negative controls (*Neg*, buffer + AIR, no enzyme). If the assay is to be used for the analysis of inhibition of N^5-CAIR synthetase, the assay is conducted as above except that the compound to be tested is added. For inhibition studies, each plate contained multiple wells (3–8) for the positive controls (*Pos*, buffer + enzyme + AIR + DMSO (if applicable)) and multiple wells (3–8) of negative controls (*Neg*, buffer + AIR + DMSO (if applicable), no enzyme) (Note 4).

2. AIR (10 µM) is added to the appropriate wells followed by the addition of 168 ng of N^5-CAIR synthetase. The reaction is incubated at room temperature for 10 min.

3. The reaction is quenched by the addition of 30 µL of the malachite green reagent to each well and the plate is incubated at room temperature for at least 15 min to allow for the formation of the phosphate molybdenum complex.

4. The absorbance at 620 nm (*Assay*) of each well is determined using a microplate reader.

5. The absorbance due to phosphate production is calculated by *Assay-Neg*. To convert this value into the concentration of phosphate produced, a standard curve must be generated (Note 5). To calculate the percent inhibition, the data is fit to Eq. 2.

$$\%Inhibition = \left(1 - \frac{Assay - Neg}{Pos - Neg}\right) \times 100 \tag{2}$$

7.6.4 *Notes*

1. This assay can be conducted in 384-well plate or with a cuvette and a regular spectrophotometer.

2. High quality ATP with minimal phosphate contamination is needed for this assay.

3. Phosphate contamination is very common and will impact the utility of this assay. It is important to utilize pure water and reagents and to store reagents in plastic containers. Laboratory detergents frequently have high levels of phosphate and glassware often has phosphate bound to it unless extreme measures are taken. Water and key reagents should be tested for phosphate levels before they are used in the assay.

4. If the compound is dissolved in DMSO, the appropriate level of DMSO should be used in the controls. DMSO up to 5 % by volume can be used in the assay without interference.
5. The assay kit comes with a premixed phosphate solution that can be used to generate the standard curve. This kit has a useful range of 0.02 μM to 40 μM.

7.7 Isatin-fluorescein enzyme activity assay (Sharma, Sharma, Streeter, & Firestine, 2023) (Note 1)

7.7.1 Equipment
96–well white flat-bottom fluorescence plates (non–binding) (Corning Cat. # 3600).
Fluorescence Microplate Reader (Biotek Synergy 2) with excitation/emission wavelengths of 485 nm/528 nm.

7.7.2 Reagents
Isatin–fluorescein (Sharma, Sharma, Streeter, & Firestine, 2023) dissolved in DMSO Buffer (25 mM Tris, pH 8 +/- 0.01 % Triton).
CAIR.
N^5-CAIR mutase or AIR carboxylase (Firestine & Davisson, 1994; Skerlova et al., 2020) (Note 2).
1 M $ZnCl_2$ dissolved in water.

7.7.3 Procedure
1. In a non–sterile 96–well plate, buffer and 34 μM CAIR is added to a final volume of 100 μL. Each plate contained multiple wells (3–8) of negative controls (*Neg*, buffer + CAIR, no enzyme). If the assay is to be used for

Carboxylation in *de novo* purine biosynthesis

the analysis of inhibition of N^5-CAIR mutase, the assay is conducted as above except that the inhibitor to be tested is added. For inhibition studies, each plate contained multiple wells (3-8) for the positive controls (*Pos*, buffer + enzyme + CAIR + DMSO (if applicable)) and multiple wells (3-8) of negative controls (*Neg*, buffer + CAIR + DMSO (if applicable), no enzyme).

2. The reaction is initiated by adding 104 ng N^5-CAIR mutase to each well and the reaction is run for 10 min (Note 3).

3. At 10 min, the reaction is quenched by the addition $ZnCl_2$ to a final concentration of 300 µM followed by the addition of isatin-fluorescein to a final concentration of 30 µM (Note 4,5).

4. The development of the AIR-isatin-fluorescein complex is monitored by an increase in fluorescence using an excitation wavelength of 485 and emission wavelength of 528 via a multimode plate reader. Maximum fluorescence (*Assay*) is typically achieved after 1 h of incubation at room temperature (Note 6).

5. The AIR produced is proportional to the fluorescent signal at 1 hr, which is calculated by *Assay-Neg*. To determine the actual concentration of AIR produced, a standard curve of AIR versus fluorescence must be generated. The percent inhibition is calculated according to Eq. 2.

$$\%Inhibition = \left(1 - \frac{Assay - Neg}{Pos - Neg}\right) \times 100 \qquad (2)$$

7.7.4 Notes

1. This assay is predominately used for screening or as an orthogonal assay to verify inhibition. It is not suitable for detailed kinetic analysis.

2. Either enzyme can be used in this assay. While the decarboxylation of CAIR by N^5-CAIR mutase generates N^5-CAIR, this compound is chemically unstable and will non-enzymatically produce AIR. Decarboxylation of CAIR by AIR carboxylase will give AIR directly.

3. The time for the enzymatic reaction can be varied depending on the concentration of enzyme added. However, care should be taken that the concentration of AIR produced is linear over the time selected for the experiment.

4. $ZnCl_2$ is added to quench the reaction (Sullivan et al., 2014). This works by forming a strong complex with CAIR preventing it from being a substrate for the enzyme. The zinc complex also prevents a reaction between CAIR and isatin-fluorescein.

5. Zinc and the isatin–fluorescein can be combined into one solution to minimize the number of pipetting steps
6. The amount of time it takes to reach equilibrium for the AIR–isatin-fluorescein complex is dependent on the amount of AIR in solution. Lower concentrations of AIR will take less time. However, the AIR-isatin-fluorescein complex is stable for several hours making it convenient to monitor the fluorescence at 1 h incubation.

7.8 Intrinsic Tryptophan Fluorescence Assay for N^5-CAIR Mutase (Constantine et al., 2006; Lei et al., 2016)

7.8.1 Equipment
Fluorimeter (Horiba FluoroMax).
1 mL, 1 cm quartz fluorescence cuvettes (Starna, Cat. # 9F-Q-10).

7.8.2 Reagents
N^5-CAIR mutase.
Buffer (25 mM Tris, pH 8 + 0.01 % Triton).
Compound to be tested.

7.8.3 Procedure
1. In a fluorescence cuvette, 1 mL of buffer is added, and the cuvette is placed into a fluorimeter at room temperature. The spectrophotometer is set to an excitation wavelength to 295 nm (1 nm slit width) and the emission spectrum is scanned from 310 and 500 nm to determine the background.
2. To the cuvette, 1 μL of N^5-CAIR mutase (final concentration 21 μg/mL) is added and the fluorescence spectrum is recorded using the settings outlined in step 1. The fluorescence intensity (Fo) at 340 nm is determined and corrected for dilution.
3. To determine compound binding, various concentrations of the compound to be tested are added, individual fluorescence spectra are determined, and F is calculated for each concentration of compound.
4. The change in fluorescence is calculated using the equation $(Fo\text{-}F)/Fo$. To calculate the K_d, a plot of $(Fo\text{-}F)/Fo$ versus ligand concentration is generated and the resulting data is fit to a single rectangular hyperbolic Eq. (3) where $\Delta Fmax$ is the maximum value of $(Fo\text{-}F)/F0$, and L is the ligand concentration.

$$\frac{(Fo - F)}{Fo} = \frac{\Delta F_{max} * L}{K_d + L} \tag{3}$$

7.9 Thermal melting assay for N⁵-CAIR mutase

7.9.1 Equipment

96-well USA Scientific TempPlate® 96-well no-skirt 0.2 mL PCR plates (Note 1).
Plate sealing film (USA Scientific TempPlate® RT select optical film).
Centrifuge (Eppendorf 5810 R).
Stratagene Mx3005P QPCR system.

7.9.2 Reagents

Buffer (25 mM Tris, pH 8).
5,000X Sypro Orange Protein Gel Stain (Invitrogen Cat. # S6650).
N⁵-CAIR mutase.
Compound to be tested.

7.9.3 Procedure

1. In a non-sterile 96-well PCR compatible plate, buffer, 5 µM N⁵-CAIR mutase, 10x SYPRO gel stain, and 0.5 µL DMSO is added to a final volume of 10 µL (Note 2, 3).
2. Each well is mixed, and the entire plate is sealed with film. The microplate is centrifuged at 15xg for 1 min to eliminate bubbles.
3. The plate is placed into a QPCR instrument, and the protocol is selected to use the 492 nm emission wavelength filter and the 610 nm emission wavelength filter. The experiment measured fluorescent changes for temperatures ranged from 25°C to 95°C with a ramp rate of 1 °C per minute. The typical measurement takes approximately 40 min to complete.
4. Raw data is imported into Excel to normalize starting fluorescence for each well (Note 4). The normalized data is imported into GraphPad Prism. Each melting curve is fit to the Boltzmann-sigmoidal Eq. (4) to determine the melting temperature (T_m). In this equation, Y = fluorescence intensity; X = temperature; *Top* and *Bottom* represent maximal fluorescence at top of curve and baseline fluorescence at bottom of curve, respectively; T_m = protein melting temperature; *Slope* represents steepness of curve. For *E. coli* N⁵-CAIR mutase, T_m is 61-62 °C.

$$Y = Bottom + \frac{Top - Bottom}{1 + \exp\left(T_m - \frac{X}{Slope}\right)} \tag{4}$$

If this method is being used to determine binding affinity, a graph of T_m versus compound concentration is generated and K_d is determined using a four-parameter non-linear regression (Eq. 5) for a dose response relationship. In this equation, Y is the response, *Top* and *Bottom* represent top and bottom of plateau in the same units as Y respectively, K_d is the dissociation constant in the same units as X and *Hill slope* is unitless and describes the steepness of the curve.

$$Y = Bottom + \frac{Top - Bottom}{\left(1 + \frac{K_d}{X}\right)^{(HillSlope)}} \tag{5}$$

7.9.4 Notes

1. This assay has been conducted in 384-well format, with a $10\,\mu L$ final volume. For a small-scale experiment, individual $0.2\,mL$ PCR tubes with clear, flat caps can be used instead of an entire plate.
2. Increasing DMSO concentration in each well can impact the protein stability. No change in melting temperature was seen with 8 % DMSO present in the buffer; however, a significant decrease ($\sim 2\,°C$) in melting temperature is seen at 12 % and higher DMSO concentrations.
3. For positive control wells, NAIR or CAIR can be used as binding ligands.
4. Often, there is a decrease in signal after the maximum change in fluorescence is observed. This decrease is due to precipitation of the unfolded protein-dye complex. Removing this region from the graph before applying the Boltzmann equation will result in a better fit to the data.

8. Summary and conclusions

The de novo purine biosynthetic pathway has a single carbon-carbon bond formation, namely in the carboxylation of AIR into CAIR. Interestingly, how this transformation is accomplished is species specific. In vertebrates and some microbes in high CO_2 environments, AIR carboxylase directly carboxylates AIR with CO_2 to give CAIR. In contrast, in bacteria, yeast, fungi and plants, two steps are needed to convert AIR into CAIR. In the first, N^5-CAIR synthetase converts AIR, ATP and bicarbonate into phosphate and the unstable carbamate, N^5-CAIR. In this reaction, CO_2 is added to the exocyclic amine of AIR. In the second step, N^5-CAIR mutase catalyzes the rearrangement of CO_2 from N5 to C4 to give CAIR. This divergence in approaches towards CAIR synthesis likely reflects differences in the CO_2/HCO_3^- environment of the organism.

Carboxylation in de novo purine biosynthesis

Vertebrates have a well-regulated and consistent concentration of CO_2, whereas microbes and plants are at the mercy of their surrounding environment. Thus, they have adapted to utilize the abundant bicarbonate in the carboxylation of AIR at the expense of 1 molecule of ATP.

The enzymes outlined in this chapter pose several interesting questions. One is the reason for the substrate selectivity of AIR carboxylase and N^5-CAIR mutase. These enzymes are evolutionarily and structurally conserved and produce the same product (CAIR), yet each is specific for its own reaction. What amino acids dictate substrate selectivity for each enzyme is currently unknown and of biochemical interest. Secondly, these enzymes represent interesting targets for drug discovery. N^5-CAIR mutase is found only in microbes and AIR carboxylase has been shown to be critical for many cancers. What are the critical design features needed to create selective and potent inhibitors of these enzymes? Although there have been several approaches towards addressing this question, drug discovery efforts have so far had limited success.

This chapter provides an overview of these interesting enzymes, gives methods to produce these enzymes, discusses the synthesis of the substrates, and supplies a collection of assays that can be used to address the questions outlined above. The easy-to-use protocols outlined in the chapter will help researchers to determine enzyme kinetics, examine mutants of these enzymes, and evaluate compounds as inhibitors.

References

Alenin, V., Kostikova, T., & Domkin, V. (1987). Chemical synthesis of N1-substituted 5-aminoimidazoles and the formation of N-carboxylation products in aqueous-solutions of potassium bicarbonate. *Zhurnal Obshchei Khimii, 57*(3), 692–701.

Bartoschek, S., Vorholt, J. A., Thauer, R. K., Geierstanger, B. H., & Griesinger, C. (2000). N-carboxymethanofuran (carbamate) formation from methanofuran and co2 in methanogenic archaea. Thermodynamics and kinetics of the spontaneous reaction. *European Journal of Biochemistry, 267*(11), 3130–3138.

Belfon, K. K. J., Sharma, N., Zigweid, R., Bolejack, M., Davies, D., Edwards, T. E., ... French, J. B. (2023). Structure-guided discovery of N(5)-CAIR mutase inhibitors. *Biochemistry, 62*(17), 2587–2596.

Bhat, B., Groziak, M. P., & Leonard, N. J. (1990). Nonenzymatic synthesis and properties of 5-aminoimidazole ribonucleotide (AIR). Synthesis of specifically ^{15}N-labeled 5-aminoimidazole ribonucleoside (AIRs) derivatives. *Journal of the American Chemical Society, 112*(12), 4891–4897.

Brown, A. M., Hoopes, S. L., White, R. H., & Sarisky, C. A. (2011). Purine biosynthesis in archaea: Variations on a theme. *Biology Direct, 6*, 63.

Cao, W., & De La Cruz, E. M. (2013). Quantitative full time course analysis of nonlinear enzyme cycling kinetics. *Scientific Reports, 3*, 2658.

Christopherson, R. I., Lyons, S. D., & Wilson, P. K. (2002). Inhibitors of de novo nucleotide biosynthesis as drugs. *Accounts of Chemical Research, 35*(11), 961–971.

Chua, S. M., & Fraser, J. A. (2020). Surveying purine biosynthesis across the domains of life unveils promising drug targets in pathogens. *Immunology & Cell Biology, 98*(10), 819–831.

Constantine, C. Z. (2009). *An investigation of acetobacter aceti N5-carboxyaminoimidazole ribonucleotide mutase and its purE-purK operon.* St. Louis: Washington University in.

Constantine, C. Z., Starks, C. M., Mill, C. P., Ransome, A. E., Karpowicz, S. J., Francois, J. A., ... Kappock, T. J. (2006). Biochemical and structural studies of N5-carboxyaminoimidazole ribonucleotide mutase from the acidophilic bacterium *Acetobacter aceti. Biochemistry, 45*(27), 8193–8208.

Dewal, M. B., & Firestine, S. M. (2013). Site-directed mutagenesis of catalytic residues in N (5)-carboxyaminoimidazole ribonucleotide synthetase. *Biochemistry, 52*(37), 6559–6567.

Fawaz, M. V., Topper, M. E., & Firestine, S. M. (2011). The ATP-grasp enzymes. *Bioorganic Chemistry, 39*(5–6), 185–191.

Firestine, S. M. (1995). *Biochemical and mechanistic characterization of Gallus gallus 5-aminoimidazole ribonucleotide carboxylase* Purdue University.

Firestine, S. M., & Davisson, V. J. (1993). A tight binding inhibitor of 5-aminoimidazole ribonucleotide carboxylase. *Journal of Medicinal Chemistry, 36*(22), 3484–3486.

Firestine, S. M., & Davisson, V. J. (1994). Carboxylases in de novo purine biosynthesis. Characterization of the Gallus gallus bifunctional enzyme. *Biochemistry, 33*(39), 11917–11926.

Firestine, S. M., Paritala, H., McDonnell, J. E., Thoden, J. B., & Holden, H. M. (2009). Identification of inhibitors of N5-carboxyaminoimidazole ribonucleotide synthetase by high-throughput screening. *Bioorganic & Medicinal Chemistry, 17*(9), 3317–3323.

Firestine, S. M., Poon, S. W., Mueller, E. J., Stubbe, J., & Davisson, V. J. (1994). Reactions catalyzed by 5-aminoimidazole ribonucleotide carboxylases from Escherichia coli and Gallus gallus: A case for divergent catalytic mechanisms. *Biochemistry, 33*(39), 11927–11934.

Firestine, S. M., Wu, W., Youn, H., & Davisson, V. J. (2009). Interrogating the mechanism of a tight binding inhibitor of AIR carboxylase. *Bioorganic & Medicinal Chemistry, 17*(2), 794–803.

Firestine, S. M., Misialek, S., Toffaletti, D. L., Klem, T. J., Perfect, J. R., & Davission, V. J. (1998). Biochemical role of the Cryptococcus neoformans ADE2 protein in fungal de novo purine biosynthesis. *Archives of Biochemistry and Biophysics, 351*(1), 123–134.

Gedeon, A., Karimova, G., Ayoub, N., Dairou, J., Gianetto, Q. G., Vichier-Guerre, S., ... Munier-Lehmann, H. (2023). Interaction network among de novo purine nucleotide biosynthesis enzymes in Escherichia coli. *FEBS Journal, 290*, 3165–3184.

Huang, N., Xu, C., Deng, L., Li, X., Bian, Z., Zhang, Y., ... Sun, F. (2020). PAICS contributes to gastric carcinogenesis and participates in DNA damage response by interacting with histone deacetylase 1/2. *Cell Death & Disease, 11*(7), 507.

Huang, Z., Xie, N., Illes, P., Di Virgilio, F., Ulrich, H., Semyanov, A., ... Tang, Y. (2021). From purines to purinergic signalling: Molecular functions and human diseases. *Signal Transduction and Targeted Therapy, 6*(1), 162.

Kappock, T. J., Ealick, S. E., & Stubbe, J. (2000). Modular evolution of the purine biosynthetic pathway. *Current Opinion in Chemical Biology, 4*(5), 567–572.

Kim, A., Wolf, N. M., Zhu, T., Johnson, M. E., Deng, J., Cook, J. L., & Fung, L. W. (2015). Identification of bacillus anthracis PurE inhibitors with antimicrobial activity. *Bioorganic & Medicinal Chemistry, 23*(7), 1492–1499.

Lei, H., Jones, C., Zhu, T., Patel, K., Wolf, N. M., Fung, L. W., ... Johnson, M. E. (2016). Identification of B. anthracis N(5)-carboxyaminoimidazole ribonucleotide mutase (PurE) active site binding compounds via fragment library screening. *Bioorganic & Medicinal Chemistry, 24*(4), 596–605.

Li, S. X., Tong, Y. P., Xie, X. C., Wang, Q. H., Zhou, H. N., Han, Y., ... Bi, R. C. (2007). Octameric structure of the human bifunctional enzyme PAICS in purine biosynthesis. *Journal of Molecular Biology, 366*(5), 1603–1614.

Lin, Q. (2018). *Design, synthesis and biological evaluation of pyrido [2, 3-d] pyrimidines as inhibitors of N^5-CAIR synthetase*. Wayne State University.

Li, X., Zheng, Q. C., Zhang, J. J., & Zhang, H. X. (2011). Theoretical study on the mechanism of rearrangement reaction catalyzed by N^5-carboxyaminoimidazole ribonucleotide mutase. *Computational and Theoretical Chemistry, 964*, 77–82.

Litchfield, G., & Shaw, G. (1971). Purines, pyrimidines, and imidazoles. Part xxxviii. A kinetics study of the decarboxylation of 5-amino-1-β-d-ribofuranosylimidazole-4-carboxylic acid 5′-phosphate and related compounds. *Journal of the Chemical Society B: Physical Organic*, 1474–1484.

Lukens, L. N., & Buchanan, J. M. (1959). Biosynthesis of the purines. xxiv. The enzymatic synthesis of 5-amino-1-ribosyl-4-imidazolecarboxylic acid 5′-phosphate from 5-amino-1-ribosylimidazole 5′-phosphate and carbon dioxide. *Journal of Biological Chemistry, 234*(7), 1799–1805.

Marolewski, A., Smith, J. M., & Benkovic, S. J. (1994). Cloning and characterization of a new purine biosynthetic enzyme: A non-folate glycinamide ribonucleotide transformylase from E. coli. *Biochemistry, 33*(9), 2531–2537.

Meng, M., Chen, Y., Jia, J., Li, L., & Yang, S. (2018). Knockdown of PAICS inhibits malignant proliferation of human breast cancer cell lines. *Biological Research, 51*(1), 24.

Meyer, E., Kappock, T. J., Osuji, C., & Stubbe, J. (1999). Evidence for the direct transfer of the carboxylate of N5-carboxyaminoimidazole ribonucleotide (N^5-CAIR) to generate 4-carboxy-5-aminoimidazole ribonucleotide catalyzed by Escherichia coli PurE, an N^5-CAIR mutase. *Biochemistry, 38*(10), 3012–3018.

Meyer, E., Leonard, N. J., Bhat, B., Stubbe, J., & Smith, J. M. (1992). Purification and characterization of the purE, purK, and purC gene products: Identification of a previously unrecognized energy requirement in the purine biosynthetic pathway. *Biochemistry, 31*(21), 5022–5032.

Moffatt, B. A., & Ashihara, H. (2002). Purine and pyrimidine nucleotide synthesis and metabolism. *Arabidopsis Book, 1*, e0018.

Mueller, E. J., Meyer, E., Rudolph, J., Davisson, V. J., & Stubbe, J. (1994). N5-carboxyaminoimidazole ribonucleotide: Evidence for a new intermediate and two new enzymatic activities in the de novo purine biosynthetic pathway of Escherichia coli. *Biochemistry, 33*(8), 2269–2278.

Murray, A. W. (1971). The biological significance of purine salvage. *Annual Review of Biochemistry, 40*, 811–826.

Offensperger, F., Tin, G., Duran-Frigola, M., Hahn, E., Dobner, S., Ende, C. W. A., ... Winter, G. E. (2024). Large-scale chemoproteomics expedites ligand discovery and predicts ligand behavior in cells. *Science, 384*(6694), eadk5864.

Ownby, K., Xu, H., & White, R. H. (2005). A methanocaldococcus jannaschii archaeal signature gene encodes for a 5-formaminoimidazole-4-carboxamide-1-β-d-ribofuranosyl 5′-monophosphate synthetase: A new enzyme in purine biosynthesis. *Journal of Biological Chemistry, 280*(12), 10881–10887.

Pedley, A. M., & Benkovic, S. J. (2017). A new view into the regulation of purine metabolism: The purinosome. *Trends in Biochemical Sciences, 42*(2), 141–154.

Prejano, M., Skerlova, J., Stenmark, P., & Himo, F. (2022). Reaction mechanism of human PAICS elucidated by quantum chemical calculations. *Journal of the American Chemical Society, 144*(31), 14258–14268.

Ravi, G. R. R., Biswal, J., Kanagarajan, S., & Jeyakanthan, J. (2019). Exploration of N^5-CAIR mutase novel inhibitors from pyrococcus horikoshii ot3: A computational study. *Journal of Computational Biology, 26*(5), 457–472.

Samant, S., Lee, H., Ghassemi, M., Chen, J., Cook, J. L., Mankin, A. S., & Neyfakh, A. A. (2008). Nucleotide biosynthesis is critical for growth of bacteria in human blood. *PLoS Pathog, 4*(2), e37.

Sauers, C. K., Jencks, W. P., & Groh, S. (1975). Alcohol-bicarbonate-water system. Structure-reactivity studies on the equilibriums for formation of alkyl monocarbonates and on the rates of their decomposition in aqueous alkali. *Journal of the American Chemical Society, 97*(19), 5546–5553.

Sharma, M. F., Sharma, S. K., Streeter, C. C., & Firestine, S. M. (2023). A fluorescence-based assay for N(5)-carboxyaminoimidazole ribonucleotide mutase. *ChemBioChem, 24*(18), e202300347.

Skerlova, J., Unterlass, J., Gottmann, M., Marttila, P., Homan, E., Helleday, T., ... Stenmark, P. (2020). Crystal structures of human PAICS reveal substrate and product binding of an emerging cancer target. *Journal of Biological Chemistry, 295*(33), 11656–11668.

Streeter, C. C., Lin, Q., & Firestine, S. M. (2019). Isatins inhibit N(5)-CAIR synthetase by a substrate depletion mechanism. *Biochemistry, 58*(17), 2260–2268.

Sullivan, K. L., Huma, L. C., Mullins, E. A., Johnson, M. E., & Kappock, T. J. (2014). Metal stopping reagents facilitate discontinuous activity assays of the de novo purine biosynthesis enzyme pure. *Analytical Biochemistry, 452*, 43–45.

Thoden, J. B., Holden, H. M., & Firestine, S. M. (2008). Structural analysis of the active site geometry of N^5-carboxyaminoimidazole ribonucleotide synthetase from Escherichia coli. *Biochemistry, 47*(50), 13346–13353.

Tong, L. (2013). Structure and function of biotin-dependent carboxylases. *Cellular and Molecular Life Sciences, 70*(5), 863–891.

Tranchimand, S., Starks, C. M., Mathews, I. I., Hockings, S. C., & Kappock, T. J. (2011). *Treponema denticola PurE Is a Bacterial AIR Carboxylase. Biochemistry, 50* 4623-3637.

Zhang, Y., Morar, M., & Ealick, S. E. (2008). Structural biology of the purine biosynthetic pathway. *Cellular and Molecular Life Sciences, 65*(23), 3699–3724.

Zhao, M. (2014). *Functional studies of 5-carboxyaminoimidazole ribonucleotide synthetase and mutase from thermotoga maritima*. Wayne State University.

Printed in the United States
by Baker & Taylor Publisher Services